Graduate Texts in Mathematics 56

Graduate Texts in Mathematics

continued after Index

William S. Massey

Algebraic Topology: An Introduction

Springer-Verlag
New York Heidelberg Berlin

Dr. William S. Massey
Department of Mathematics
Yale University
New Haven, Connecticut 06520
USA

AMS Subject Classification 55-01

Library of Congress Cataloging in Publication Data

Massey, William S
Algebraic topology, an introduction.
(Graduate texts in mathematics ; 56)
Includes bibliographies.
1. Algebraic topology. I. Title. II. Series.
QA612.M37 1977 514'.2 77-22206

First three printings published by
Harcourt, Brace & World, Inc.

Printed in the United States of America

9 8 7 (Corrected Seventh Printing, 1987)

ISBN 0-387-90271-6 Springer-Verlag New York

ISBN 3-540-90271-6 Springer-Verlag Berlin Heidelberg

To Ethel

Preface to the New Printing

I have taken advantage of the opportunity afforded by this new printing to correct some minor errors in the text, to add some additional exercises, and to include references to some of the more recent books and papers on algebraic topology. Other than this, the main body of the text is unchanged. It is my intention to publish in this same Springer-Verlag series a sequel to this book on singular homology theory and related topics.

W. S. MASSEY

New Haven, Connecticut
May, 1977

Preface

This textbook is designed to introduce advanced undergraduate or beginning graduate students to algebraic topology as painlessly as possible. The principal topics treated are 2-dimensional manifolds, the fundamental group, and covering spaces, plus the group theory needed in these topics. The only prerequisites are some group theory, such as that normally contained in an undergraduate algebra course on the junior-senior level, and a one-semester undergraduate course in general topology.

The topics discussed in this book are "standard" in the sense that several well-known textbooks and treatises devote a few sections or a chapter to them. This, I believe, is the first textbook giving a straightforward treatment of these topics, stripped of all unnecessary definitions, terminology, etc., and with numerous examples and exercises, thus making them intelligible to advanced undergraduate students.

The subject matter is used in several branches of mathematics other than algebraic topology, such as differential geometry, the theory of Lie groups, the theory of Riemann surfaces, or knot theory. In the development of the theory, there is a nice interplay between algebra and topology which causes each to reinforce interpretations of the other. Such an interplay between different topics of mathematics breaks down the often artificial subdivision of mathematics into different "branches" and emphasizes the essential unity of all mathematics.

Undoubtedly some experts will be shocked that a textbook purporting to be an introduction to algebraic topology does not even mention homology theory. It is certainly true that homology and cohomology theory form the core of algebraic topology. However, it is difficult to motivate the student who is learning these subjects for the first time, and their systematic treatment requires the patient development of a great deal of machinery. Only after several months of classroom lectures and study can interesting applications be given which show that the development of all the machinery was worthwhile. For these reasons, I believe that it is easier for the student to understand and appreciate homology theory after he has studied the fundamental group and allied topics presented in this book.

To those with a strictly logical mind, Chapter I, which discusses 2-dimensional manifolds, will perhaps seem the least rigorous part of the book. There certainly would be no real problem in giving a strictly rigorous treatment of this subject matter. However, such a treatment would be rather dull and tedious, with long-winded proofs of facts that are visually obvious. Moreover, the results of Chapter I are not basic to the main theorems in the rest of the book; rather, they furnish examples, illustrations, and applications of the results of the later chapters.

Chapter II gives the definition and basic properties of the fundamental group and the homomorphism induced by a continuous map. General methods for determining the structure of the fundamental group of a space are developed later, in Chapter IV, after certain essential group-theoretic notions have been introduced in Chapter III.

In Chapters III and IV the characterization of certain mathematical structures as the solutions of "universal mapping problems" is emphasized for two different reasons. First, it seems that the most efficient method of determining the structure of the fundamental group of a wide variety of spaces is by use of the Seifert-Van Kampen theorem (Chapter IV); the best formulation of this essential theorem involves the notion of a universal mapping problem. Second, this method of characterizing various mathematical structures as solutions to universal mapping problems seems to be one of the truly unifying mathematical principles to have emerged since 1945, and it should be brought into the mathematics curriculum as early as possible.

Chapter V contains a rather thorough discussion of covering spaces. The relationship between covering spaces and the fundamental group is emphasized throughout.

In Chapters VI and VII are given topological proofs of several well-known theorems of group theory, especially the Nielsen-Schreier theorem on subgroups of a free group, the Kurosh theorem on subgroups of a free product, and the Grushko theorem on the decomposition of a finitely generated group as a free product. These theorems belong to a section of group theory whose original development was largely motivated by combinatorial topology. I believe that the proofs of these theorems using the fundamental groups and covering spaces of certain low-dimensional complexes are more easily comprehended than the purely algebraic proofs. I hope the unified treatment of these theorems by these essentially geometric methods will make this section of group theory less formidable and more readily accessible.

Chapter VIII is rather brief and of a strictly descriptive nature; no theorems are proved. Its purpose is to help the student make the transition to the study of more advanced topics in algebraic topology.

Although triangulations of 2-manifolds are used in Chapter I, and the CW-complexes of J. H. C. Whitehead are introduced in the last chapter, there is no systematic treatment of simplicial complexes in this book. This may surprise some readers in view of the fact that many treatises on algebraic topology start off with just such a discussion. However, it is difficult to see how it could have materially simplified the exposition. Moreover, it is my personal opinion that any such discussion must of necessity be rather dull. One of the tendencies of algebraic topology during the last fifteen years or so has been the replacement of simplicial complexes by CW-complexes as the main object of study.

The sections listed below are not absolutely necessary to the further developments of the theory, and they can be omitted completely or given less emphasis in a briefer course or on a first reading of the book:

Chapter I, Sections 9–13.
Chapter II, Sections 7 and 8.
Chapter III, Section 7.
Chapter IV, Section 6.
Chapter V, Sections 10–12.
Chapter VI, Section 8.
Chapter VII, Sections 5 and 6.

Also, a briefer course could be built around the material in the first five chapters, omitting the same sections.

This book has developed from lectures given at Yale University to both graduate and undergraduate students over a period of several years. It is a pleasure to acknowledge my indebtedness to these students. Their questions, criticisms, and suggestions have given me many insights. I am also deeply indebted to my colleagues for many discussions of the ideas presented in this book. Most of the theorems and definitions in this book may be found in well-known textbooks or articles in mathematical journals. In this regard, special mention must be made of the following German textbooks: B. Kerekjarto, *Topologie* (Springer, 1923); K. Reidemeister, *Einführung in die Kombinatorische Topologie* (Teubner, 1932), H. Seifert and W. Threlfall, *Lehrbuch der Topologie* (Vieweg, 1934). In many cases I have tried to indicate the person or persons to whom I thought an idea or theorem should be credited. However, in a subject such as this, whose development spans most of the past century and which has been the joint work of many mathematicians in many countries, it is inevitable that I have committed some errors in assigning credit. To those whose names have been inadvertently omitted, I apologize; I trust that they will be understanding.

W. S. MASSEY

New Haven, Connecticut

Note to the Student

Prerequisites This book assumes that the student knows enough group theory to understand such standard terms as group, subgroup, normal subgroup, homomorphism, quotient group, coset, abelian group, and cyclic group. Moreover, it is hoped that he has seen enough examples and has worked enough exercises to have some feeling for the true significance of these concepts. An appendix on permutation and transformation groups is supplied for the benefit of those who are unfamiliar with this topic. Most of the additional topics needed in group theory are developed in the text, especially in Chapter III.

The necessary background in point set topology can be obtained from a one-semester undergraduate course in the subject. Because most textbooks for such a course either treat the subject very briefly or omit it entirely, a short discussion of quotient spaces is appended. No knowledge of any branch of algebra other than group theory is needed; in particular, nothing is used from the theory of rings, fields, modules, or vector spaces.

Terminology and notation Since most terminology and notation is standard in contemporary mathematics books on this level, little explanation is needed. In group theory, all groups (with a few standard exceptions, such as the additive group of integers) are written multiplicatively, not additively. A homomorphism from one group to another is called an *epimorphism* if it is onto, a *monomorphism* if it is one-to-one (i.e., the kernel contains only the identity), and an *isomorphism* if it is both one-to-one and onto. A diagram of groups and homomorphisms, such as

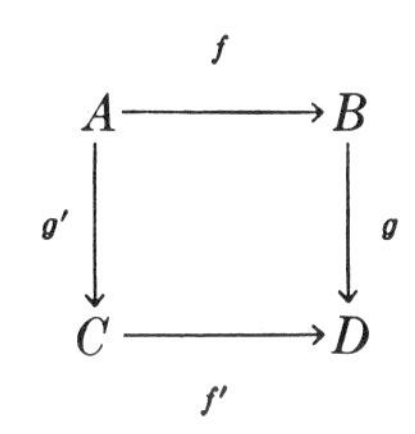

is said to be *commutative* if all possible homomorphisms from one group to another in the diagram are equal. In the above diagram, there are two homomorphisms from group A to group D, namely, gf (i.e., f followed by g) and $f'g'$. Thus, requiring that this diagram be commutative is equivalent to requiring that $gf = f'g'$. Note that the requirement that a diagram be commutative has nothing to do with whether or not any of the groups involved is commutative or abelian. For example, the above diagram could be commutative even if A, B, C, and D were non-abelian groups.

In set theory, the notation

$$\prod_{i \in I} S_i$$

denotes the product (or cartesian product) of the family of sets S_i, $i \in I$. An element x of the cartesian product is a function that assigns to each index $i \in I$ an element $x_i \in S_i$. The element $x_i \in S_i$ is also called the *coordinate* of the element x corresponding to the index $i \in I$.

If A is a subset of B, then there is a uniquely defined *inclusion map* of A into B: It assigns to each element $x \in A$ the element x itself. In symbols, if $i : A \to B$ denotes the inclusion map, then $i(x) = x$ for any $x \in A$. If C is another set and $f : B \to C$ is any function from B to C, then $f \mid A$ denotes the *restriction* of f to the subset A; i.e., for any $a \in A$, $(f \mid A)(a) = f(a) \in C$.

The following notation is fixed throughout the book:

$\mathbf{Z}$ = set of all integers, positive and negative.

$\mathbf{Q}$ = set of all rational numbers.

$\mathbf{R}$ = set of all real numbers.

$\mathbf{C}$ = set of all complex numbers.

The notation $\mathbf{R}^n$ (respectively, $\mathbf{C}^n$) for any integer $n > 0$ denotes the set of all n-tuples $(x_1, \ldots, x_n)$ of real (respectively, complex) numbers; $\mathbf{R}^n$ is the *Euclidean n-space* and has its usual topology. If $x = (x_1, \ldots, x_n)$ is a point of $\mathbf{R}^n$, then the *norm* or *absolute value* of x, denoted by $|x|$, is defined as usual:

$$|x| = \left(\sum_{i=1}^{n} x_i^2\right)^{1/2}.$$

With this notation, we define the following standard subsets of Euclidean n-space for any $n > 0$:

$$E^n = \{x \in \mathbf{R}^n : |x| \leqq 1\},$$

$$U^n = \{x \in \mathbf{R}^n : |x| < 1\},$$

$$S^{n-1} = \{x \in \mathbf{R}^n : |x| = 1\}.$$

These spaces are called the *closed n-dimensional disc* or *ball*, the *open n-dimensional disc* or *ball*, and the $(n-1)$*-dimensional sphere*, respectively. Each is topologized as a subset of $\mathbf{R}^n$. The same names are sometimes applied to any topological space homeomorphic to one of the spaces just mentioned.

If a and b are real numbers such that $a < b$, then the following standard notation is used for the open and closed intervals with a and b as end points:

$$(a, b) = \{x \in \mathbf{R} : a < x < b\},$$

$$[a, b] = \{x \in \mathbf{R} : a \leqq x \leqq b\},$$

$$(a, b] = \{x \in \mathbf{R} : a < x \leqq b\}.$$

We say two spaces are of the same *topological type* if they are homeomorphic.

References A reference to Theorem or Lemma III. 8.4 indicates Theorem or Lemma 4 in Section 8 of Chapter III; if the reference is simply to Theorem 8.4, then the theorem is in Section 8 of the same chapter in which the reference occurs.

At the end of each chapter is a brief bibliography. Numbers in square brackets in the text refer to items in the bibliography.

On studying this book The exercises and examples are an integral part of the text; without them it would be much more difficult to gain an understanding of the subject. Many assertions are made without proof, and the details of certain proofs are omitted. Regard the filling in of the missing details as an exercise that tests whether you really understand the ideas involved.

Remember that the path from ignorance to knowledge in any subject is not straight and true, but is almost always rather zigzagged. One seems to learn things by a method of successive approximations to the truth. Thus, the first attempt to master some of the more difficult theorems in this book is not likely to be completely successful. However, do not give up. Rather, proceed with the study of the exercises and examples and some of the later material, confident that your perseverance will be rewarded with a deeper understanding of the ideas involved.

Contents

CHAPTER TWO

The Fundamental Group 55

CHAPTER THREE

Free Groups and Free Products of Groups 85

CHAPTER FOUR

Seifert and Van Kampen Theorem on the Fundamental Group of the Union of Two Spaces. Applications 113

Algebraic Topology: An Introduction

CHAPTER ONE

Two-Dimensional Manifolds

1 Introduction

The topological concept of a surface or 2-dimensional manifold is a mathematical abstraction of the familiar concept of a surface made of paper, sheet metal, plastic, or some other thin material. A surface or 2-dimensional manifold is a topological space with the same local properties as the familiar plane of Euclidean geometry. An intelligent bug crawling on a surface could not distinguish it from a plane if he had a limited range of visibility.

The natural, higher dimensional analog of a surface is an n-dimensional manifold, which is a topological space with the same local properties as Euclidean n-space. Because they occur frequently and have application in many other branches of mathematics, manifolds are certainly one of the most important classes of topological spaces. Although we define and give some examples of n-dimensional manifolds for any positive integer n, we devote most of this chapter to the case $n = 2$. Because there is a classification theorem for compact 2-manifolds, our knowledge of 2-dimensional manifolds is incomparably more complete than our knowledge of the higher dimensional cases. This classification theorem gives a simple procedure for obtaining all possible compact 2-manifolds. Moreover, there are simple computable invariants which enable us to decide whether or not any two compact 2-manifolds are homeomorphic. This may be considered an ideal theorem. Much research in topology has been directed toward the development of analogous classification theorems for other situations. Unfortunately, no such theorem is known for compact 3-manifolds, and logicians have shown that we cannot even hope for such a complete result for n-manifolds, $n \geqq 4$. Nevertheless, the theory of higher dimensional manifolds is currently a very active field of mathematical research, and will probably continue to be so for a long time to come.

We shall use the material developed in this chapter, especially in Sections 1–8, later in the book.

2 Definition and examples of n-manifolds

Assume n is a positive integer. An *n-dimensional manifold* is a Hausdorff space (i.e., a space that satisfies the T_2 separation axiom) such that each point has an open neighborhood homeomorphic to the open n-dimensional disc U^n ($= \{x \in \mathbf{R}^n : |x| < 1\}$). Usually we shall say "n-manifold" for short.

Examples

2.1 Euclidean n-space $\mathbf{R}^n$ is obviously an n-dimensional manifold. We can easily prove that the unit n-dimensional sphere

$$S^n = \{x \in \mathbf{R}^{n+1} : |x| = 1\}$$

is an n-manifold. For the point $x = (1, 0, \ldots, 0)$, the set $\{(x_1, \ldots, x_{n+1}) \in S^n : x_1 > 0\}$ is a neighborhood with the required properties, as we see by orthogonal projection on the hyperplane in $\mathbf{R}^{n+1}$ defined by $x_1 = 0$. For any other point $x \in S^n$, there is a rotation carrying x into the point $(1, 0, \ldots, 0)$. Such a rotation is a homeomorphism of S^n onto itself; hence, x also has the required kind of neighborhood.

2.2 If M^n is any n-dimensional manifold, then any open subset of M^n is also an n-dimensional manifold. The proof is immediate.

2.3 If M is an m-dimensional manifold and N is an n-dimensional manifold, then the product space $M \times N$ is an $(m + n)$-dimensional manifold. This follows from the fact that $U^m \times U^n$ is homeomorphic to U^{m+n}. To prove this, note that, for any positive integer k, U^k is homeomorphic to $\mathbf{R}^k$, and $\mathbf{R}^m \times \mathbf{R}^n$ is homeomorphic to $\mathbf{R}^{m+n}$.

In addition to the 2-sphere S^2, the reader can easily give examples of many other subsets of Euclidean 3-space $\mathbf{R}^3$, which are 2-manifolds, e.g., surfaces of revolution, etc.

As these examples show, an n-manifold may be either connected or disconnected, compact or noncompact. In any case, an n-manifold is always locally compact.

What is not so obvious is that a connected manifold need not satisfy the second axiom of countability (i.e., it need not have a countable base). The simplest example is the "long line."[1] Such manifolds are usually regarded as pathological, and we shall restrict our attention to manifolds with a countable base.

Note that in our definition we required that a manifold satisfy the Hausdorff separation axiom. We must make this requirement explicit

---

[1] See *General Topology* by J. L. Kelley. Princeton, N.J.: Van Nostrand, 1955. Exercise L, p. 164.

in the definition because it is *not* a consequence of the other conditions imposed on a manifold. We leave it to the reader to construct examples of non-Hausdorff spaces, such that each point has an open neighborhood homeomorphic to U^n for $n = 1$ or 2.

3 Orientable vs. nonorientable manifolds

Connected n-manifolds for $n > 1$ are divided into two kinds: orientable and nonorientable. We will try to make the distinction clear without striving for mathematical precision.

Consider first the case where $n = 2$. We can prescribe in various ways an orientation for the Euclidean plane $\mathbf{R}^2$ or, more generally, for a small region in the plane. For example, we could designate which of the two possible kinds of coordinate systems in the plane is to be considered a right-handed coordinate system and which is to be considered a left-handed coordinate system. Another way would be to prescribe which direction of rotation in the plane about a point is to be considered the positive direction and which is to be considered the negative direction. Let us imagine an intelligent bug or some 2-dimensional being constrained to move in the plane; once he decides on a choice of orientation at any point in the plane, he can carry this choice with him as he moves about. If two such bugs agree on an orientation at a given point in the plane, and one of them travels on a long trip to some distant point in the plane and eventually returns to his starting point, both bugs will still agree on their choice of orientation.

Similar considerations apply to any connected 2-dimensional manifold because each point has a neighborhood homeomorphic to a neighborhood of a point in the plane. Here our two hypothetical bugs agree on a choice of orientation at a given point. It is possible, however, that after one of them returns from a long trip to some distant point on the manifold, they may find they are no longer in agreement. This phenomenon can occur even though both were meticulously careful about keeping an accurate check of the positive orientation.

The simplest example of a 2-dimensional manifold exhibiting this phenomenon is the well-known Möbius strip. As the reader probably knows, we construct a model of a Möbius strip by taking a long, narrow rectangular strip of paper and gluing the ends together with a half twist (see Figure 1.1). Mathematically, a Möbius strip is a topological space that is described as follows. Let X denote the following rectangle in the plane:

$$X = \{(x, y) \in \mathbf{R}^2 : -10 \leqq x \leqq +10, \ -1 < y < +1\}.$$

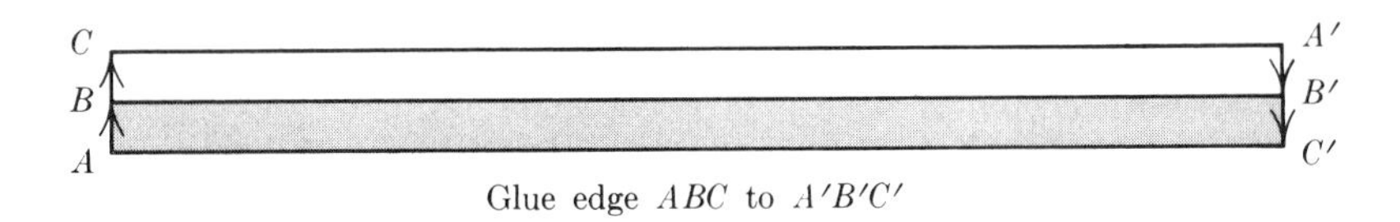

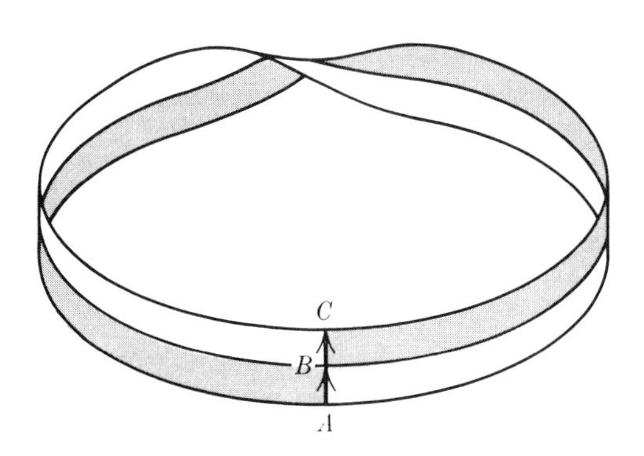

FIGURE 1.1 Constructing a Möbius strip.

We then form a quotient space of X by identifying the points $(10, y)$ and $(-10, -y)$ for $-1 < y < +1$. (See Appendix A for information on quotient spaces.) Note that the two boundaries of the rectangle corresponding to $y = +1$ and $y = -1$ were omitted. This omission is crucial; otherwise the result would not be a manifold (it would be a "manifold with boundary," a concept we will take up later in this chapter). Alternatively, we could specify a certain subset of $\mathbf{R}^3$ which is homeomorphic to the quotient space just described.

However we define the Möbius strip, the center line of the rectangular strip becomes a circle after the gluing or identification of the two ends. We leave it to the reader to verify that if our imaginary bug started out at any point on this circle with a definite choice of orientation and carried this orientation with him around the circle once, he would come back to his initial point with his original orientation reversed. We will call such a path in a manifold an *orientation-reversing* path. A closed path that does not have this property will be called an *orientation-preserving* path. For example, any closed path in the plane is orientation preserving.

A connected 2-manifold is defined to be *orientable* if every closed path is orientation preserving; a connected 2-manifold is *nonorientable* if there is at least one orientation-reversing path.

We now consider the orientability of 3-manifolds. We can specify an orientation of Euclidean 3-space or a small region thereof by designating which type of coordinate system is to be considered right handed and which type is to be considered left handed. An alternative method would be to specify which type of helix or screw thread is to be designated as right handed and which kind is to be left handed. We can now describe a closed path in a 3-manifold as *orientation preserving* or *orientation reversing*, depending on whether or not a traveler who traverses the path comes back to his initial point with his initial choice of right and left unchanged. If our universe were nonorientable, then an astronaut who made a journey along some orientation-reversing path would return to earth with the right and left sides of his body interchanged: His heart would now be on the right side of his chest, etc.

There is a 3-dimensional generalization of the Möbius strip which furnishes a particularly simple example of a nonorientable 3-manifold. Let

$$X = \{(x, y, z) \in \mathbf{R}^3 : -10 \leqq x \leqq +10,\ -1 < y < +1,\ -1 < z < +1\}.$$

Form a quotient space of X by identifying the points $(10, y, z)$ and $(-10, -y, z)$ for $-1 < y < +1$ and $-1 < z < +1$. This space may also be considered the product of an ordinary 2-dimensional Möbius strip with the open interval $\{z \in \mathbf{R} : -1 < z < +1\}$. In any case, the segment $-10 \leqq x \leqq +10$ of the x axis becomes a circle under the identification, and we leave it to the reader to convince himself that this circle is an orientation-reversing path in the resulting 3-manifold.

To make analogous definitions for n-dimensional manifolds, we must first be able to distinguish between two kinds of coordinate systems in Euclidean n-space. This distinction can be made as follows. If we have given two coordinate systems, then any point x will have coordinates $(x_1, \ldots, x_n)$ and $(x_1', \ldots, x_n')$ in the two systems, and these coordinates will be related by equations of the following type:

$$x_i' = \sum_{j=1}^{n} a_{ij}x_j + b_i, \qquad i = 1, 2, \ldots, n. \tag{1.3-1}$$

Here the a_{ij}'s and b_i's are real numbers that do not depend on the choice of the point x. Furthermore, it is well known that the determinant of the a_{ij}'s,

$$\begin{vmatrix} a_{11} & a_{12} & \cdots & a_{1n} \\ a_{21} & a_{22} & \cdots & a_{2n} \\ \vdots & \vdots & & \vdots \\ a_{n1} & a_{n2} & \cdots & a_{nn} \end{vmatrix},$$

is nonzero. We call these two coordinate systems *of the same class* if this determinant is >0. From standard properties of the determinant of a system of linear equations such as (1.3-1), it follows that the relation being "of the same class" is an equivalence relation between coordinate systems in $\mathbf{R}^n$, and that there are exactly two equivalence classes. To choose an orientation of $\mathbf{R}^n$ is to choose one of these two equivalence classes of coordinate systems as the preferred class. We may designate such a preferred coordinate system by some adjective such as "positive" or "right handed."

Once the preferred class of coordinate systems is chosen, an orientation-preserving or an orientation-reversing path in a connected n-dimensional manifold is defined in essentially the same way as for 2- and 3-dimensional manifolds. The only difference is that we do not have much geometric intuition to guide us in the higher dimensional cases. In a complete mathematical development of the subject it is necessary to go into much more detail to achieve mathematical rigor.

In any case, it is possible to define the concepts of orientability and nonorientability for the case of connected n-dimensional manifolds. Euclidean n-space $\mathbf{R}^n$ and the n-sphere S^n are examples of orientable n-manifolds. We can easily define an n-dimensional generalization of the Möbius strip, which is a nonorientable n-dimensional manifold. It is homeomorphic to the product of an ordinary 2-dimensional Möbius strip and an $(n - 2)$-dimensional open disc U^{n-2}.

In the remainder of this chapter, we shall be mainly concerned with 2-dimensional manifolds; hence, we shall not go any further into these topics.

4 Examples of compact, connected 2-manifolds

To save words, from now on we shall refer to a connected 2-manifold as a *surface*. The simplest example of a compact surface is the 2-sphere S^2; another important example is the *torus*. A torus may be roughly described as any surface homeomorphic to the surface of a doughnut or of a solid ring. It may be defined more precisely as

(a) Any topological space homeomorphic to the product of two circles, $S^1 \times S^1$.

(b) Any topological space homeomorphic to the following subset of $\mathbf{R}^3$:

$$\{(x, y, z) \in \mathbf{R}^3 : [(x^2 + y^2)^{1/2} - 2]^2 + z^2 = 1\}.$$

[This is the set obtained by rotating the circle $(x - 2)^2 + z^2 = 1$ in the xz plane about the z axis.]

(c) Let X denote the unit square in the plane $\mathbf{R}^2$:

$$\{(x, y) \in \mathbf{R}^2 : 0 \leqq x \leqq 1, 0 \leqq y \leqq 1\}.$$

Then, a torus is any space homeomorphic to the quotient space of X obtained by identifying opposite sides of the square X according to the following rules. The points $(0, y)$ and $(1, y)$ are to be identified for $0 \leqq y \leqq 1$, and the points $(x, 0)$ and $(x, 1)$ are to be identified for $0 \leqq x \leqq 1$.

We will find it convenient to indicate symbolically how such identifications are to be made by a diagram such as Figure 1.2. Sides that are to be identified are labeled with the same letter of the alphabet, and the identifications should be made so that the directions indicated by the arrows agree.

We leave it to the reader to prove that the topological spaces described in (a), (b), and (c) are actually homeomorphic. The reader should also convince himself that a torus is orientable.

Our next example of a compact surface is the *real projective plane* (referred to as the *projective plane* for short). It is a compact, nonorientable surface. Because it is not homeomorphic to any subset of Euclidean 3-space, the projective plane is much more difficult to visualize than the 2-sphere or the torus.

Definition The quotient space of the 2-sphere S^2 obtained by identifying every pair of diametrically opposite points is called a *projective plane*. We shall also refer to any space homeomorphic to this quotient space as a projective plane.

For readers who have studied projective geometry, we shall explain why this surface is called the real projective plane. Such a reader will recall that, in the study of projective plane geometry, a point has "homogeneous" coordinates (x_0, x_1, x_2), where x_0, x_1, and x_2 are real numbers, at least one of which is $\neq 0$. The term "homogeneous" means (x_0, x_1, x_2) and (x'_0, x'_1, x'_2) represent the same point if and only if there exists a real number λ (of necessity $\neq 0$) such that

$$x_i = \lambda x'_i, \qquad i = 0, 1, 2.$$

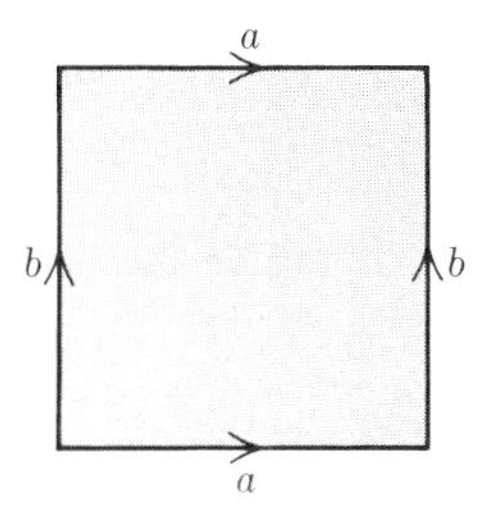

FIGURE 1.2 Construction of a torus.

If we interpret (x_0, x_1, x_2) as the ordinary Euclidean coordinates of a point in $\mathbf{R}^3$, then we see that (x_0, x_1, x_2) and (x_0', x_1', x_2') represent the same point in the projective plane if and only if they are on the same line through the origin. Thus, we may reinterpret a point of the projective plane as a line through the origin in $\mathbf{R}^3$. The next question is, how shall we topologize the set of all lines through the origin in $\mathbf{R}^3$? Perhaps the easiest way is to note that each line through the origin in $\mathbf{R}^3$ intersects the unit sphere S^2 in a pair of diametrically opposite points. This leads to the above definition.

Let $H = \{(x, y, z) \in S^2 : z \geqq 0\}$ denote the closed upper hemisphere of S^2. It is clear that, of each diametrically opposite pair of points in S^2, at least one point lies in H. If both points lie in H, then they are on the equator, which is the boundary of H. Thus, we could also define the projective plane[2] as the quotient space of H obtained by identifying diametrically opposite points on the boundary of H. As H is obviously homeomorphic to the closed unit disc E^2 in the plane,

$$E^2 = \{(x, y) \in \mathbf{R}^2 : x^2 + y^2 \leqq 1\},$$

the quotient space of E^2 obtained by identifying diametrically opposite points on the boundary is a projective plane. For E^2 we could substitute any homeomorphic space, e.g., a square. Thus, a projective plane is obtained by identifying the opposite sides of a square as indicated in Figure 1.3. The reader should compare this with the construction of a torus in Figure 1.2.

The projective plane is easily seen to be nonorientable; in fact, it contains a subset homeomorphic to a Möbius strip.

We shall now describe how to give many additional examples of compact surfaces by forming what are called connected sums. Let S_1 and S_2 be disjoint surfaces. Their *connected sum*, denoted by $S_1 \# S_2$, is

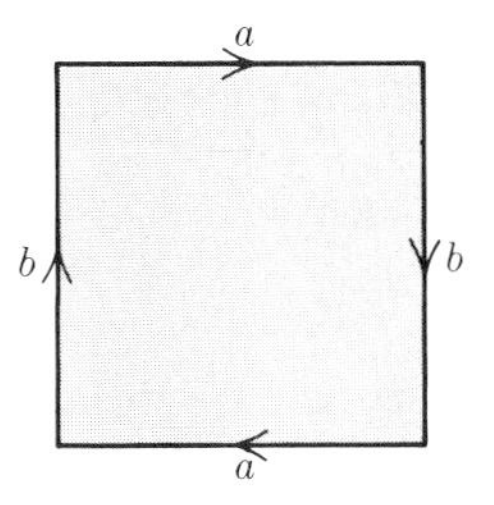

FIGURE 1.3 Construction of a projective plane from a square.

[2] For a rigorous justification of this assertion, we must use Proposition 4.2 in Appendix A, which is applicable because the natural map from S^2 to the projective plane is a closed map, and H is a closed subset of S^2.

formed by cutting a small circular hole in each surface, and then gluing the two surfaces together along the boundaries of the holes. To be precise, we choose subsets $D_1 \subset S_1$ and $D_2 \subset S_2$ such that D_1 and D_2 are closed discs (i.e., homeomorphic to E^2). Let S_i' denote the complement of the interior of D_i in S_i for $i = 1$ and 2. Choose a homeomorphism h of the boundary circle of D_1 onto the boundary of D_2. Then $S_1 \# S_2$ is the quotient space of $S_1' \cup S_2'$ obtained by identifying the points x and $h(x)$ for all points x in the boundary of D_1. It is clear that $S_1 \# S_2$ is a surface. It seems plausible, and can be proved rigorously, that the topological type of $S_1 \# S_2$ does not depend on the choice of the discs D_1 and D_2 or the choice of the homeomorphism h.

Examples

4.1 If S_2 is a 2-sphere, then $S_1 \# S_2$ is homeomorphic to S_1.

4.2 If S_1 and S_2 are both tori, then $S_1 \# S_2$ is homeomorphic to the surface of a block that has two holes drilled through it. (It is assumed, of course, that the holes are not so close together that their boundaries touch or intersect.)

4.3 If S_1 and S_2 are projective planes, then $S^1 \# S^2$ is a "Klein Bottle," i.e., homeomorphic to the surface obtained by identifying the opposite sides of a square as shown in Figure 1.4. We may prove this by the "cut and paste" technique, as follows. If S_i is a projective plane, and D_i is a closed disc such that $D_i \subset S_i$, then S_i', the complement of the interior of D_i, is homeomorphic to a Möbius strip (including the boundary). In fact, if we think of S_i as the space obtained by identification of the diametrically opposite points on the boundary of the unit disc E^2 in $\mathbf{R}^2$, then we can choose D_i to be the image of the set $\{(x, y) \in E^2 : |y| \geqq \frac{1}{2}\}$ under the identification, and the truth of the assertion is clear. From this it follows that $S_1 \# S_2$ is obtained by gluing together two Möbius strips along their boundaries. On the other hand, Figure 1.5 shows how to cut a Klein Bottle so as to obtain two Möbius strips. We cut along the lines AB' and BA'; under the identification, this cut becomes a circle.

We will now consider some properties of this operation of forming connected sums.

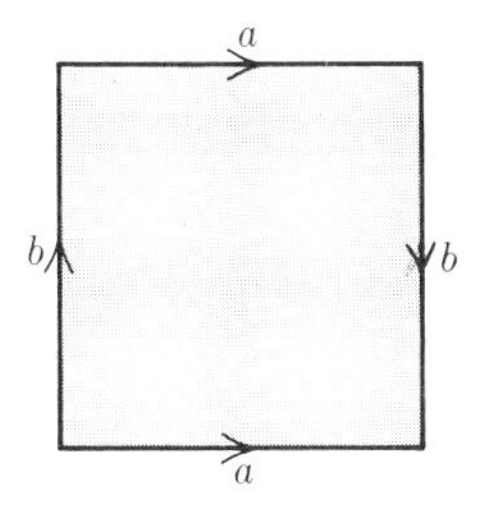

FIGURE 1.4 Construction of a Klein bottle from a square.

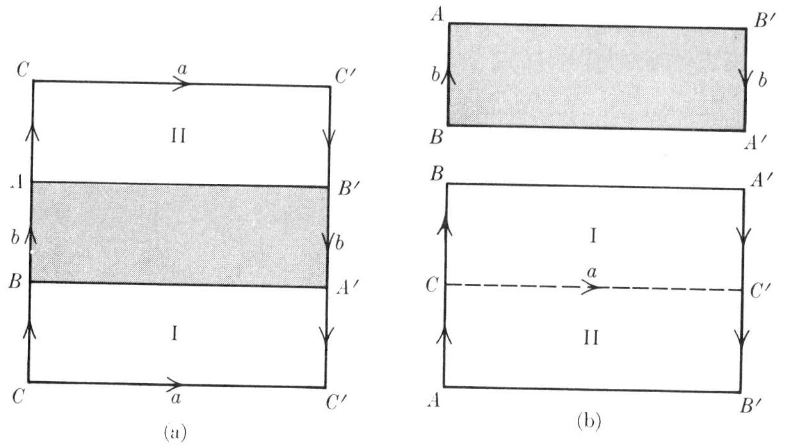

FIGURE 1.5 The Klein bottle is the union of two Möbius strips.

It is clear from our definitions that there is no distinction between $S_1 \# S_2$ and $S_2 \# S_1$; i.e., the operation is commutative. It is not difficult to see that the manifolds $(S_1 \# S_2) \# S_3$ and $S_1 \# (S_2 \# S_3)$ are homeomorphic. Thus, we see that the connected sum is a commutative, associative operation on the set of homeomorphism types of compact surfaces. Moreover, Example 4.1 shows the sphere is a unit or neutral element for this operation. We must not jump to the conclusion that the set of homeomorphism classes of compact surfaces forms a group under this operation: There are no inverses. It only forms what is called a semigroup.

The connected sum of two orientable manifolds is again orientable. On the other hand, if either S_1 or S_2 is nonorientable, then so is $S_1 \# S_2$.

5 Statement of the classification theorem for compact surfaces

In the preceding section we have seen how examples of compact surfaces can be constructed by forming connected sums of various numbers of tori and/or projective planes. Our main theorem asserts that these examples exhaust all the possibilities. In fact, it is even a slightly stronger statement, in that we do not need to consider surfaces that are connected sums of both tori and projective planes.

Theorem 5.1 *Any compact surface is either homeomorphic to a sphere, or to a connected sum of tori, or to a connected sum of projective planes.*

As preparation for the proof, we shall describe what might be called a "canonical form" for a connected sum of tori or projective planes.

Recall our description of a torus as a square with the opposite sides identified (see Figure 1.2). We can obtain an analogous description of the connected sum of two tori as follows. Represent each of the tori T_1 and T_2 as a square with opposite sides identified as shown in Figure 1.6(a). Note that all four vertices of each square are identified to a single point of the corresponding torus. To form their connected sum, we must first cut out a circular hole in each torus, and we can do this in any way that

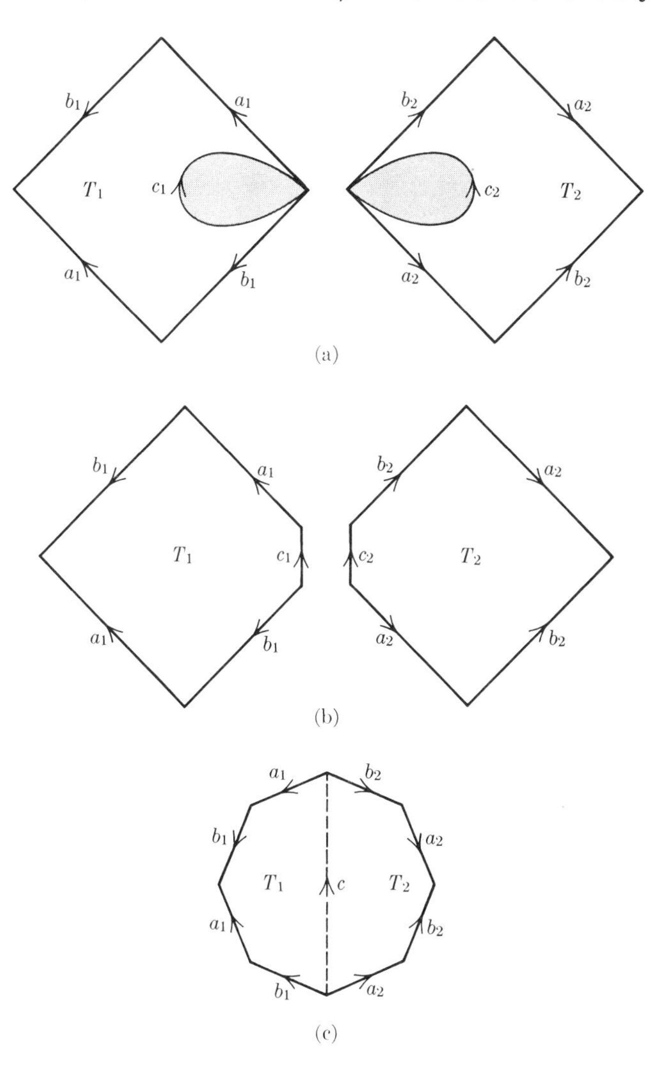

FIGURE 1.6 (a) Two disjoint tori, T_1 and T_2. (b) Disjoint tori with holes cut out. (c) After gluing together.

we wish. It is convenient to cut out the regions shaded in the diagrams. The boundaries of the holes are labeled c_1 and c_2, and they are to be identified as indicated by the arrows. We can also represent the complement of the holes in the two tori by the pentagons shown in Figure 1.6(b), because the indicated edge identifications imply that the two end points of the segment c_i are to be identified, $i = 1, 2$. We now identify the segments c_1 and c_2; the result is the octagon in Figure 1.6(c), in which the sides are to be identified in pairs, as indicated. Note that all eight vertices of this octagon are to be identified to a single point in $T_1 \# T_2$.

This octagon with the edges identified in pairs is our desired "canonical form" for the connected sum of two tori. By repeating this process, we can show that the connected sum of three tori is the quotient space of the 12-gon shown in Figure 1.7, where the edges are to be identified in pairs as indicated. It should now be clear how to prove by induction that the connected sum of n tori is homeomorphic to the quotient space of a $4n$-gon whose edges are to be identified in pairs according to a scheme, the precise description of which is left to the reader.

Next, we must consider the analogous procedure for the connected sum of projective planes. We have considered the projective plane as the quotient space of a circular disc; diametrically opposite points on the boundary are to be identified. By choosing a pair of diametrically opposite points on the boundary as vertices, the circumference of the disc is divided into two segments. Thus, we can regard the projective plane as obtained from a 2-gon by identification of the two edges; see Figure 1.8.

Figure 1.9 shows how to obtain a representation of the connected sum of two projective planes as a square with the edges identified in pairs.

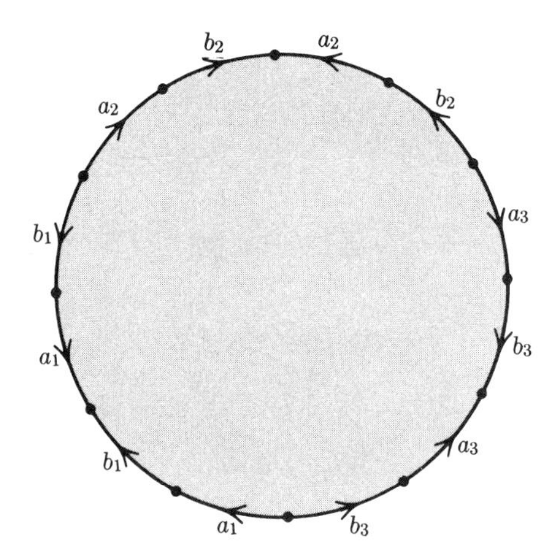

FIGURE 1.7 The connected sum of 3 tori is obtained by identifying the edges of a 12-gon in pairs as shown.

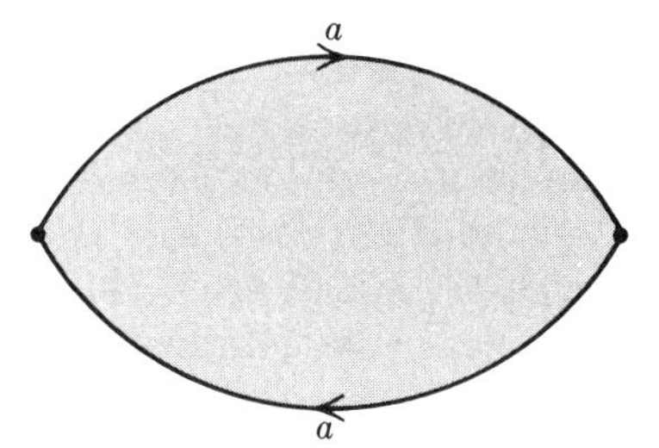

FIGURE 1.8 The projective plane is obtained by identifying opposite edges of a 2-gon.

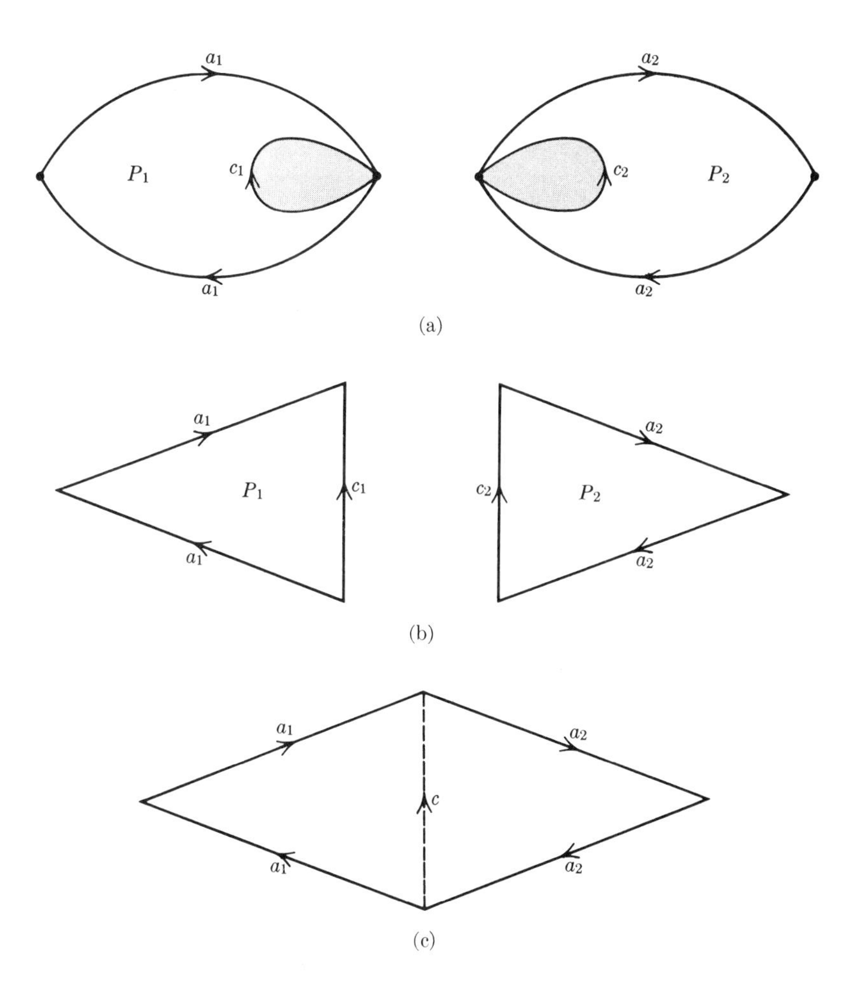

FIGURE 1.9 (a) Two disjoint projective planes, P_1 and P_2. (b) Disjoint projective planes with holes cut out. (c) After gluing together.

The method is basically the same as that used to obtain a representation of the connected sum of two tori as a quotient space of an octagon (Figure 1.6). By repeating this process, we see that the connected sum of three projective planes is the quotient space of a hexagon with the sides identified in pairs as indicated in Figure 1.10. By a rather obvious induction, we can prove that, for any positive integer n, the connected sum of n projective planes is the quotient space of a $2n$-gon with the sides identified in pairs according to a certain scheme. Note that all the vertices of this polygon are identified to one point.

It remains to represent the sphere as the quotient space of a polygon with the sides identified in pairs. We can do this as shown in Figure 1.11. We can think of a sphere with a zipper on it, like a purse; when the zipper is opened, the purse can be flattened out.

Thus, we have shown how each of the compact surfaces mentioned in Theorem 5.1 can be considered as the quotient space of a polygon with

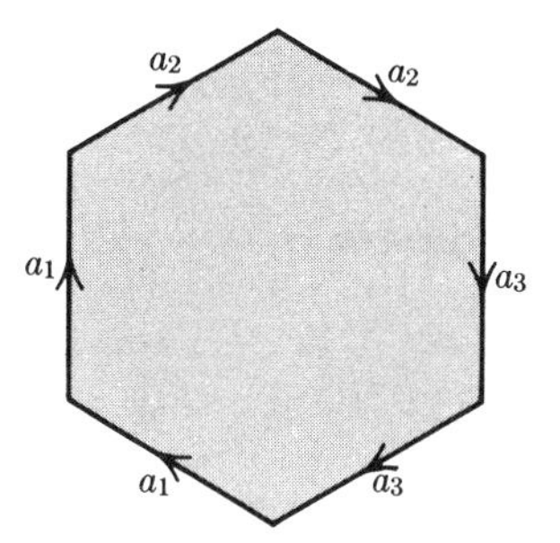

FIGURE 1.10 Construction of the connected sum of three projective planes by identifying the sides of a hexagon in pairs.

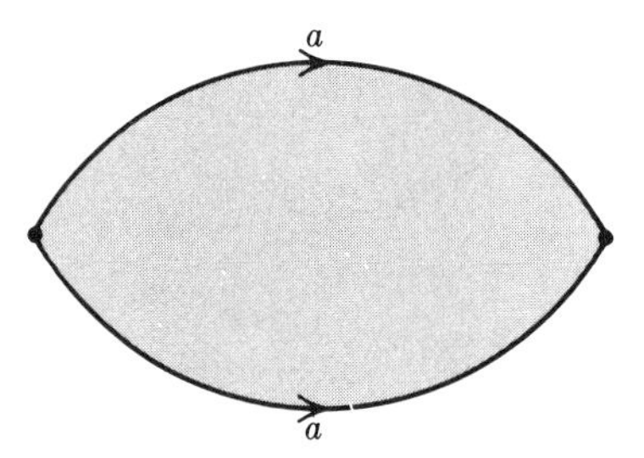

FIGURE 1.11 The sphere is a quotient space of a 2-gon with edges identified as shown.

the edges identified in pairs. We now introduce a rather obvious and convenient method of indicating precisely which paired edges are to be identified in such a polygon. Consider the diagram which indicates how the edges are identified; starting at a definite vertex, proceed around the boundary of the polygon, recording the letters assigned to the different sides in succession. If the arrow on a side points in the *same* direction that we are going around the boundary, then we write the letter for that side with no exponent (or the exponent $+1$). On the other hand, if the arrow points in the *opposite* direction, then we write the letter for that side with the exponent -1. For example, in Figures 1.7 and 1.10 the identifications are precisely indicated by the symbols

$$a_1b_1a_1^{-1}b_1^{-1}a_2b_2a_2^{-1}b_2^{-1}a_3b_3a_3^{-1}b_3^{-1} \quad \text{and} \quad a_1a_1a_2a_2a_3a_3.$$

In each case we started at the bottom vertex of the diagram and read clockwise around the boundary. It is clear that such a symbol unambiguously describes the identifications; on the other hand, in writing the symbol corresponding to a given diagram, we can start at any vertex, and proceed either clockwise or counterclockwise around the boundary.

We summarize our results by writing the symbols corresponding to each of the surfaces mentioned in Theorem 5.1.

(a) The sphere: aa^{-1}.

(b) The connected sum of n tori:

$$a_1b_1a_1^{-1}b_1^{-1}a_2b_2a_2^{-1}b_2^{-1} \ \ldots \ a_nb_na_n^{-1}b_n^{-1}.$$

(c) The connected sum of n projective planes:

$$a_1a_1a_2a_2 \ \ldots \ a_na_n.$$

Exercise

5.1 Let P be a polygon with an even number of sides. Suppose that the sides are identified in pairs in accordance with any symbol whatsoever. Prove that the quotient space is a compact surface.

6 Triangulations of compact surfaces

To prove Theorem 5.1, we must assume that the given surface is triangulated, i.e., divided up into triangles which fit together nicely. We can easily visualize the surface of the earth divided into triangular regions, and such a subdivision is very useful in the study of compact surfaces in general.

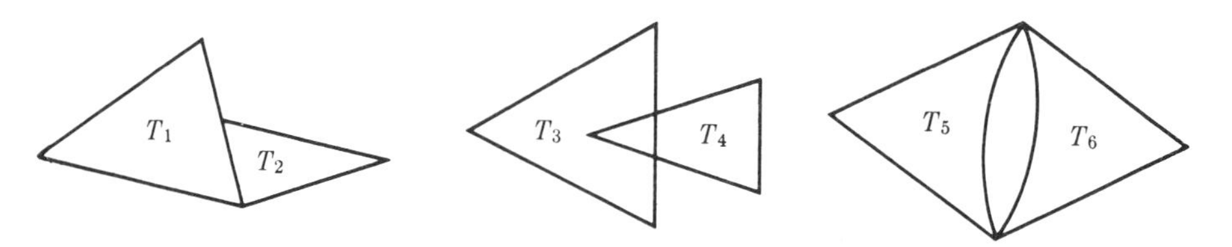

FIGURE 1.12 Some types of intersection forbidden in a triangulation.

Definition A *triangulation* of a compact surface S consists of a finite family of closed subsets $\{T_1, T_2, \ldots, T_n\}$ that cover S, and a family of homeomorphisms $\varphi_i : T_i' \to T_i$, $i = 1, \ldots, n$, where each T_i' is a triangle in the plane $\mathbf{R}^2$ (i.e., a compact subset of $\mathbf{R}^2$ bounded by three distinct straight lines). The subsets T_i are called "triangles." The subsets of T_i that are the images of the vertices and edges of the triangle T_i' under φ_i are also called "vertices" and "edges," respectively. Finally, it is required that any two distinct triangles, T_i and T_j, either be disjoint, have a single vertex in common, or have one entire edge in common.

Perhaps the conditions in the definition are clarified by Figure 1.12, which shows three *unallowable* types of intersection of triangles.

Given any compact surface S, it seems plausible that there should exist a triangulation of S. A rigorous proof of this fact (first given by T. Radó in 1925) requires the use of a strong form of the Jordan curve theorem. Although it is not difficult, the proof is tedious, and we will not repeat it here.

We can regard a triangulated surface as having been constructed by gluing together the various triangles in a certain way, much as we put together a jigsaw puzzle or build a wall of bricks. Because two different triangles cannot have the same vertices we can specify completely a triangulation of a surface by numbering the vertices, and then listing which triples of vertices are vertices of a triangle. Such a list of triangles completely determines the surface together with the given triangulation up to homeomorphism.

Examples

6.1 The surface of an ordinary tetrahedron in Euclidean 3-space is homeomorphic to the sphere S^2; moreover, the four triangles satisfy all the conditions for a triangulation of S^2. In this case there are four vertices, and every triple of vertices is the set of vertices of a triangle. No other triangulation of any surface can have this property.

6.2 In Figure 1.13 we show a triangulation of the projective plane, considered as the space obtained by identifying diametrically opposite points on the bound-

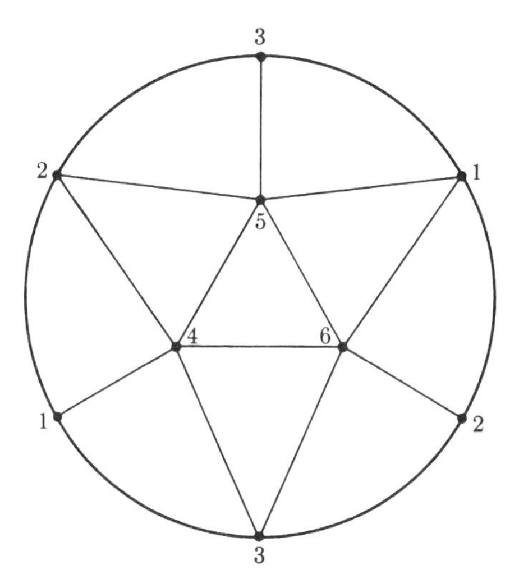

FIGURE 1.13 A triangulation of the projective plane.

ary of a disc. The vertices are numbered from 1 to 6, and there are the following 10 triangles:

124	245
235	135
156	126
236	346
134	456

6.3 In Figure 1.14 we show a triangulation of a torus, regarded as a square

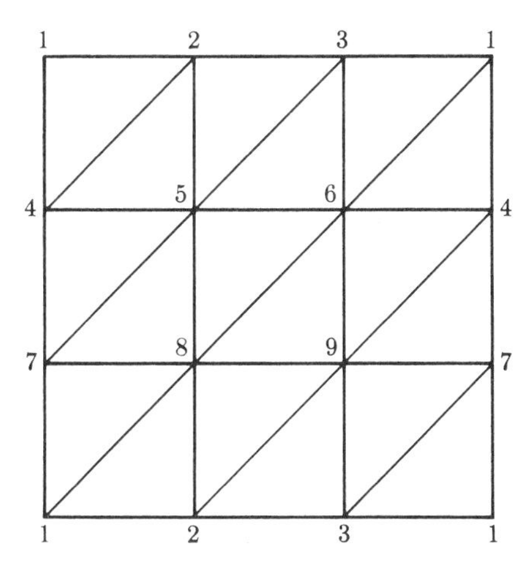

FIGURE 1.14 A triangulation of a torus.

with the opposite sides identified. There are 9 vertices, and the following 18 triangles:

124	245	235
356	361	146
457	578	658
689	649	479
187	128	289
239	379	137

We conclude our discussion of triangulations by noting that any triangulation of a compact surface satisfies the following two conditions:

(1) Each edge is an edge of exactly two triangles.

(2) Let v be a vertex of a triangulation. Then we may arrange the set of all triangles with v as a vertex in cyclic order, T_0, T_1, T_2, $\ldots$, T_{n-1}, $T_n = T_0$, such that T_i and T_{i+1} have an edge in common for $0 \leqq i \leqq n - 1$.

The truth of (1) follows from the fact that each point on the edge in question must have an open neighborhood homeomorphic to the open disc U^2. If an edge were an edge of only one triangle or more than two triangles, this would not be possible. The rigorous proof of this last assertion would take us rather far afield; however, its plausibility cannot be disputed.

Condition (2) can be demonstrated as follows. The fact that the set of all the triangles with v as a vertex can be divided into several disjoint subsets, such that the triangles in each subset can be arranged in cyclic order as described, is an easy consequence of condition (1). However, if there were more than one such subset, then the requirement that v have a neighborhood homeomorphic to U^2 would be violated. We shall not attempt a rigorous proof of this last assertion.

7 Proof of Theorem 5.1

Let S be a compact surface. We shall demonstrate Theorem 5.1 by proving that S is homeomorphic to a polygon with the edges identified in pairs as indicated by one of the symbols listed at the end of Section 5.

First step. From the discussion in the preceding section, we may assume that S is triangulated. Denote the number of triangles by n. We assert that we can number the triangles T_1, T_2, $\ldots$, T_n, so that the triangle T_i has an edge e_i in common with at least one of the triangles T_1, $\ldots$, T_{i-1}, $2 \leqq i \leqq n$. To prove this assertion, label any of the triangles T_1; for T_2 choose any triangle that has an edge in common with T_1, for T_3 choose any triangle that has an edge in common with T_1 or T_2, etc. If at any stage we could not continue this process, then we would have two sets of triangles $\{T_1, \ldots, T_k\}$, and $\{T_{k+1}, \ldots, T_n\}$ such that no triangle in the first set would have an edge or vertex in common with any triangle of the second set. But this would give a partition of S into two disjoint nonempty closed sets, contrary to the assumption that S was connected.

We now use this ordering of the triangles, $T_1, T_2, \ldots, T_n$, together with the choice of edges $e_2, e_3, \ldots, e_n$, to construct a "model" of the surface S in the Euclidean plane; this model will be a polygon whose sides are to be identified in pairs. Recall that for each triangle T_i there exists an ordinary Euclidean triangle T_i' in $\mathbf{R}^2$ and a homeomorphism φ_i of T_i' onto T_i. We can assume that the triangles $T_1', T_2', \ldots, T_n'$ are pairwise disjoint; if they are not, we can translate some of them to various other parts of the plane $\mathbf{R}^2$. Let

$$T' = \bigcup_{i=1}^{n} T_i';$$

then T' is a compact subset of $\mathbf{R}^2$. Define a map $\varphi : T' \to S$ by $\varphi \mid T_i' = \varphi_i$; the map φ is obviously continuous and onto. Because T' is compact and S is a Hausdorff space, φ is a closed map, and hence S has the quotient topology determined by φ (see Section 1 of Appendix A). This is a rigorous mathematical statement of our intuitive idea that S is obtained by gluing the triangles $T_1, T_2, \ldots$ together along the appropriate edges.

The polygon we desire will be constructed as a quotient space of T'. Consider any of the edges e_i, $2 \leqq i \leqq n$. By assumption, e_i is an edge of the triangle T_i and one other triangle T_j, for which $1 \leqq j < i$. Therefore, $\varphi^{-1}(e_i)$ consists of an edge of the triangle T_i' and an edge of the triangle T_j'. We identify these two edges of the triangles T_i' and T_j' by identifying points which map onto the same point of e_i (speaking intuitively, we glue together the triangles T_i' and T_j'). We make these identifications for each of the edges $e_2, e_3, \ldots, e_n$. Let D denote the resulting quotient space of T'. It is clear that the map $\varphi : T' \to S$ induces a map ψ of D onto S, and that S has the quotient topology induced by ψ (because D is compact and S is Hausdorff, ψ is a closed map).

We now assert that topologically D is a closed disc. The proof depends on two facts:

(a) Let E_1 and E_2 be disjoint spaces, which topologically are closed discs (i.e., they are homeomorphic to E^2). Let A_1 and A_2 be subsets of the boundary of E_1 and E_2, respectively, which are homeomorphic to the closed interval $[0, 1]$, and let $h : A_1 \to A_2$ be a definite homeomorphism. Form a quotient space of $E_1 \cup E_2$

by identifying points that correspond under h. Then, topologically, the quotient space is also a closed disc. The reader may either take this very plausible fact for granted, or construct a proof using the type of argument given in II.4. Intuitively, it means that if we glue two discs together along a common segment of their boundaries, the result is again a disc.

(b) In forming the quotient space D of T', we may either make all the identifications at once, or make the identifications corresponding to e_2, then those corresponding to e_3, etc., in succession. This is a consequence of Lemma 2.4 of Appendix A [see application (a) of this lemma].

We now use these facts to prove that D is a disc as follows. T_1' and T_2' are topologically equivalent to discs. Therefore, the quotient space of $T_1' \cup T_2'$ obtained by identifying points of $\varphi^{-1}(e_2)$ is again a disc by (a). Form a quotient space of this disc and T_3' by making the identifications corresponding to the edge e_3, etc.

It is clear that S is obtained from D by identifying certain paired edges on the boundary of D.

Example

7.1 Figure 1.15 shows an easily visualized example. The surface of a cube has been triangulated by dividing each face by a diagonal into two triangles.

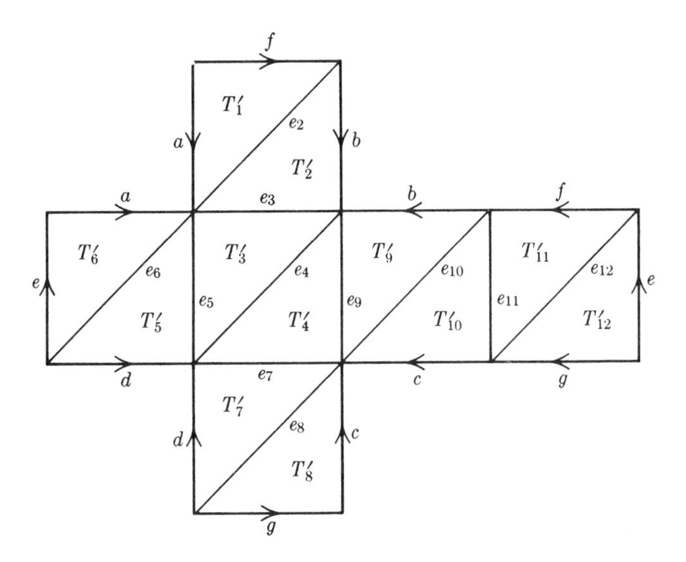

FIGURE 1.15 Example illustrating the first step of the proof of Theorem 5.1.

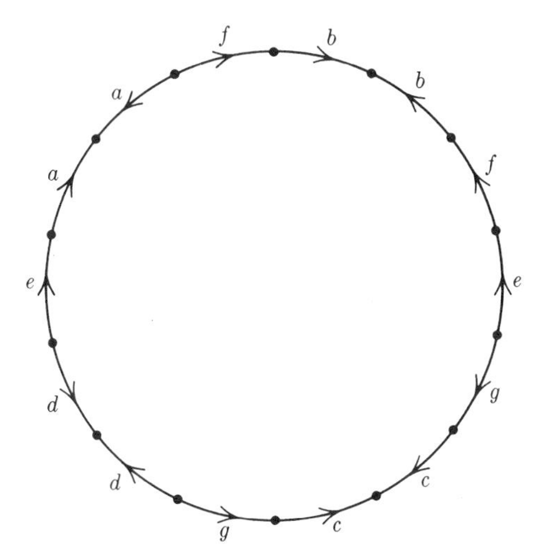

FIGURE 1.16 Simplified version of polygon shown in Figure 1.15.

The resulting disc D might look like the diagram, depending, of course, on how the triangles were enumerated, and how the edges $e_2, \ldots, e_{12}$ were chosen. The edges of D that are to be identified are labeled in the usual way. At this stage, we can forget about the edges $e_2, e_3, \ldots, e_{12}$. Thus, instead of the polygon in Figure 1.15, we could work equally well with the one in Figure 1.16.

Exercises

Carry out the above process for each of the surfaces whose triangulations are given below. (NOTE: these examples will be used later.)

7.1	124	236	134	246		
	367	347	469	459		
	698	678	457	259		
	289	578	358	125		
	238	135				
7.2	123	234	341	412		
7.3	123	234	345	451	512	
	136	246	356	416	526	
7.4	124	235	346	457	561	672
	713	134	245	356	467	571
	126	237				
7.5	123	256	341	451		
	156	268	357	468		
	167	275	374	476		
	172	283	385	485		

Second step. Elimination of adjacent edges of the first kind. We have now obtained a polygon D whose edges have to be identified in pairs to obtain the given surface S. These identifications may be indicated by the appropriate symbol; e.g., in Figure 1.16, the identifications are described by

$$aa^{-1}fbb^{-1}f^{-1}e^{-1}gcc^{-1}g^{-1}dd^{-1}e.$$

If the letter designating a certain pair of edges occurs with *both* exponents, $+1$ and -1, in the symbol, then we will call that pair of edges a pair of the *first kind;* otherwise, the pair is of the *second kind.* For example, in Figure 1.16, all seven pairs are of the first kind.

We wish to show that an adjacent pair of edges of the first kind can be eliminated, provided there are at least four edges in all. This is easily seen from the sequence of diagrams in Figure 1.17. We can continue this process until all such pairs are eliminated, or until we obtain a polygon with only two sides. In the latter case, this polygon, whose symbol will be aa or aa^{-1}, must be a projective plane or a sphere, and we have completed the proof. Otherwise, we proceed as follows.

Third step. Transformation to a polygon such that all vertices must be identified to a single vertex. Although the edges of our polygon must be

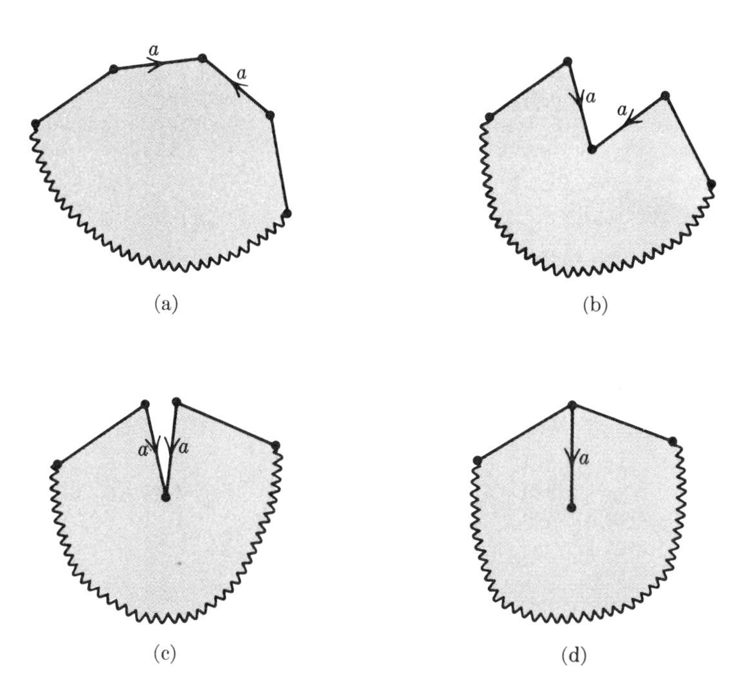

FIGURE 1.17 Elimination of an adjacent pair of edges of the first kind.

identified in pairs, the vertices may be identified in sets of one, two, three, four, ... Let us call two vertices of the polygon *equivalent* if and only if they are to be identified. For example, the reader can easily verify that in Figure 1.16 there are eight different equivalence classes of vertices. Some equivalence classes contain only one vertex, whereas other classes contain two or three vertices.

Assume we have carried out step two as far as possible. We wish to prove we can transform our polygon into another polygon with all its vertices belonging to one equivalence class.

Suppose there are at least two different equivalence classes of vertices. Then, the polygon must have an adjacent pair of vertices which are nonequivalent. Label these vertices P and Q. Figure 1.18 shows how to proceed. As P and Q are nonequivalent, and we have carried out step two, it follows that sides a and b are *not* to be identified. Make a cut along the line labeled c, from the vertex labeled Q to the other vertex of the edge a (i.e., to the vertex of edge a, which is distinct from P). Then, glue the two edges labeled a together. A new polygon with one less vertex in the equivalence class of P and one more vertex in the equivalence class of Q results. If possible, perform step two again. Then carry out step three to reduce the number of vertices in the equivalence class of P still further, then do step two again. Continue alternately doing step three and step two until the equivalence class of P is eliminated entirely. If more than one equivalence class of vertices remains, we can repeat this procedure to reduce the number by one. If we continue in this manner, we ultimately obtain a polygon such that all the vertices are to be identified to a single vertex.

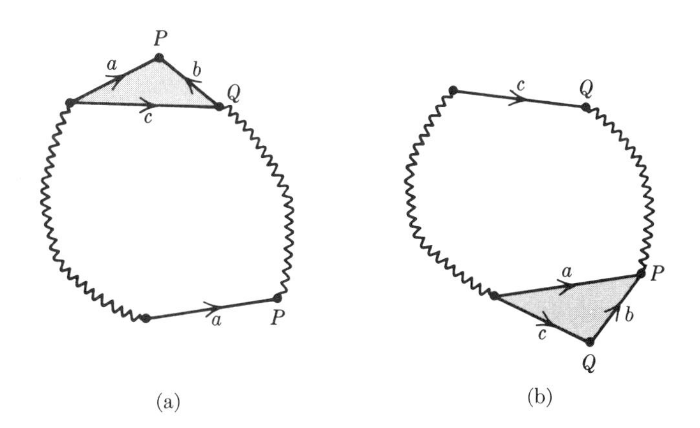

FIGURE 1.18 Third step in the proof of Theorem 5.1.

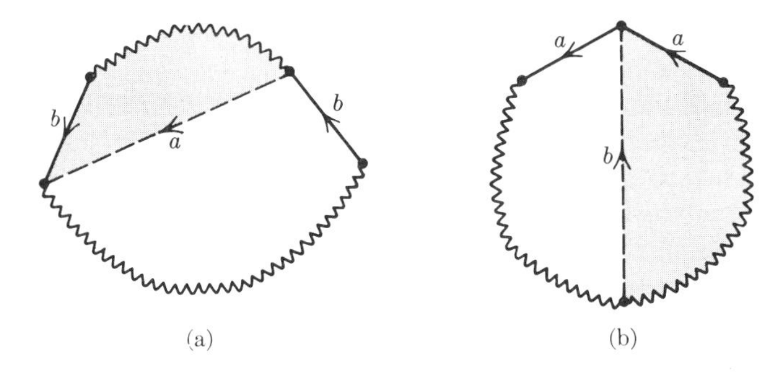

FIGURE 1.19 Fourth step in the proof of Theorem 5.1.

Fourth step. How to make any pair of edges of the second kind adjacent. We wish to show that our surface can be transformed so that any pair of edges of the second kind are adjacent to each other. Suppose we have a pair of edges of the second kind which are nonadjacent, as in Figure 1.19(a). Cut along the dotted line labeled a and paste together along b. As shown in Figure 1.19(b), the two edges are now adjacent.

Continue this process until all pairs of edges of the second kind are adjacent. If there are no pairs of the first kind, we are finished, because the symbol of the polygon must then be of the form $a_1a_1a_2a_2 \ldots a_na_n$, and hence S is the connected sum of n projective planes.

Assume to the contrary that at this stage there is at least one pair of edges of the first kind, each of which is labeled with the letter c. Then we assert that there is at least one other pair of edges of the first kind such that these two pairs separate one another; i.e., edges from the two pairs occur alternately as we proceed around the boundary of the polygon (hence, the symbol must be of the form $c \ldots d \ldots c^{-1} \ldots d^{-1} \ldots$, where the dots denote the possible occurrence of other letters).

To prove this assertion, assume that the edges labeled c are not separated by any other pair of the first kind. Then our polygon has the appearance indicated in Figure 1.20. Here A and B each designate a whole sequence of edges. The important point is that any edge in A must be identified with another edge in A, and similarly for B. No edge in A is to be identified with an edge in B. But this contradicts the fact that the initial and final vertices of either edge labeled "c" are to be identified, in view of step number three.

Fifth step. Pairs of the first kind. Suppose, then, that we have two pairs of the first kind which separate each other as described (see Figure 1.21). We shall show that we can transform the polygon so that the four sides in question are consecutive around the perimeter of the polygon.

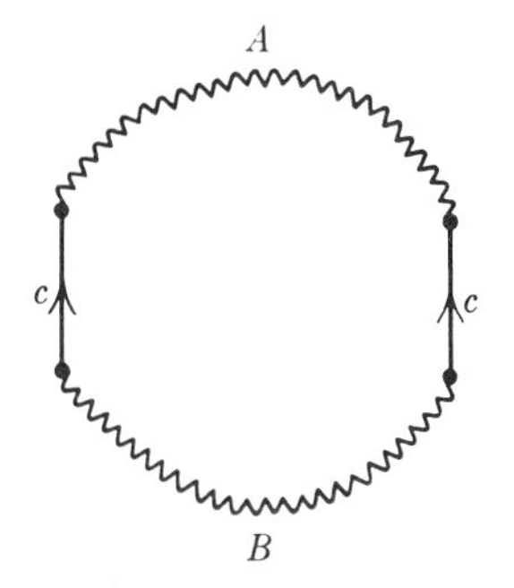

FIGURE 1.20 A pair of edges of the first kind.

First, cut along c and paste together along b to obtain Figure 1.21(b). Then, cut along d and paste together along a to obtain (c), as desired.

Continue this process until all pairs of the first kind are in adjacent groups of four, as $cdc^{-1}d^{-1}$ in Figure 1.21(c). If there are no pairs of the

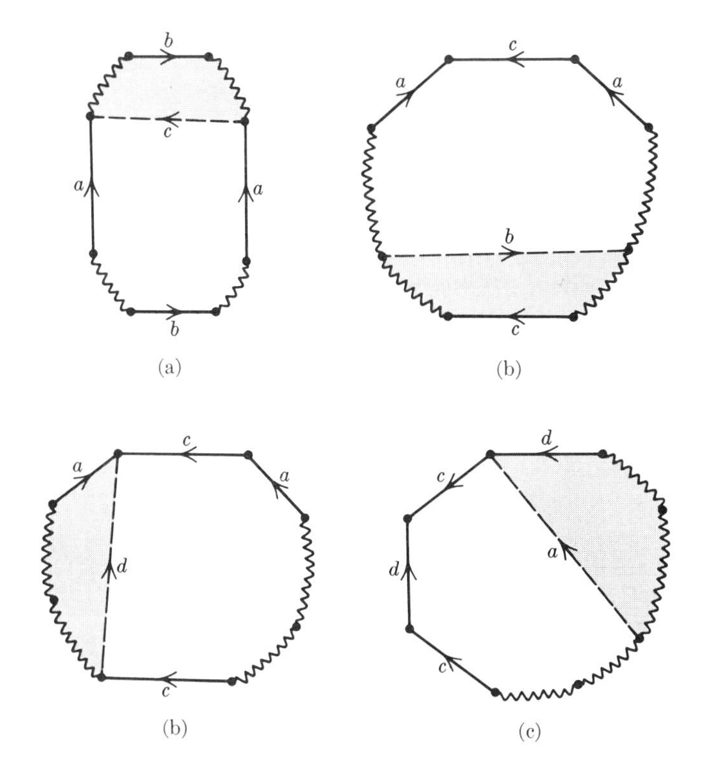

FIGURE 1.21 Fifth step in the proof of Theorem 5.1.

second kind, this leads to the desired result because, in that case, the symbol must be of the form

$$a_1b_1a_1^{-1}b_1^{-1}a_2b_2a_2^{-1}b_2^{-1} \dots a_nb_na_n^{-1}b_n^{-1}$$

and the surface is the connected sum of n tori.

It remains to treat the case in which there are pairs of both the first and second kind at this stage. The key to the situation is the following rather surprising lemma:

Lemma 7.1 *The connected sum of a torus and a projective plane is homeomorphic to the connected sum of three projective planes.*

PROOF: We have remarked that the connected sum of two projective planes is homeomorphic to a Klein Bottle (see Example 4.3). Thus, we must prove that the connected sum of a projective plane and a torus is homeomorphic to the connected sum of a projective plane and a Klein Bottle. To do this it will be convenient to give an alternative construction for a connected sum of any surface S with a torus or a Klein Bottle. We can represent the torus and Klein Bottle as rectangles with opposite sides identified as shown in Figure 1.22. To form the connected sum, we first cut out the disc that is shaded in the diagrams, cut a similar hole in S, and glue the boundary of the hole in the torus or Klein Bottle to the boundary of the hole in S. However, instead of gluing on the entire torus or Klein Bottle in one step, we may do it in two stages: First, glue on the part of the torus or Klein Bottle that is the image of the rectangle $ABB'A'$ under the identification, and then glue on the rest of the torus or Klein Bottle. In the first stage we form the connected sum of S with an open tube or cylinder. Such an open tube or cylinder is homeomorphic to a sphere with two holes cut in it, and forming the connected sum of S with a sphere does not change anything. Thus, the space resulting from the first stage is homeomorphic to the original surface S with two holes

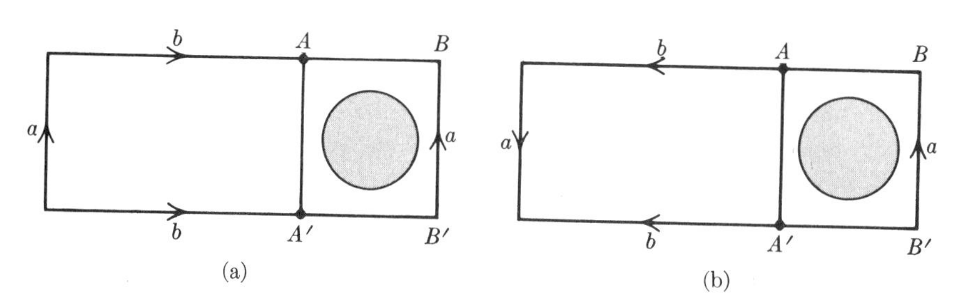

FIGURE 1.22 (a) Torus with hole. (b) Klein bottle with hole.

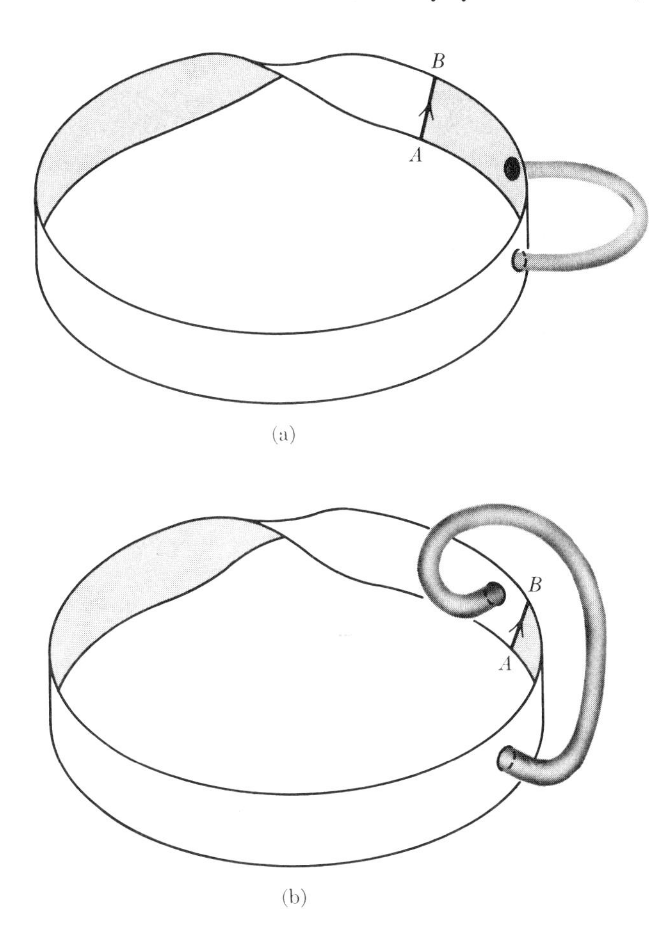

FIGURE 1.23 (a) Connected sum of a Möbius strip and a torus. (b) Connected sum of a Möbius strip and a Klein bottle.

cut in it. In the second stage we then connect the boundaries of these two holes with a tube that is the remainder of the torus or Klein Bottle. The difference between the two cases depends on whether we connect the boundaries so they will have the same or opposite orientations. This is illustrated in Figure 1.23, where S is a Möbius strip.

We now assert that the two spaces shown in Figure 1.23(a) and (b) (i.e., the connected sum of a Möbius strip with a torus and a Klein Bottle, respectively) are homeomorphic. To see this, imagine that we cut each of these topological spaces along the lines AB. In each case, the result is the connected sum of a rectangle and a torus, with the two ends of the

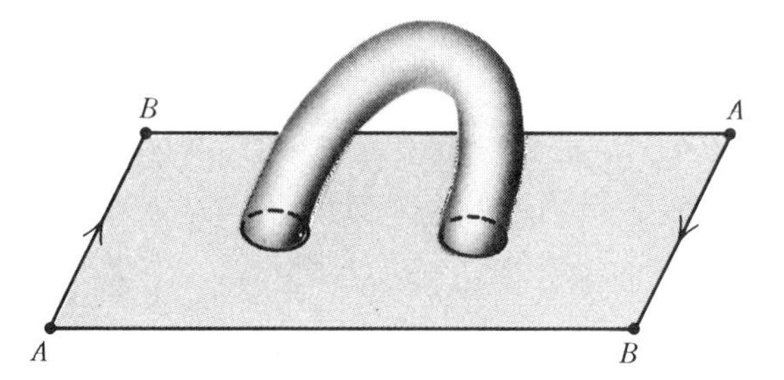

FIGURE 1.24 The result of cutting the spaces shown in Figure 1.23 along the line AB.

rectangle to be identified with a twist, as shown in Figure 1.24. Hence, the two spaces are homeomorphic.

As stated previously, we obtain the projective plane by gluing the boundary of a disc to the boundary of a Möbius strip. As the spaces shown in Figure 1.23 are homeomorphic, so are the spaces obtained by gluing a disc on the boundary of each. Thus, the connected sum of a projective plane and a torus is homeomorphic to the connected sum of a projective plane and a Klein Bottle, as was to be proved.

It should be clear that this lemma takes care of the remaining case. For, assume that after the fifth step has been completed, the polygon has m pairs ($m > 0$) of the second kind such that the two edges of each pair are adjacent, and n quadruples ($n > 0$) of sides, each quadruple consisting of two pairs of the first kind which separate each other. Then, the surface is the connected sum of m projective planes and n tori, which by the lemma is homeomorphic to the connected sum of $m + 2n$ projective planes. This completes the proof of Theorem 5.1.

Exercises

7.6 Carry out each of the above steps for the examples given in Exercises 7.1–7.5.

It is clear that we can also work the process described above backwards; whenever there are three pairs of the second kind, we can replace them by one pair of the second kind and two pairs of the first kind. Alternatively, we can apply Lemma 7.1 to any connected sum of which three or more of the summands are projective planes. The following alternative form of Theorem 5.1, which may be preferable in some cases, results.

Theorem 7.2 *Any compact, orientable surface is homeomorphic to a sphere or a connected sum of tori. Any compact, nonorientable surface is homeomorphic to the connected sum of either a projective plane or Klein Bottle and a compact, orientable surface.*

8 The Euler characteristic of a surface

Although we have shown that any compact surface is homeomorphic to a sphere, a sum of tori, or a sum of projective planes, we do not know that all these are topologically different. It is conceivable that there exist positive integers m and n, $m \neq n$, such that the sum of m tori is homeomorphic to the sum of n tori. To show that this cannot happen, we introduce a numerical invariant called the *Euler characteristic.*

First, we define the Euler characteristic of a triangulated surface. Let M be a compact surface with triangulation $\{T_1, \ldots, T_n\}$. Let

v = total number of vertices of M,

e = total number of edges of M,

t = total number of triangles (in this case, $t = n$).

Then,

$$\chi(M) = v - e + t$$

is called the *Euler characteristic* of M.

Examples

8.1 Figure 1.25 suggests uniform methods of triangulating the sphere, torus, and projective plane so that we may make the number of triangles as large as we please. Using such triangulations, the reader should verify that the Euler characteristics of the sphere, torus, and projective plane are 2, 0, and 1, respectively. He should also verify that the Euler characteristics are independent of the number of vertical and horizontal dividing lines in the diagrams for the sphere and torus, and of the number of radial lines or concentric circles in the case of the diagram for the projective plane.

Consideration of these and other examples suggests that $\chi(M)$ depends only on M, not on the triangulation chosen. We wish to suggest a method of proving this. To do this, we shall allow subdivisions of M into arbitrary polygons, not just triangles. These polygons may have any number

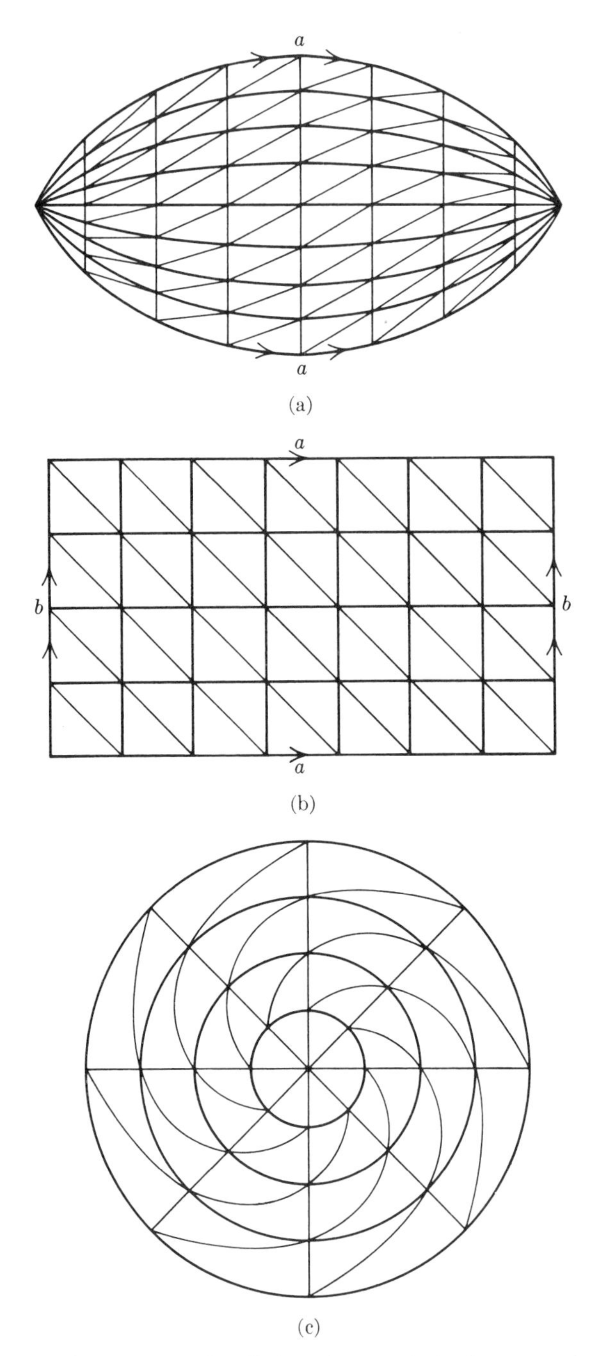

FIGURE 1.25 Computing the Euler characteristic from a triangulation. (a) Sphere. (b) Torus. (c) Projective plane.

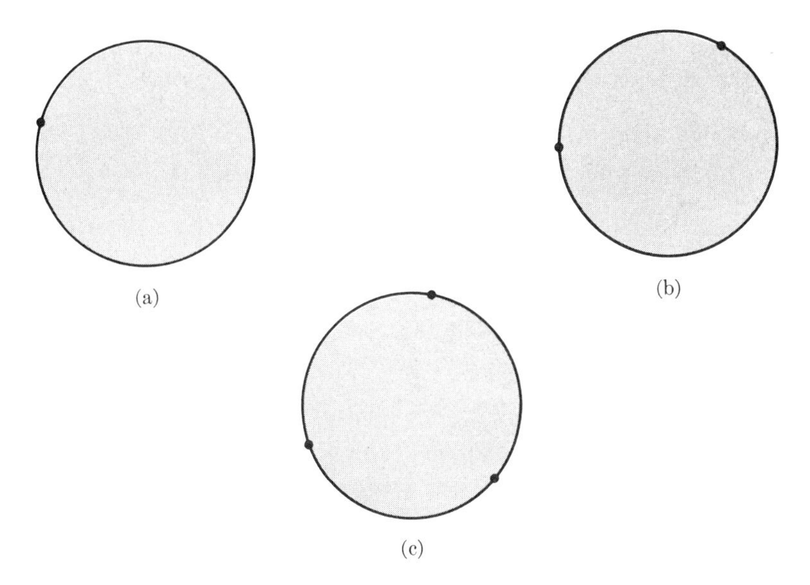

FIGURE 1.26 (a) A 1-gon. (b) A 2-gon. (c) A 3-gon.

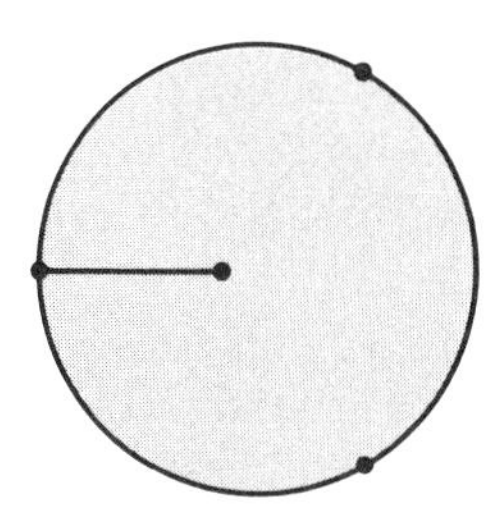

FIGURE 1.27 An allowable kind of edge.

n of sides and vertices, $n \geqq 1$ (see Figure 1.26). We shall also allow for the possibility of edges that do not subdivide a region, as in Figure 1.27. In any case, the interior of each polygonal region is required to be homeomorphic to an open disc, and each edge is required to be homeomorphic to an open interval of the real line, once the vertices are removed (the closure of each edge shall be homeomorphic to a closed interval or a circle). Finally, the number of vertices, edges, and polygonal regions will be finite. As before, we define the Euler characteristic of such a subdivision of a compact surface M to be

$$\chi(M) = (\text{no. of vertices}) - (\text{no. of edges}) + (\text{no. of regions}).$$

It is now easily shown that the Euler characteristic is invariant under the following processes:

(a) Subdividing an edge by adding a new vertex at an interior point (or, inversely, if only two edges meet at a given vertex, we can amalgamate the two edges into one and eliminate the vertex).
(b) Subdividing an n-gon, $n \geqq 1$, by connecting two of the vertices by a new edge (or, inversely, amalgamating two regions into one by removing an edge).
(c) Introducing a new edge and vertex running into a region, as shown in Figure 1.27 (or, inversely, eliminating such an edge and vertex).

The invariance of the Euler characteristic would now follow if it could be shown that we could get from any one triangulation (or subdivision) to any other by a finite sequence of "moves" of types (a), (b), and (c). Suppose we have two triangulations

$$\mathfrak{T} = \{T_1, T_2, \ldots, T_m\}$$

$$\mathfrak{T}' = \{T_1', T_2', \ldots, T_n'\}$$

of a given surface. If the intersection of any edge of the triangulation $\mathfrak{T}$ with any edge of the triangulation $\mathfrak{T}'$ consists of a finite number of points and a finite number of closed intervals, then it is easily seen that we can get from the triangulation $\mathfrak{T}$ to the triangulation $\mathfrak{T}'$ in a finite number of such moves; the details are left to the reader. However, it may happen that an edge of $\mathfrak{T}$ intersects an edge of $\mathfrak{T}'$ in an infinite number of points, like the following two curves in the xy plane:

$$\{(x, y) : y = 0 \quad \text{and} \quad -1 \leqq x \leqq +1\},$$

$$\{(x, y) : y = x \sin \frac{1}{x} \quad \text{and} \quad 0 < |x| \leqq 1\} \cup \{(0, 0)\}.$$

If this is the case, it is clearly impossible to get from the triangulation $\mathfrak{T}$ to the triangulation $\mathfrak{T}'$ by any finite number of moves. It appears plausible that we could always avoid such a situation by "moving" one of the edges slightly. This is true, and can be proved rigorously. However, we do not attempt such a proof here, for several reasons: (a) The details are tedious and involved. (b) The Euler characteristic can be defined for more general spaces than surfaces and its invariance can be proven by the use of homology theory. In these more general circumstances, the type of proof we have suggested is not possible. (c) We shall use the Euler characteristic to distinguish between compact surfaces. We shall achieve this purpose with complete rigor in a later chapter by the use of the fundamental group.

Proposition 8.1 *Let S_1 and S_2 be compact surfaces. The Euler characteristics of S_1 and S_2 and their connected sum, $S_1 \# S_2$, are related by the formula*

$$\chi(S_1 \# S_2) = \chi(S_1) + \chi(S_2) - 2.$$

PROOF: The proof is very simple; assume S_1 and S_2 are triangulated. Form their connected sum by removing from each the interior of a triangle, and then identifying edges and vertices of the boundaries of the removed triangles. The formula then follows by counting vertices, edges, and triangles before and after the formation of the connected sum.

Using this theorem, and an obvious induction, starting from the known results for the sphere, torus, and projective plane, we obtain the following values for the Euler characteristics of the various possible compact surfaces:

Surface	*Euler characteristic*
Sphere	2
Connected sum of n tori	$2 - 2n$
Connected sum of n projective planes	$2 - n$
Connected sum of projective plane and n tori	$1 - 2n$
Connected sum of Klein Bottle and n tori	$-2n$

Note that the Euler characteristic of an orientable surface is always even, whereas for a nonorientable surface it may be either odd or even.

Assuming the topological invariance of the Euler characteristic and Theorem 5.1, we have the following important result:

Theorem 8.2 *Let S_1 and S_2 be compact surfaces. Then, S_1 and S_2 are homeomorphic if and only if their Euler characteristics are equal and both are orientable or both are nonorientable.*

This is a topological theorem par excellence; it reduces the classification problem for compact surfaces to the determination of the orientability and Euler characteristic, both problems usually readily soluble. Moreover, Theorem 5.1 makes clear what are all possible compact surfaces.

Such a complete classification of any class of topological spaces is very rare. No corresponding theorem is known for compact 3-manifolds, and for 4-manifolds it has been proven (roughly speaking) that no such result is possible.

We close this section by giving some standard terminology. A surface that is the connected sum of n tori or n projective planes is said to be of *genus* n, whereas a sphere is of *genus* 0. The following relation holds between the genus g and the Euler characteristic χ of a compact surface:

$$g = \begin{cases} \frac{1}{2}(2 - \chi) & \text{in the orientable case,} \\ 2 - \chi & \text{in the nonorientable case.} \end{cases}$$

Exercises

8.1 For over 2000 years it has been known that there are only five regular polyhedra, namely, the regular tetrahedron, cube, octahedron, dodecahedron, and icosahedron. Prove this by considering subdivisions of the sphere into n-gons (n fixed) such that exactly m edges meet at each vertex (m fixed, $m, n \geqq 3$). Use the fact that $\chi(S^2) = 2$.

8.2 For any *triangulation* of a compact surface, show that

$$3t = 2e$$

$$e = 3(v - \chi)$$

$$v \geqq \tfrac{1}{2}(7 + \sqrt{49 - 24\chi}).$$

In the case of the sphere, projective plane, and torus, what are the minimum values of the numbers v, e, and t? (Here, t, e, and v denote the number of triangles, edges, and vertices, respectively.)

8.3 In how many pieces do n great circles, no three of which pass through a common point, dissect a sphere?

8.4 (a) The sides of a regular octagon are identified in pairs in such a way as to obtain a compact surface. Prove that the Euler characteristic of this surface is $\geqq -2$.

(b) Prove that any surface (orientable or nonorientable) of Euler characteristic $\geqq -2$ can be obtained by suitably identifying in pairs the sides of a regular octagon.

8.5 Prove that it is not possible to subdivide the surface of a sphere into regions, each of which has 6 sides (i.e., it is a hexagon) and such that distinct regions have no more than one side in common.

8.6 Let S_1 be a surface that is the sum of m tori, $m \geqq 1$, and let S_2 be a surface that is the sum of n projective planes, $n \geqq 1$. Suppose two holes are cut in each of these surfaces, and the two surfaces are then glued together along the boundaries of the holes. What surface is obtained by this process?

8.7 What surface is represented by a regular 10-gon with edges identified in pairs, as indicated by the symbol $abcdec^{-1}da^{-1}b^{-1}e^{-1}$? (HINT: How are the vertices identified around the boundary?)

8.8 What surface is represented by a $2n$-gon with the edges identified in pairs according to the symbol

$$a_1a_2 \ldots a_na_1^{-1}a_2^{-1} \ldots a_{n-1}^{-1}a_n?$$

8.9 What surface is represented by a $2n$-gon with the edges identified in pairs according to the symbol

$$a_1a_2 \ldots a_na_1^{-1}a_2^{-1} \ldots a_{n-1}^{-1}a_n^{-1}?$$

(HINT: The cases where n is odd and where n is even are different.)

Remark: The results of Exercises 8.8 and 8.9 together give an alternative "normal form" for the representation of a compact surface as a quotient space of polygon.

9 Manifolds with boundary

The concept of a manifold with boundary is a slight generalization of that of a manifold.

Definition An *n-dimensional manifold with boundary* is a Hausdorff space such that each point has an open neighborhood homeomorphic either to the open disc U^n or to the space

$$\{(x_1, x_2, \ldots, x_n) \in U^n : x_1 \geqq 0\}.$$

The set of all points that have an open neighborhood homeomorphic to U^n is called the *interior* of the manifold, and the set of those points p that have an open neighborhood V such that there exists a homeomorphism h of V onto $\{x \in U^n : x_1 \geqq 0\}$ with $h(p) = (0, 0, \ldots, 0)$ is called the *boundary* of the manifold.

Examples

9.1 The closed disc or ball

$$E^n = \{x \in \mathbf{R}^n : |x| \leqq 1\}$$

is an n-dimensional manifold with boundary. The sphere S^{n-1} is the boundary, and the open disc U^n is the interior.

9.2 Another example is the "half-space," $\{x \in \mathbf{R}^n : x_1 \geqq 0\}$.

9.3 The Möbius strip, as it is usually defined, is a 2-dimensional manifold with boundary.

9.4 Other examples of 2-dimensional manifolds with boundary may be obtained by removing a collection of small, open discs from a 2-dimensional manifold.

It is quite plausible and can be proved rigorously that the set of boundary points and the set of interior points are mutually disjoint. It is readily seen that the set of interior points is an open everywhere dense subset; hence, the set of boundary points is a closed set. The set of boundary points of an n-dimensional manifold is an $(n - 1)$-dimensional manifold. The interior is a noncompact n-manifold.

The reader should note that the terms "interior" and "boundary" were used in the preceding paragraphs in a sense different from that which is usual in point set topology. However, this will seldom lead to any confusion.

Examples show that a manifold with boundary may be compact or noncompact, connected or not connected. A noncompact manifold with boundary may or may not have a countable basis of open sets. In any case, it is always locally compact. We should note that the boundary of a connected manifold may be disconnected; also, the boundary of a noncompact manifold may be compact.

The concepts of orientability and nonorientability apply to manifolds with boundary exactly as in the case of manifolds. For example, a Möbius strip is a nonorientable manifold with boundary, whereas the cylinder

$$\{(x, y, z) \in \mathbf{R}^3 : x^2 + y^2 = 1, 0 \leqq z \leqq 1\}$$

is an orientable manifold with boundary.

The orientability of a manifold with boundary depends essentially on the orientability of the interior considered as a noncompact manifold. We should note that each boundary component of an n-manifold is an $(n - 1)$-manifold, which may be either orientable or nonorientable. Both cases may actually occur. It can be shown, though, that every boundary component of an orientable manifold must be orientable. On the other hand, a nonorientable manifold may have both orientable and nonorientable boundary components. For example, if P^2 denotes the projective plane and I denotes the closed unit interval, then $P^2 \times I$ is a nonorientable 3-manifold with boundary. The boundary consists of two components, $P^2 \times \{0\}$ and $P^2 \times \{1\}$. If we remove a small open 3-dimensional disc from the interior of $P^2 \times I$, then we obtain a manifold with boundary such that the boundary has three components: $P^2 \times \{0\}$, $P^2 \times \{1\}$, and a 2-sphere which is the boundary of the removed disc. Thus, two of the boundary components are nonorientable, and one of them is orientable.

Exercises

9.1 Prove that the product of a manifold and a manifold with boundary is a manifold with boundary. What is the boundary of the product?

9.2 Let P be a polygon. Assume that certain paired edges of P are identified, but not all edges of P are included among these pairs. Prove that the resulting quotient space is a compact, connected 2-manifold with boundary.

Remark on Terminology: In view of our definitions, every n-manifold (as defined in Section 1) is an n-manifold with boundary (as defined in this section). For convenience, from now on we use the following convention: When referring to a manifold with boundary, we shall mean the boundary is nonempty; otherwise we shall use the single word "manifold." Because our main interest will be in the 2-dimensional case, we shall call a connected 2-dimensional manifold with (nonempty) boundary a *bordered surface*. The word "surface" alone will continue to mean one without boundary.

10 The classification of compact, connected 2-manifolds with boundary

We have alluded to the fact that, if we select a finite number of disjoint closed discs in a compact surface and remove their interiors, we obtain a bordered surface. The number of boundary components is equal to the number of discs chosen.

Conversely, assume that M is a compact, bordered surface and that the boundary has k components $k \geqq 1$. Each boundary component is a compact, connected 1-manifold, i.e., a circle. It is clear that we obtain a compact surface M^* if we take k closed discs and glue the boundary of the ith disc to the ith component of the boundary of M. The topological type of the resulting surface M^* obviously depends only on the topological type of M. What is not so obvious is that a sort of converse statement is true: The topological type of the bordered surface M depends only on the number of its boundary components and the topological type of the surface M^* obtained by gluing a disc onto each boundary component.

We can state this in another way: If we start with a compact surface M^* and construct a bordered surface by removing the interiors of k closed discs, which are pairwise disjoint, then the location of the discs that are to be removed does not matter. The resulting manifold with boundary will be topologically the same no matter how the position of the discs is chosen. We will state this result formally as follows:

Theorem 10.1 *Let M_1 and M_2 be compact bordered surfaces; assume that their boundaries have the same number of components. Then, M_1 and M_2 are homeomorphic if and only if the surfaces M_1^* and M_2^* (obtained by gluing a disc to each boundary component) are homeomorphic.*

PROOF: We will now outline a proof of the "if" part of this theorem. It depends heavily on the classification theorem for compact surfaces. As in the demonstration of Theorem 5.1, the proof is made by showing that M_1 and M_2 are homeomorphic to a polygon with certain paired edges identified, a so-called "normal form." First, we shall explain the normal forms in detail.

(*a*) *Normal form for a sphere with k holes.* A sphere is represented by a 2-sided polygon whose edges are identified according to the symbol aa^{-1}. Cut k holes in such a polygon as shown in Figure 1.28(a) for the case where $k = 4$. Then, from a vertex on the boundary make cuts $c_1, c_2, \ldots, c_k$ to the corresponding boundary components $B_1, B_2, \ldots, B_k$. Open up each cut to obtain the

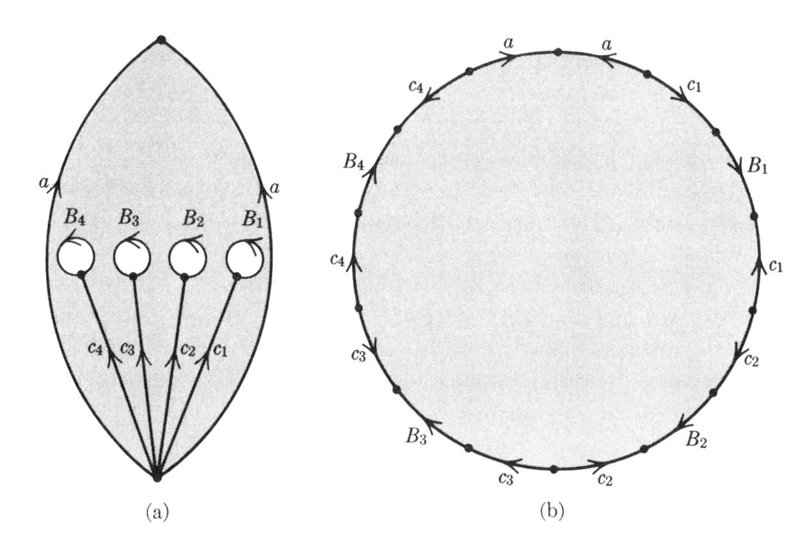

FIGURE 1.28 A sphere with four holes.

polygon shown in Figure 1.28(b). In general, we obtain a polygon whose edges are identified in accordance with the symbol

$$aa^{-1}c_1B_1c_1^{-1}c_2B_2c_2^{-1} \ldots c_kB_kc_k^{-1}.$$

(*b*) *Normal form for the connected sum of n tori with k holes.* The diagrams in Figure 1.29(a) and (b) show how to proceed when $n = 2$ and $k = 4$. It is entirely analogous to the case of a sphere with holes cut in it. The result is a polygon with $4n + 3k$ sides, which must be identified in accordance with the following symbol:

$$a_1b_1a_1^{-1}b_1^{-1} \ldots a_nb_na_n^{-1}b_n^{-1}c_1B_1c_1^{-1} \ldots c_kB_kc_k^{-1}.$$

(*c*) *Normal form for the connected sum of n projective planes with k holes.* We leave it to the reader to see that in this case we obtain a polygon with $2n + 3k$ sides, which are identified by the symbol

$$a_1a_1 \ldots a_na_nc_1B_1c_1^{-1} \ldots c_kB_kc_k^{-1}.$$

Note that in the above constructions we were careful to cut the holes in a straight line so that it was clear that we could make the cuts $c_1, \ldots, c_k$ in such a way that they would be disjoint except for one end point.

Next, we consider triangulations of compact bordered surfaces. The definition is exactly the same as that given in Section 6 for the case of compact surfaces. There is, however, one difference between the two cases which should be emphasized: In the case of a triangulation of a

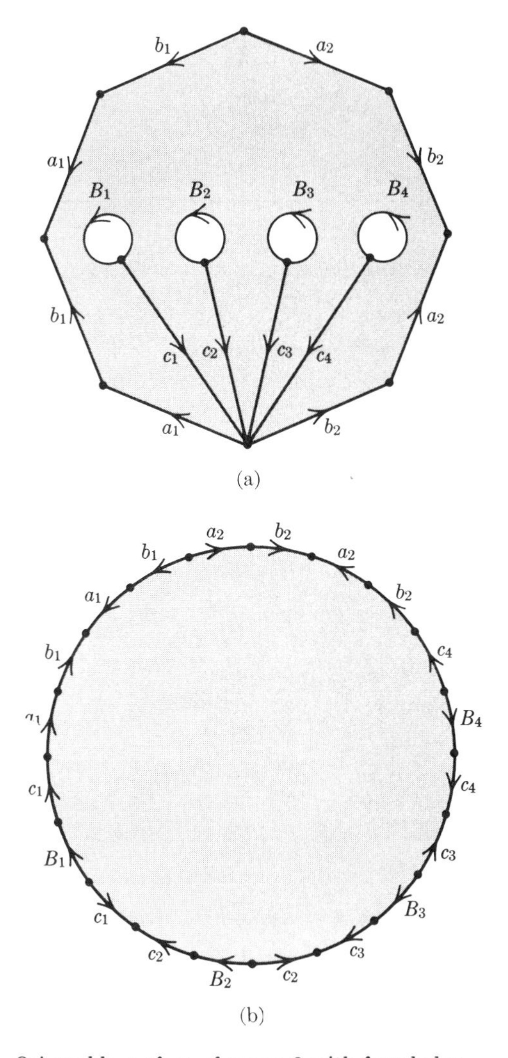

FIGURE 1.29 Orientable surface of genus 2 with four holes.

surface, every edge is an edge of exactly two triangles. However, if a bordered surface is triangulated, some edges will be edges of only one triangle. Such edges will be contained in the boundary. It is a theorem, which we will assume without proof, that every compact bordered surface may be triangulated (for a proof, see Ahlfors and Sario [1], Chapter I, Section 8).

Let M be a compact bordered surface, with a given triangulation. We assert that we may assume the triangulation satisfies the following

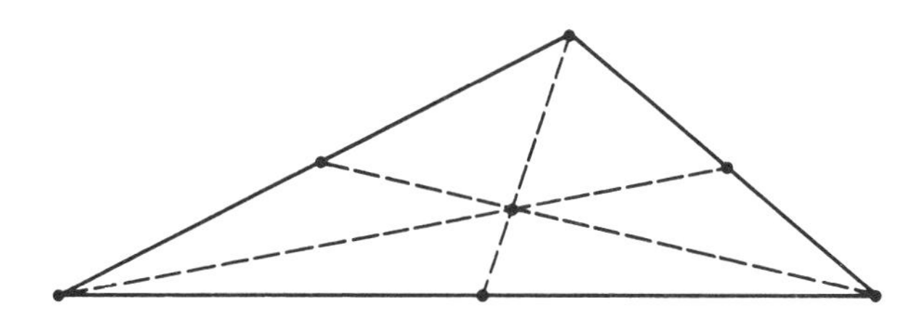

FIGURE 1.30 Barycentric subdivision of a triangle.

conditions: *No edge has both vertices contained in the boundary unless the entire edge is contained in the boundary, and no triangle has more than one edge contained in the boundary.* For, if this condition does not hold, we can achieve it by subdividing each edge into two edges and each triangle into six triangles, as shown in Figure 1.30. This process is called *barycentric subdivision.* By barycentrically subdividing once more if necessary, we may assume that our triangulation satisfies the following even stronger condition: *Let T_i and T_j be triangles each of which has one edge contained in the boundary. Then, T_i and T_j are disjoint, or else have one vertex in common, which is a vertex of the boundary.*

Let $B_1, \ldots, B_k$ denote the components of the boundary. If T is a triangle that meets one of the components B_i, then T has exactly two edges which have one vertex in B_i but do not lie in B_i. Similarly, if e is an edge that has a vertex in B_i but does lie in B_i, then e is an edge of two triangles, both of which meet B_i. It follows that the edges and triangles that meet B_i but do not lie in B_i can be arranged in one or more cycles of alternating edges and triangles,

$$T_1, e_1, T_2, e_2, \ldots, T_n, e_n, T_{n+1} = T_1,$$

such that each e_j is an edge of T_j and T_{j+1}, whereas each T_j has e_{j-1} and e_j as edges. An easy argument shows that there can only be one such cycle corresponding to each boundary component B_i. From the conditions imposed on the triangulation of M, it is clear that the union of the triangles $T_1, T_2, \ldots, T_n$ that meet B_i is homeomorphic to a polygonal region in the plane with a hole in it; Figure 1.31 illustrates how such a region might look when $n = 17$. There will be one such polygonal region P_i for each boundary component B_i, $1 \leqq i \leqq k$.

Let $T_1, \ldots, T_l$ denote the remaining triangles of the given triangulation of M not contained in any of the polygons P_i, $1 \leqq i \leqq k$. Using these k polygons and l triangles, we now perform the process described in the first step of the proof of Theorem 5.1 (as described in Section 7). A single polygon in the plane, which has k holes in its interior, and such that the exterior edges of this polygon are to be identified in pairs [see, e.g., Figure 1.29(a)] results.

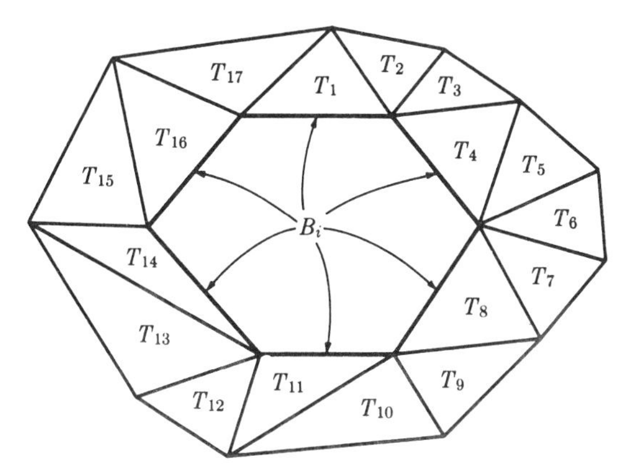

FIGURE 1.31 Triangles near the boundary component B_i.

We can now apply the remaining six steps of the proof of Theorem 5.1 to this polygon with holes. There is one proviso, however. Each of the steps requires certain processes of cutting and pasting together again. It is assumed that these cuts are made so as to avoid all of the holes. It is clear that this can always be done. It is also clear that the number of holes remains unchanged throughout all these steps.

As a result we obtain one of the three types of polygons shown in Figure 1.32. For convenience, we have taken $k = 4$ in each drawing.

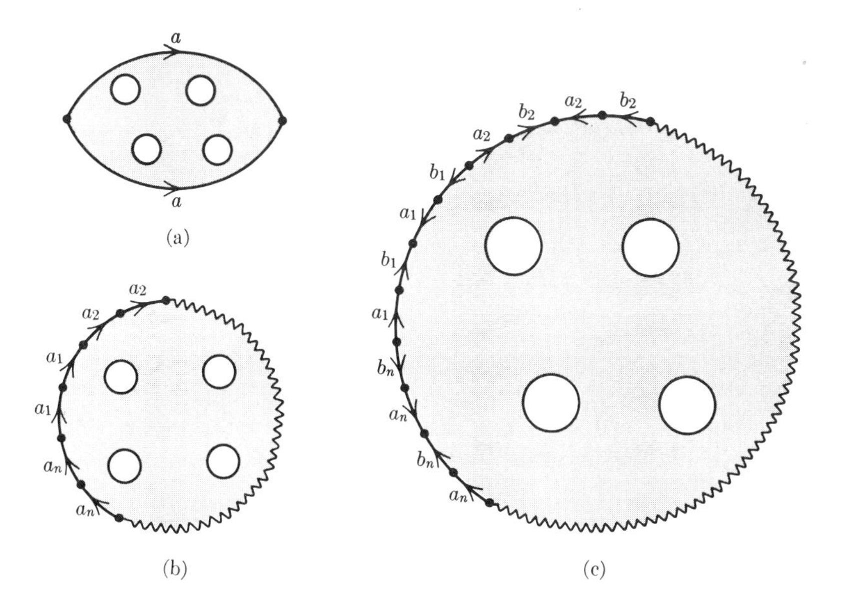

FIGURE 1.32 Possible types of bordered surfaces with $k = 4$.

Diagram (a) corresponds to a sphere with holes in it, (b) corresponds to a connected sum of projective planes with holes, and (c) corresponds to a connected sum of tori with holes. In each case, all vertices around the edge of the polygon are to be identified to a single vertex.

To complete the proof, we must now make cuts $c_1, c_2, \ldots, c_k$ from the initial vertex on the boundary of the polygon to the boundary of each of the holes, and open up each of the cuts to obtain a polygon in the desired normal form. Of course, we must be sure that any two cuts are pairwise disjoint except for the initial vertex. If k is very large (e.g., $k = 10^{10}$) and the holes are arranged in some peculiar way, it may not be immediately obvious how to proceed. We can get around this difficulty by an inductive procedure, as follows: Make a cut from the initial vertex to the *nearest* hole. Open up the cut thus made to obtain a new polygon with three more sides and one less hole. Again, make a cut from the initial vertex to the nearest hole in this new polygon, and open up the cut to obtain a polygon with three more sides and one less hole. Repeat the process k times until the necessary number of cuts has been made. It is clear that we obtain a polygon in one of the three possible normal forms, and thus the proof of the theorem is complete.

11 The Euler characteristic of a bordered surface

The Euler characteristic of a triangulated bordered surface is defined exactly the same as in the case of a surface without boundary. We can give the same type of argument as in Section 8 to show that it is independent of the triangulation. With the use of the Euler characteristic, we can now give a complete set of invariants for the classification of compact bordered surfaces:

Theorem 11.1 *Two compact bordered surfaces are homeomorphic if and only if they have the same number of boundary components, they are both orientable or nonorientable, and they have the same Euler characteristic.*

PROOF: Let M be a compact, connected 2-manifold, with or without boundary. Assume that M is given a definite triangulation, and that we form a new bordered surface M' by removing the interior of one triangle, which is contained entirely in the interior of M. It is clear that the boundary of M' has one more component than the boundary of M, and that

$$\chi(M') = \chi(M) - 1;$$

i.e., the Euler characteristic is reduced by one.

It follows that, if we start with a triangulated surface M^* (without boundary), and remove the interiors of k pairwise disjoint triangles, we obtain a bordered surface, and

$$\chi(M) = \chi(M^*) - k.$$

According to Theorem 5.1, we obtain in this way every bordered surface M whose boundary has k components. Thus, we see that the Euler characteristic of M uniquely determines that of M^* and vice versa. It is also clear that M and M^* are both orientable or nonorientable. The theorem now follows from Theorems 5.1 and 10.1. Q.E.D.

Definition The *genus* of a compact bordered surface M is defined to be the genus of the compact surface M^* obtained by attaching a disc to each boundary component of M.

Exercises

11.1 Prove that the Euler characteristic of a compact bordered surface having k boundary components is $\leqq 2 - k$.

11.2 Give a formula for the genus of a compact bordered surface in terms of its Euler characteristic and the number of components of the boundary (treat the orientable and nonorientable cases separately).

11.3 Make a table listing all compact surfaces M, with or without boundary, such that $-2 \leqq \chi(M) \leqq +2$.

12 Models of compact bordered surfaces in Euclidean 3-space

We can construct a variety of concrete models of bordered surfaces in the following manner. Take a disc and several long rectangular strips made of paper; then paste both ends of each strip to the boundary of the disc. We can paste the ends of the strips around the boundary in various orders, and if we desire, we can give some of the strips a half-twist. It is understood, of course, that the ends of the strips do not overlap on the boundary of the disc. Figures 1.34 through 1.36 illustrate the procedure.

We now assert that models of *all compact bordered surfaces can be constructed in this way.* The relatively simple proof is as follows. If M is any compact bordered surface, and M' is constructed from M by gluing the two ends of a rectangular strip to the boundary of M in any way whatsoever, then

$$\chi(M') = \chi(M) - 1.$$

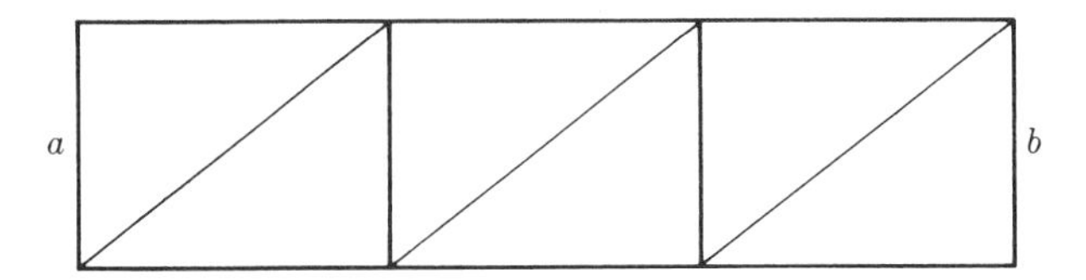

FIGURE 1.33 A triangulated strip.

We can verify this by assuming that M has been suitably triangulated, and that the strip has been triangulated as shown in Figure 1.33. We identify the edges a and b with two edges of the triangulation of the boundary of M, and count vertices, edges, and triangles before and after the identification.

Now, we shall show how to construct any compact, orientable bordered surface whose boundary has k components, $k \geqq 1$. First, paste $k - 1$ strips to the boundary (without twisting) as shown in Figure 1.34 for $k = 4$. An orientable bordered surface of Euler characteristic $2 - k$ whose boundary has k components results. Note that the Euler characteristic is the maximum possible for the given number of boundary components.

We next attach pairs of strips so as to keep the number of boundary components the same and reduce the Euler characteristic to the desired value as shown in Figure 1.35. Here we have attached two more strips to the model shown in Figure 1.34, in such a way as to reduce the Euler characteristic by 2, but to keep the number of boundary components fixed. We can repeatedly attach such "crossed" pairs of strips and reduce the Euler characteristic by any even integer.

To construct a nonorientable bordered surface whose boundary has k components, we begin in the same way: Attach $(k - 1)$ strips (as shown in Figure 1.34) to obtain an orientable surface of Euler characteristic $2 - k$ whose boundary has k components. If we attach a strip with a half-twist as shown in Figure 1.36 we reduce the Euler characteristic by one, keep the number of boundary components the same, and have a nonorientable bordered surface. By attaching more such half-twisted strips, we can reduce the Euler characteristic by any desired amount.

From these indications, it should be clear how to construct a model of any compact bordered surface by this method. This shows that any such surface, whether orientable or not, can be imbedded homeomorphically in Euclidean 3-space (recall that the corresponding statement is *not* true for compact 2-manifolds without boundary).

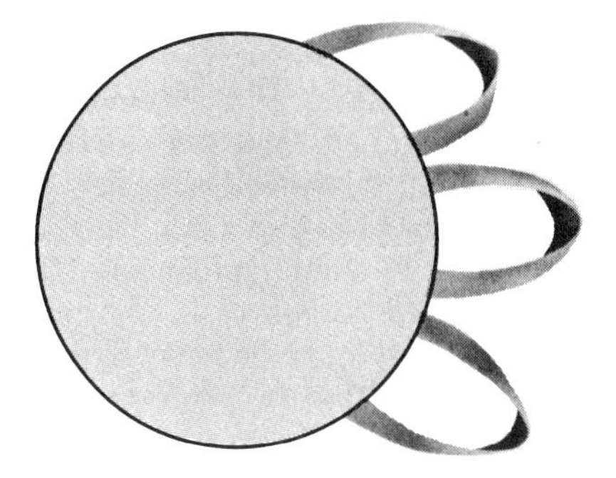

FIGURE 1.34 Method of pasting strips to a disc.

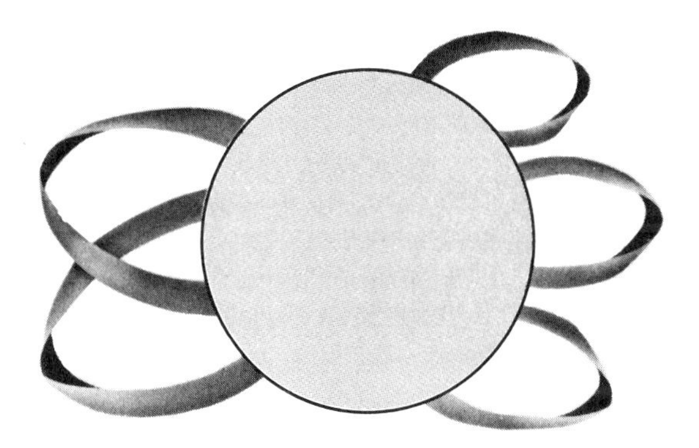

FIGURE 1.35 Method of pasting strips to a disc (orientable case, higher genus).

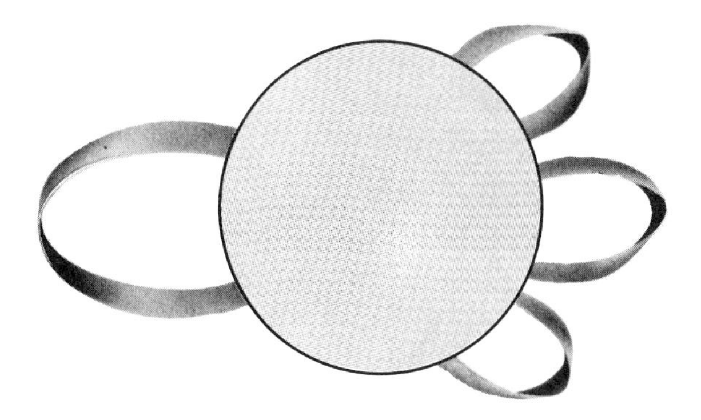

FIGURE 1.36 Method of pasting strips to a disc (nonorientable case).

Exercises

12.1 Which compact bordered surfaces are homeomorphic to a subset of the plane $\mathbf{R}^2$? Give your answer in terms of the Euler characteristic, number of boundary components, and orientability.

12.2 Tell which compact bordered surfaces can be constructed as follows: Choose a closed disc D, and a finite number of smaller closed discs which are pairwise disjoint and contained in the interior of D. Cut holes in D by removing the interiors of these smaller discs. Join the boundaries of certain pairs of holes by a tube. These tubes may be attached in two different ways, as shown in Figure 1.37.

12.3 If we go through the preceding construction, starting with a Möbius strip instead of a disc, which compact bordered surfaces can be constructed?

12.4 Let M_1 and M_2 be compact bordered surfaces. Form a new bordered surface, $M_1 \# M_2$, called the *boundary connected sum*, as follows: Choose a subset e_i of the boundary of M_i such that e_i is homeomorphic to the closed interval $[0, 1]$, $i = 1, 2$. Glue M_1 and M_2 together by gluing e_i to e_2; i.e., choose a homeomorphism of e_1 onto e_2 and form a quotient space by identifying points of e_1 and e_2 which correspond under the chosen homeomorphism.

(a) Express the Euler characteristic of $M_1 \# M_2$ in terms of that of M_1 and M_2.
(b) How many components has the boundary of $M_1 \# M_2$?
(c) Prove that any compact bordered surface can be built up as an iterated boundary connected sum of copies of the following four bordered surfaces: (a) closed disc, (b) annulus (i.e., disc with a single hole), (c) Möbius strip, and (d) torus with a single hole.

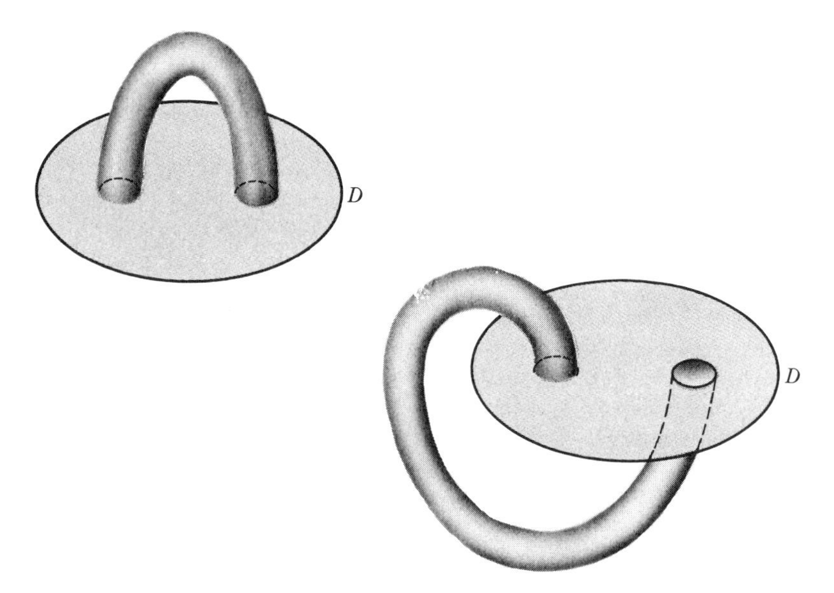

FIGURE 1.37 Methods of attaching a tube to a disc D with two holes.

13 Remarks on noncompact surfaces

Because there are so many different examples of noncompact surfaces, and many of the theorems are rather complicated, we shall only give a brief introduction to this topic.

First, we can divide noncompact surfaces into two broad classes: those that have a countable basis for their topology, and those that do not. The standard example of a connected surface that does not have a countable basis of open sets is due to Prüfer (see Radó [9]; this example is also reproduced in the following books: R. Nevanlinna, *Uniformisierung.* Berlin-Göttingen-Heidelberg: Springer-Verlag, 1953, p. 51, and G. Springer, *Introduction to Riemann Surfaces.* Reading: Addison-Wesley, 1957, p. 56). Such surfaces are usually regarded as pathological, and ignored; in most work on the subject, it is assumed that there is a countable basis of open sets. A theorem of Radó [9] asserts that a surface can be triangulated if and only if it has a countable basis for its topology (the proof is reproduced in the book of Ahlfors and Sario [1]). Triangulation of a noncompact surface means the same as triangulation of a compact surface, except that the number of triangles is infinite, and it is further required that each point have a neighborhood that meets only finitely many triangles.

The existence of triangulations for surfaces having a countable basis is very important, and many of the known results in the subject are only proved by using this fact. For the remainder of this chapter we shall only consider such surfaces.[3]

We now give some examples of noncompact surfaces:

(a) Any open subset of a compact surface. Already this gives a bewildering variety of examples. Consider, for example, the complement of any finite subset, or more generally, of any compact, totally disconnected subset (e.g., a Cantor set) of a surface.

(b) The surface of a ladder of infinite length, with an infinite number of rungs.

(c) Consider the following three families of parallel lines in Euclidean 3-space: lines parallel to the x axis through the points with integral coordinates in the yz plane; lines parallel to the y axis through the points with integral coordinates in the xz plane; and lines parallel to the z axis through the points with integral coordinates in the xy plane. Imagine all these lines "thickened" slightly, so that they are like solid rods; then the surface of the resulting solid is a noncompact surface.

---

[3] It can be shown that a surface is metrizable if and only if it has a countable basis of open sets. Similarly, paracompactness is equivalent to the existence of a countable basis.

(d) We can vary the construction of the preceding two examples by taking some other connected collection of lines and curves in $\mathbf{R}^3$, and then thickening each line and curve slightly. The surface of the resulting solid is often a 2-manifold.

(e) The process of forming the connected sum is also applicable to noncompact surfaces, only the possibilities are much greater now because we can form infinite connected sums. For example, we may start with the Euclidean plane $\mathbf{R}^2$ and remove a small circular hole about each point with integral coordinates. Then, we fill in each hole with a Möbius strip, gluing the boundary of the Möbius strip to the boundary of the hole. It is obvious how this procedure could be varied: Instead of starting with the Euclidean plane, we could start with some other noncompact surface. Instead of a Möbius strip, we could use some other bordered surface with a connected boundary.

(f) Let M be a compact surface, other than a sphere or projective plane, and let $(\tilde{M}, p)$ be a covering space of M (see Chapter 5) corresponding to a subgroup of $\pi(M)$ of infinite index. Then $\tilde{M}$ is a noncompact surface. Alternatively, if M is noncompact, then so is *any* covering space of M.

By combining these different methods of construction, still more examples can be given. In any case, it is clear that the possibilities are enormous.

Because there is a classification theorem for compact surfaces, it is natural to inquire whether or not there is a satisfactory classification theorem for noncompact manifolds. Here the answer depends on our interpretation of the word "satisfactory"; there *is* a classification theorem, but it does not seem to be easily applicable to problems that arise in the subject. Although it would take us too far afield to give all the details, we can explain the idea behind this theorem.

Let M be a noncompact surface. As usual, by a *compactification* of M we mean a compact Hausdorff space X, which contains M as an open, everywhere dense subspace. Two compactifications, X and Y, are regarded as equivalent if there exists a homeomorphism h of X onto Y such that $h \mid M$ is the identity map. Let us give some examples:

(1) Because M is locally compact, we can always form the Alexandroff 1-point compactification. This is the unique "minimal" compactification because only one point is added. Because M is completely regular, we can also form the Stone-Cech compactification. In a certain sense, this is a maximal compactification.

(2) Supose that M is an open subset of a compact surface M'. Then $\overline{M}$, the closure of M in M', is a compactification of M.

(3) Let M' be a compact bordered surface, and let M be its interior. Then M' is a compactification of M.

To state our next theorem, we need one more definition. Let X be a topological space and let A be a subspace. A is said to be *nonseparating* on X if, for any open connected subset U of X, $U - A$ is connected. For example, any finite subset of a surface M is nonseparating on M; any curve in $\mathbf{R}^3$ is nonseparating on $\mathbf{R}^3$.

Theorem 13.1 *Let M be a noncompact surface. There exists a compactification M^* of M, which has the following three properties:*

(1) *M^* is locally connected.*
(2) *$\beta(M) = M^* - M$ is totally disconnected.*
(3) *$\beta(M)$ is nonseparating on M^*.*

Moreover, any two compactifications of M having these three properties are equivalent.

The proof of the existence and uniqueness of M^* is rather long; the reader is referred to Ahlfors and Sario [1], Chapter I, Section 6, for the proof and references to earlier work on this subject.

Example

13.1 Let X be a compact, connected 2-manifold, and let A be a closed, totally disconnected subset of X. For example, A could be a finite subset, or A could be homeomorphic to a closed subset of the Cantor set. Let $M = X - A$; then it is plausible, and can be proved rigorously, that X is a compactification of M having the three properties stated in Theorem 13.1. Hence, we may take $M^* = X$ and $\beta(M) = A$. In general, however, M^* will *not* be a surface.

The space $\beta(M)$ is called the *ideal boundary* or *set of ends* of M; its points are called *boundary components* or *ends*. It is a theorem that $\beta(M)$ is a compact metric space.

We may get some hint as to how $\beta(M)$ is constructed by considering the above example. Each point $x \in A = \beta(M)$ has arbitrarily small open connected neighborhoods U such that U is an open 2-dimensional disc and the boundary of U is a circle that does not meet A. Consider the subsets $U \cap M$ for all such neighborhoods U of x in X. This is a family of subsets of M which can be characterized *intrinsically* (i.e., without reference to $M^* = X$) by a few simple properties. Thus, a point of the ideal boundary $\beta(M)$ corresponds to each family of open connected subsets of M having the required properties. In the example, $U \cap M$ is homeomorphic to a disc with points removed, and the boundary of $U \cap M$ is a circle; in general, these two properties will not hold; $U \cap M$ may or may not be homeomorphic to a subset of a disc, and it may or may not be orientable. In general, we call a point x of $\beta(M)$ *planar* if $U \cap M$ is homeomorphic to a subset of the plane for all sufficiently small open neighborhoods U of x; similarly, x is called *orientable* if $U \cap M$ is orientable for all sufficiently small U. Let $\beta'(M)$ denote the subset of orientable points of $\beta(M)$, and let $\beta''(M)$ denote the subset of planar points of $\beta(M)$. It can be shown that $\beta(M)$ is a compact, metric space which is totally disconnected, and both $\beta'(M)$ and $\beta''(M)$ are open subsets of $\beta(M)$. Obviously, $\beta'(M) \supset \beta''(M)$.

Because $\beta(M)$ and the subspaces $\beta'(M)$ and $\beta''(M)$ are defined intrinsically, they are topological invariants of the manifold M. The surprising thing is that together with a few other simple properties they characterize M. First, we have to describe the remaining properties needed.

Definitions (a) A noncompact surface M is of *finite genus* if there exists a compact bordered surface $A \subset M$ such that $M - A$ is homeomorphic to a subset of the plane $\mathbf{R}^2$. In this case, the genus of M is defined to the genus of A. In the contrary case, M is of *infinite genus*. (See Section 8 for the definition of the genus of a bordered surface.)

(b) A noncompact, nonorientable surface M is *finitely nonorientable* if there exists a compact subset $A \subset M$ such that $M - A$ is orientable; in the contrary case, M is *infinitely nonorientable.*

Clearly, a nonorientable surface of finite genus is finitely nonorientable, but not conversely.

(c) Finitely nonorientable surfaces are said to be of *even* or *odd nonorientability type* depending on whether every sufficiently large compact subset A, which is a bordered surface, is of odd or even genus. (This definition makes sense because the connected sum of a projective plane and a torus is homeomorphic to the connected sum of three projective planes; more generally, if we "add" an orientable surface to a nonorientable surface, the genus is unchanged mod 2.)

The properties of a surface, which we have just defined, are obviously topologically invariant.

Theorem 13.2 *Let M_1 and M_2 be noncompact surfaces, which have the same genus and orientability type (in accordance with definitions (a), (b), and (c) just given). Then M_1 and M_2 are homeomorphic if and only if there is a homeomorphism of $\beta(M_1)$ onto $\beta(M_2)$, such that $\beta'(M_1)$ and $\beta''(M_1)$ are mapped onto $\beta'(M_2)$ and $\beta''(M_2)$, respectively.*

This theorem is originally due to Kerekjarto; for a proof, we refer the reader to I. Richards [10], who recently completed the theorem as follows:

Theorem 13.3 *Let X be a totally disconnected compact metric space, and let U and V be open subsets of X such that $U \supset V$. Then there exists a* *noncompact surface M, such that $\beta(M)$ is homeomorphic to X under a homeomorphism which maps $\beta'(M)$ onto U and $\beta''(M)$ onto V, respectively.*

It is not hard to show that M may have any prescribed genus and orientability type compatible with the requirement that $\beta'(M)$ and $\beta''(M)$ be homeomorphic to U and V, respectively.

Finally, we consider some miscellaneous properties of noncompact 2-manifolds.

In the preceding section we proved that every compact bordered surface is homeomorphic to a subset of Euclidean 3-space. The same thing is true of noncompact surfaces.

Theorem 13.4 *Every noncompact 2-manifold is homeomorphic to a subset of* $\mathbf{R}^3$.

Note that it is possible for a given noncompact 2-manifold to be homeomorphic both to a closed and a nonclosed subset of $\mathbf{R}^3$: For example, the xy plane in $\mathbf{R}^3$ and the open unit disc in the xy plane are homeomorphic. It can be proved, however, that a nonorientable surface is *not* homeomorphic to any *closed* subset of $\mathbf{R}^3$.

This theorem is a special case of a theorem of M. Hirsch ([8], Theorem 4.6). Hirsch's work depends on results of J. H. C. Whitehead [11]. In particular, J. H. C. Whitehead proved that every triangulated noncompact surface M has a *spine*, i.e., a closed subset $L \subset M$ such that L is a union of edges of the triangulation, and there are arbitrarily small open neighborhoods U of L, which are homeomorphic to M. Moreover, for each such neighborhood U it is required that there exists a smaller neighborhood V of L such that points of V are left fixed under such a homeomorphism. In some cases the existence of a spine is almost obvious. For example, in the case of an open Möbius strip, the center circle is a spine. In other cases, the existence of a spine may not be so plausible.

NOTES

Definition of the connected sum of two manifolds

The definition of the connected sum given in Section 4 is adequate for 2-dimensional manifolds, but more care is necessary when we define the connected sum of two orientable n-manifolds for $n > 2$. We must worry about whether the homeomorphism h in our definition preserves or reverses orientations. The essential reason for this difference is that any orientable surface admits an orientation-reversing self-homeomorphism, whereas there exist orientable manifolds in higher dimensions which do not admit such a self-homeomorphism. Seifert and Threlfall [6], p. 280, give an example of a 3-dimensional manifold with this property. The complex projective plane is a 4-dimensional manifold having the property in question.

Triangulation of manifolds

In the early days of topology, it was apparently taken for granted that all surfaces and all higher dimensional manifolds could be triangulated. The first rigorous proof that surfaces can be triangulated was published by Tibor Radó in a paper on Riemann surfaces [9]. Radó pointed out the necessity of assuming the surface has a countable basis for its topology, and gave an example (due to Prüfer) of a surface that does not have such a countable basis. Radó's proof, given in Chapter I of the text by Ahlfors and Sario [1], makes essential use of a strong form of the Jordan Curve Theorem. The triangulability of 3-manifolds was proved by E. Moise ("Affine Structures in 3-manifolds, V: The triangulation theorem and Hauptvermutung." *Ann. Math.*, *56*, 1952, pp. 96–114).

Recent results of A. Casson and M. Freedman show that some 4-dimensional manifolds cannot be triangulated.

Models of nonorientable surfaces in Euclidean 3-space

No closed subset of Euclidean n-space is homeomorphic to a nonorientable $(n - 1)$-manifold. This result, first proved by the Dutch mathematician L. E. J. Brouwer in 1912, can nowadays be proved as an easy corollary of some general theorems of homology theory. This fact seriously hampers the development of our geometric intuition regarding compact, nonorientable surfaces, since they cannot be imbedded homeomorphically in Euclidean 3-space. However, it is possible to construct models of such surfaces in Euclidean 3-space provided we allow "singularities" or "self-intersections." We can even construct a mathematical theory of such models by considering the concept of *immersion* of manifolds. We say that a continuous map f of a compact n-manifold M^n into m-dimensional Euclidean space $\mathbf{R}^m$ is a *topological immersion* if each point of M^n has a neighborhood mapped homeomorphically onto its image by f. (The definition of a *differentiable immersion* is analogous; f is required to be differentiable and have a Jacobian everywhere of maximal rank.) The usual model of a Klein bottle in $\mathbf{R}^3$ is an immersion of the Klein bottle in 3-space. Werner Boy, in his thesis at the University of Göttingen in 1901 ["Über die Abbildung der projektiven Ebene auf eine im Endlichen geschlossene singularitätenfreie Fläche." *Nach. Königl. Gesell. Wiss Göttingen* (Math. Phys. Kl.), 1901, pp. 20–33. See also *Math. Annalen*, *57*, 1903, pp. 173–184], constructed immersions of the projective plane in $\mathbf{R}^3$. One of the immersions given by Boy is reproduced in Hilbert and Cohn-Vossen [3]. Since any compact, nonorientable surface is homeomorphic to the connected sum of an orientable surface and a projective plane or a Klein bottle, it is now easy to construct immersions of the remaining compact, nonorientable surfaces in $\mathbf{R}^3$.

The usual immersion of the Klein bottle in $\mathbf{R}^3$ is much nicer than any of the immersions of the projective plane given by Boy. The set of singular points for the immersion of the Klein bottle consists of a circle of double points, whereas the set of singular points for Boy's immersions of the projective plane is much more complicated. This raises the question, does there exist an immersion of the projective plane in $\mathbf{R}^3$ such that the set of singular points consists of disjoint circles of double points? The answer to this question is negative, at least in the case of differentiable immersions; for the proof, see the two papers by T. Banchoff in volume 46 (1974) of the *Proc. Amer. Math. Soc.*, pp. 402–413.

For further information on the immersion of compact surfaces in $\mathbf{R}^3$, see the interesting article entitled "Turning a Surface Inside Out" by Anthony Phillips in *Scientific American, 214,* 1966, pp. 112–120.

Bibliographical notes

The first proof of the classification theorem for compact surfaces is ascribed by some to H. R. Brahana (*Ann. Math. 23,* 1922, pp. 144–68). However, Seifert and Threlfall [6], p. 319, attribute it to Dehn and Heegard and do not even list Brahana's paper in their bibliography. The nonexistence of any algorithm for the classification of compact triangulable 4-manifolds is a result of the Russian mathematician A. A. Markov (*Proceedings of the International Congress of Mathematicians,* 1958, pp. 300–306). For the use of the Euler characteristic to prove the 5-color theorem for maps, see R. Courant and H. Robbins, *What Is Mathematics?* (New York: Oxford University Press, 1941), pp. 264–267. We also refer the student to excellent drawings in the books by Cairns [2], p. 28, and Hilbert and Cohn-Vossen [3], p. 265, illustrating how the connected sum of two or three tori can be cut open to obtain a polygon whose opposite edges are to be identified in pairs.

REFERENCES

Books

1. Ahlfors, L. V., and L. Sario. *Riemann Surfaces.* Princeton, N.J.: Princeton University Press, 1960. Chapter I.
2. Cairns, S. S. *Topology.* New York: Ronald, 1961. Chapter II.
3. Hilbert, D., and S. Cohn-Vossen. *Anschauliche Geometrie.* New York: Dover, 1944. Chapter VI. There is also an English translation by P. Nemenyi entitled *Geometry and the Imagination.* New York: Chelsea, 1956.
4. Kerekjarto, B. *Vorlesungen über Topologie.* Berlin: J. Springer, 1923. Chapters IV and V.
5. Reidemeister, K. *Einführung in die kombinatorische Topologie.* Braunschweig: Friedr. Vieweg & Sohn, 1932. Chapter V.
6. Seifert, H., and W. Threlfall. *A Textbook of Topology.* New York: Academic Press, 1980. Chapter 6.

Papers

7. Freudenthal, H. "Über die Enden topologischer Räume und Gruppen." *Math. Zeit., 33,* 1931, pp. 692–713.

8. Hirsch, M. "On Imbedding Differentiable Manifolds in Euclidean Space." *Ann. Math.*, *73*, 1961, pp. 566–571.
9. Radó, T. "Über den Begriff der Riemannschen Fläche." *Acta Litt. Sci. Szeged.*, *2*, 1925, pp. 101–121.
10. Richards, I. "On the classification of noncompact surfaces." *Trans. Amer. Math. Soc.*, *106*, 1963, pp. 259–269.
11. Whitehead, J. H. C. "The immersion of an open 3-manifold in Euclidean 3-space." *Proc. London Math. Soc.*, *11*, 1961, pp. 81–90.
12. Doyle, P. H., and D. A. Moran. "A Short Proof That Compact 2-Manifolds Can Be Triangulated." *Inventiones Math.*, *5*, 1968, pp. 160–162.

CHAPTER TWO

The Fundamental Group

1 Introduction

For any topological space X and any point $x_0 \in X$, we will define a group, called *the fundamental group* of X, and denoted by $\pi(X, x_0)$. (Actually, the choice of the point x_0 is usually of minor importance, and hence it is often omitted from the notation.) We define this group by a very simple and intuitive procedure involving the use of closed paths in X. From the definition, it will be clear that the group is a topological invariant of X; i.e., if two spaces are homeomorphic, their fundamental groups are isomorphic. This gives us the possibility of proving that two spaces are not homeomorphic by proving that their fundamental groups are nonisomorphic. For example, this method suffices to distinguish between the various compact surfaces and in many other cases.

Not only does the fundamental group give information about spaces, but it also is often useful in studying continuous maps. As we shall see, any continuous map from a space X into a space Y induces a homomorphism of the fundamental group of X into that of Y. Certain topological properties of the continuous map will be reflected in the properties of this induced homomorphism. Thus, we can prove facts about certain continuous maps by studying the induced homomorphism of the fundamental groups.

We can summarize the above two paragraphs as follows: By using the fundamental group, topological problems about spaces and continuous maps can sometimes be reduced to purely algebraic problems about groups and homomorphisms. This is the basic strategy of the entire subject of algebraic topology: to find methods of reducing topological problems to questions of pure algebra, and then hope that algebraists can solve the latter.

This chapter will only give the basic definition and properties of the fundamental group and induced homomorphism, and determine its structure for a few very simple spaces. In later chapters we shall develop more general methods for determining the fundamental groups of some more interesting spaces.

2 Basic notation and terminology

As usual, for any real numbers a and b such that $a < b$, $[a, b]$ denotes the closed interval of the real line with a and b as end points. For conciseness, we set $I = [0, 1]$. We note that, given any two closed intervals $[a, b]$ and $[c, d]$, there exist unique *linear* homeomorphisms

$$h_1, h_0 : [a, b] \to [c, d],$$

such that

$$h_0(a) = c, \qquad h_0(b) = d,$$

$$h_1(a) = d, \qquad h_1(b) = c.$$

We distinguish between these two by calling h_0 *orientation preserving* and h_1 *orientation reversing.*

A *path* or *arc* in a topological space X is a continuous map of some closed interval into X. The images of the end points of the interval are called the *end points* of the path or arc, and the path is said to *join* its end points. One of the end points is called the *initial* point, the other is called the *terminal* point (it is clear which is which).

A space X is called *arcwise connected* or *pathwise connected* if any two points of X can be joined by an arc. An arcwise connected space is connected, but the converse statement is not true. The *arc components* of X are the maximal arcwise-connected subsets of X (by analogy with the ordinary components of X). Note that the arc components of X need not be closed sets. A space is *locally arcwise connected* if each point has a basic family of arcwise-connected neighborhoods (by analogy with ordinary local connectivity).

Exercise

2.1 Prove that a space which is connected and locally arcwise connected is arcwise connected.

Definition Let $f_0, f_1 : [a, b] \to X$ be two paths in X such that $f_0(a) = f_1(a)$, $f_0(b) = f_1(b)$ (i.e., the two paths have the same initial and terminal points). We say that these two paths are *equivalent*, denoted by $f_0 \sim f_1$, if and only if there exists a continuous map

$$f : [a, b] \times I \to X,$$

such that

$$\left.\begin{aligned} f(t, 0) &= f_0(t) \\ f(t, 1) &= f_1(t) \end{aligned}\right\} t \in [a, b],$$

$$\left.\begin{aligned} f(a, s) &= f_0(a) = f_1(a) \\ f(b, s) &= f_0(b) = f_1(b) \end{aligned}\right\} s \in I.$$

Note that in the above definition we could replace I by any other closed interval if necessary. We leave it as an exercise to verify that this relation is reflexive, symmetric, and transitive.

Intuitively we say that two paths are equivalent if one can be continuously deformed into the other in the space X. During the deformation, the end points must remain fixed.

Our second basic definition is that of the *product* of two paths. The product of two paths is only defined if the terminal point of the first path is the initial point of the second path. If this condition holds, the product path is traversed by traversing the first path and then the second path, in the given order. To be precise, assume

$$f : [a, b] \to X$$

$$g : [b, c] \to X$$

are paths such that $f(b) = g(b)$ (here $a < b < c$). Then the product $f \cdot g$ is defined by

$$(f \cdot g)t = \begin{cases} f(t), & t \in [a, b] \\ g(t), & t \in [b, c]. \end{cases} \tag{2.2-1}$$

It is a map $[a, c] \to X$. In the above definition, we had the rather cumbersome requirement that the domains of f and g had to be the intervals $[a, b]$ and $[b, c]$, respectively. We can remove this requirement by changing the domain of f or g by means of an orientation-preserving linear homeomorphism. Actually, in the future we shall only be interested in equivalence classes of paths rather than the paths themselves. By "equivalence class," we mean, with respect to the equivalence relation defined above, and also with respect to the following obvious equivalence relation: If $f : [a, b] \to X$ and $g : [c, d] \to X$ are paths such that $g = fh$, where $h : [c, d] \to [a, b]$ is an *orientation-preserving* linear homeomorphism, then f and g are to be regarded as equivalent. Rather than considering paths whose domain is an arbitrary closed interval and allowing orientation-preserving linear homeomorphisms between any two such intervals, we find it technically simpler to demand that all paths be functions defined on one fixed interval, namely, the interval $I = [0, 1]$. As a result of this simplification, the simple formula for the product of two paths, (2.2-1), has to be replaced by a more complicated formula. Also, it will not be immediately obvious that the multiplication of path classes is associative. However, the reader should keep in mind that there are various alternative ways of proceeding with this subject.

3 Definition of the fundamental group of a space

From now on, by a *path* in X we mean a continuous map $I \to X$. If f and g are paths in X such that the terminal point of f is the initial point of g, then the product $f \cdot g$ is defined by

$$(f \cdot g)t = \begin{cases} f(2t), & 0 \leqq t \leqq \frac{1}{2}, \\ g(2t - 1), & \frac{1}{2} \leqq t \leqq 1. \end{cases}$$

We say two paths, f_0 and f_1, are *equivalent* ($f_0 \sim f_1$) if the condition in Section 2 is satisfied.

Lemma 3.1 *The equivalence relation and the product we have defined are compatible in the following sense: If $f_0 \sim f_1$ and $g_0 \sim g_1$, then $f_0 \cdot g_0 \sim f_1 \cdot g_1$ (it is assumed, of course, that the terminal point of f_0 is the initial point of g_0).*

The proof may be left to the reader. In proving lemmas such as this, the following fact is often useful: *Let A and B be closed subsets of the topological space X such that $X = A \cup B$. If f is a function defined on X such that the restrictions $f \mid A$ and $f \mid B$ are both continuous, then f is continuous.* The proof, which is easy, is left to the reader. In the future, we will use this fact without comment.

As a result of Lemma 3.1, the multiplication of paths defines a multiplication of equivalence classes of paths (provided the terminal point of the first path and the initial point of the second path coincide). It is this multiplication of equivalence classes with which we are primarily concerned. Note that the multiplication of paths is not associative in general, i.e., $(f \cdot g) \cdot h \neq f \cdot (g \cdot h)$ (we assume both products are defined). However, we have

Lemma 3.2 *The multiplication of equivalence classes of paths is associative.*

PROOF: It suffices to prove the following: Let f, g, and h be paths such that the terminal point of f = initial point of g, and the terminal point of g = initial point of h. Then

$$(f \cdot g) \cdot h \sim f \cdot (g \cdot h).$$

To prove this, consider the function $F : I \times I \to X$ defined by

$$F(t, s) = \begin{cases} f\left(\dfrac{4t}{1+s}\right), & 0 \leqq t \leqq \dfrac{s+1}{4}, \\[2ex] g(4t - 1 - s), & \dfrac{s+1}{4} \leqq t \leqq \dfrac{s+2}{4}, \\[2ex] h\left(1 - \dfrac{4(1-t)}{2-s}\right), & \dfrac{s+2}{4} \leqq t \leqq 1. \end{cases}$$

Then, F is continuous, $F(t, 0) = [(f \cdot g) \cdot h]t$, and $F(t, 1) = [f \cdot (g \cdot h)]t$. The motivation for the definition of F is given in Figure 2.1.

For any point $x \in X$, let us denote by ε_x the equivalence class of the constant map of I into the point x of X. This path class has the following fundamental property:

Lemma 3.3 *Let α be an equivalence class of paths with initial point x and terminal point y. Then $\varepsilon_x \cdot \alpha = \alpha$ and $\alpha \cdot \varepsilon_y = \alpha$.*

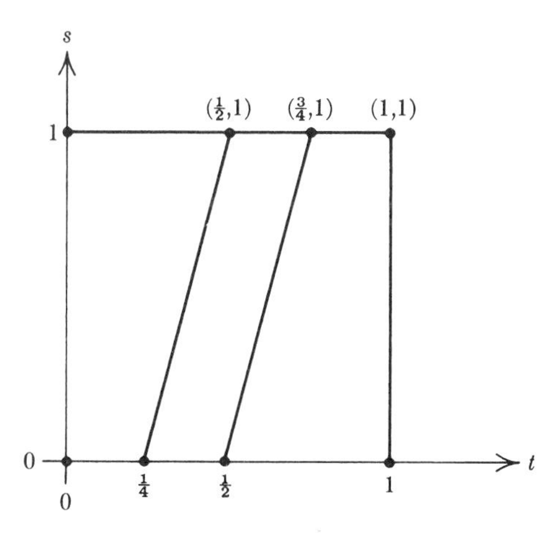

FIGURE 2.1 Proof of associativity.

PROOF: Let $e : I \to X$ be the constant map such that $e(I) = \{x\}$ and let $f : I \to X$ be a representative of the path class α. To prove the first relation, it suffices to prove that $e \cdot f \sim f$. Define $F : I \times I \to X$ by

$$F(t, s) = \begin{cases} x, & 0 \leqq t \leqq \frac{1}{2}s, \\ f\left(\dfrac{2t - s}{2 - s}\right), & \frac{1}{2}s \leqq t \leqq 1. \end{cases}$$

Then $F(t, 0) = f(t)$ and $F(t, 1) = (e \cdot f)t$ as required. The motivation for the definition of F is shown in Figure 2.2. The proof that $\alpha \cdot \varepsilon_y = \alpha$ is similar, and is left to the reader. Q.E.D.

For any path $f : I \to X$, let $\bar{f}$ denote the path defined by

$$\bar{f}(t) = f(1 - t), \qquad t \in I.$$

The path $\bar{f}$ is obtained by traversing the path f in the opposite direction.

Lemma 3.4 *Let α and $\bar{\alpha}$ denote the equivalence classes of the paths f and $\bar{f}$, respectively. Then,*

$$\alpha \cdot \bar{\alpha} = \varepsilon_x, \qquad \bar{\alpha} \cdot \alpha = \varepsilon_y,$$

where x and y are the initial and terminal points of the path f.

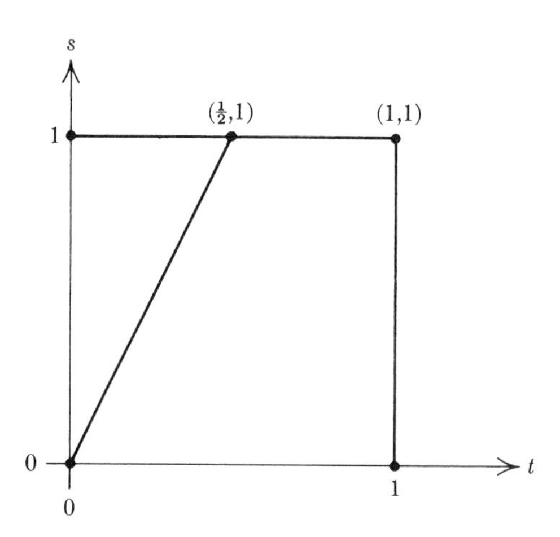

FIGURE 2.2 Proof of existence of units.

PROOF: To prove the first equation, it suffices to show that $f \cdot \bar{f} \sim e$, where e is the constant path at the point x. Therefore, we define $F : I \times I \to X$ by

$$F(t, s) = \begin{cases} f(2t), & 0 \leqq t \leqq \frac{1}{2}s, \\ f(s), & \frac{1}{2}s \leqq t \leqq 1 - \frac{1}{2}s, \\ f(2 - 2t), & 1 - \frac{1}{2}s \leqq t \leqq 1. \end{cases}$$

We then see that $F(t, 0) = x$, whereas $(f \cdot \bar{f})t = F(t, 1)$. Figure 2.3 explains the choice of the function F. We can also motivate the deformation of the path $f \cdot \bar{f}$ into the constant path e by a simple mechanical analogy. Consider the path f as an elastic "thread" in the space X from the point x to y; then $\bar{f}$ is another "thread" in the opposite direction, from y to x, and $f \cdot \bar{f}$ is represented by joining the two threads at the point y. We can now "pull in" the doubled thread to the point x, because we do not need to keep it attached to the point y.

The proof that $\bar{\alpha} \cdot \alpha = \varepsilon_y$ is similar, and is left to the reader. Q.E.D.

In view of these properties of the path class $\bar{\alpha}$, from now on we will denote it by α^{-1}. It is readily seen that the conditions of the lemma just proved characterize α^{-1} uniquely. Hence, if $f_0 \sim f_1$, then $\bar{f}_0 \sim \bar{f}_1$.

We can summarize the lemmas just proved by saying that the set of all path classes in X satisfies the axioms for a group, except that the product of two paths is not always defined.

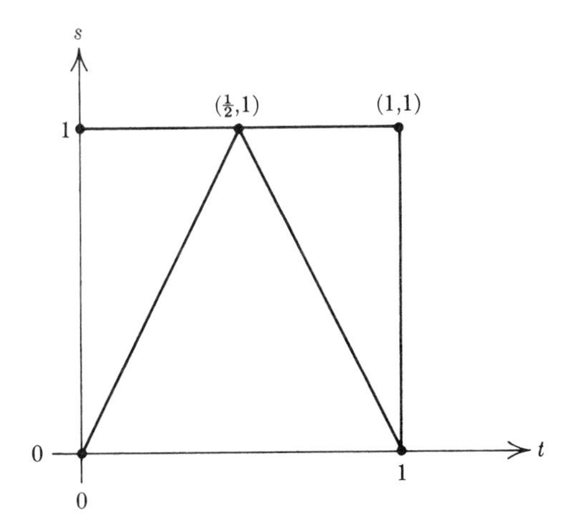

FIGURE 2.3 Proof of existence of inverses.

Definition A path, or path class, is called *closed,* or a *loop,* if the initial and terminal points are the same. The loop is said to be *based* at the common end point.

Let x be any point of X; it is readily seen that the set of all loops based at x is a group. This group is called the *fundamental group* or *Poincaré group* of X at the base point x, and denoted by $\pi(X, x)$.

Next, we will investigate the dependence of the group $\pi(X, x)$ on the base point x. Let x and y be two points in X, and let γ be a path class with initial point x and terminal point y (hence, x and y belong to the same arc component of X). Using the path γ, we define a mapping $u : \pi(X, x) \to \pi(X, y)$ by the formula $\alpha \to \gamma^{-1}\alpha\gamma$. We see immediately that this mapping is a homomorphism of $\pi(X, x)$ into $\pi(X, y)$. By using the path γ^{-1} instead of γ, we can define a homomorphism $v : \pi(X, y) \to \pi(X, x)$ in a similar manner. We immediately verify that the composed homomorphisms vu and uv are the identity maps of $\pi(X, x)$ and $\pi(X, y)$, respectively. Thus, u and v are isomorphisms, each of which is the inverse of the other. Thus, we have proved

Theorem 3.5 *If X is arcwise connected, the groups $\pi(X, x)$ and $\pi(X, y)$ are isomorphic for any two points x, $y \in X$.*

The importance of this theorem is obvious; e.g., the question as to whether or not $\pi(X, x)$ has any given group theoretic property (e.g., it is abelian, finite, nilpotent, free, etc.) is independent of the point x, and thus depends only on the space X, provided X is arcwise connected.

On the other hand, we must keep in mind that there is no *canonical* or *natural* isomorphism between $\pi(X, x)$ and $\pi(X, y)$; corresponding to each choice of a path class from x to y there will be an isomorphism, from $\pi(X, x)$ to $\pi(X, y)$, and, in general, different path classes will give rise to different isomorphisms.

Exercises

3.1 Under what conditions will two path classes, γ and γ', from x to y give rise to the same isomorphism of $\pi(X, x)$ onto $\pi(X, y)$?

3.2 Let X be an arcwise-connected space. Under what conditions is the following statement true: For any two points x, $y \in X$, all path classes from x to y give rise to the same isomorphism of $\pi(X, x)$ onto $\pi(X, y)$?

3.3 Let $f, g : I \to X$ be two paths with initial point x_0 and terminal point x_1. Prove that $f \sim g$ if and only if $f \cdot \bar{g}$ is equivalent to the constant path at x_0 ($\bar{g}$ is defined as in Lemma 3.4).

We will actually determine the structure of the fundamental group of various spaces later in this chapter and in Chapter IV.

4 The effect of a continuous mapping on the fundamental group

Let $\varphi : X \to Y$ be a continuous mapping, and let $f_0, f_1 : I \to X$ be paths in X. It is readily seen that, if f_0 and f_1 are equivalent, then so are the paths φf_0 and φf_1 represented by the composed functions. Thus, if α denotes the path class that contains f_0 and f_1, it makes sense to denote by $\varphi_*(\alpha)$ the path class that contains the paths φf_0 and φf_1. $\varphi_*(\alpha)$ is the image of the path class α in the space Y, and it is readily verified that the mapping φ_* which sends α into $\varphi_*(\alpha)$ has the following properties:

(a) If α and β are path classes in X such that $\alpha \cdot \beta$ is defined, then $\varphi_*(\alpha \cdot \beta) = (\varphi_*\alpha) \cdot (\varphi_*\beta)$.
(b) For any point $x \in X$, $\varphi_*(\varepsilon_x) = \varepsilon_{\varphi(x)}$.
(c) $\varphi_*(\alpha^{-1}) = (\varphi_*\alpha)^{-1}$.

For these reasons, we shall call φ_* a "homomorphism," or, the "homomorphism induced by φ."

If $\psi : Y \to Z$ is also a continuous map, then we can verify the following property easily:

(d) $(\psi\varphi)_* = \psi_*\varphi_*$.

Finally, if $\varphi : X \to X$ is the identity map, then

(e) $\varphi_*(\alpha) = \alpha$ for any path class α in X; i.e., φ_* is the identity homomorphism.

Note that, in view of these properties, a continuous map $\varphi : X \to Y$ induces a homomorphism $\varphi_* : \pi(X, x) \to \pi(Y, \varphi(x))$; and, if φ is a nomeomorphism, then φ_* is an isomorphism. This induced homomorphism will be extremely important in studying the fundamental group.

Caution: If φ is a one-to-one map, it does *not* follow that φ_* is one-to-one; similarly, if φ is onto, it does not follow that φ_* is onto. We shall see examples to illustrate this point later.

Exercise

4.1 Let $\varphi : X \to Y$ be a continuous map, and let γ be a class of paths in X from x_0 to x_1. Prove that the following diagram is commutative:

$$\begin{array}{ccc} \pi(X, x_0) & \xrightarrow{\varphi_*} & \pi(Y, \varphi(x_0)) \\ \downarrow u & & \downarrow v \\ \pi(X, x_1) & \xrightarrow{\varphi_*} & \pi(Y, \varphi(x_1)). \end{array}$$

Here the isomorphism u is defined by $u(\alpha) = \gamma^{-1}\alpha\gamma$, and v is defined similarly using $\varphi_*(\gamma)$ in place of γ. [NOTE: An important special case occurs if $\varphi(x_0) = \varphi(x_1)$. Then, $\varphi_*(\gamma)$ is an element of the group $\pi(Y, \varphi(x_0))$.]

To make further progress in the study of the induced homomorphism φ_*, we must introduce the important notion of *homotopy* of continuous maps.

Definition Two continuous maps $\varphi_0, \varphi_1 : X \to Y$ are *homotopic* if and only if there exists a continuous map $\varphi : X \times I \to Y$ such that, for $x \in X$,

$$\varphi(x, 0) = \varphi_0(x),$$

$$\varphi(x, 1) = \varphi_1(x).$$

If two maps φ_0 and φ_1 are homotopic, we shall denote this by $\varphi_0 \simeq \varphi_1$. We leave it to the reader to verify that this is an equivalence relation on the set of all continuous maps $X \to Y$. The equivalence classes are called *homotopy classes* of maps.

To better visualize the geometric content of the definition, let us write $\varphi_t(x) = \varphi(x, t)$ for any $(x, t) \in X \times I$. Then, for any $t \in I$,

$$\varphi_t : X \to Y$$

is a continuous map. Think of the parameter t as representing time. Then, at time $t = 0$, we have the map φ_0, and, as t varies, the map φ_t varies *continuously* so that at time $t = 1$ we have the map φ_1. For this reason a homotopy is often spoken of as a continuous deformation of a map.[1]

Definition Two maps $\varphi_0, \varphi_1 : X \to Y$ are *homotopic relative to the subset A of X* if and only if there exists a continuous map $\varphi : X \times I \to Y$ such that

$$\varphi(x, 0) = \varphi_0(x), \qquad x \in X,$$

$$\varphi(x, 1) = \varphi_1(x), \qquad x \in X,$$

$$\varphi(a, t) = \varphi_0(a) = \varphi_1(a), \qquad a \in A,\ t \in I.$$

Note that this condition implies $\varphi_0 \mid A = \varphi_1 \mid A$.

---

[1] The student who is familiar with the compact-open topology for function spaces will recognize that two maps $\varphi_0, \varphi_1 : X \to Y$ are homotopic if and only if they can be joined by an arc in the space of all continuous functions $X \to Y$ (provided X and Y satisfy certain hypotheses). Indeed, the map $t \to \varphi_t$ in the above notation is a path from φ_0 to φ_1.

Theorem 4.1 *Let $\varphi_0, \varphi_1 : X \to Y$ be maps that are homotopic relative to the subset $\{x\}$. Then*

$$\varphi_{0*} = \varphi_{1*} : \pi(X, x) \to \pi(Y, \varphi_0(x)),$$

i.e., the induced homomorphisms are the same.

PROOF: The proof is immediate.

Unfortunately, the condition that the homotopy should be relative to the base point x is too restrictive for many purposes. This condition can be omitted, but we then complicate the statement of the theorem. We shall, however, do this in Section 8.

We shall now apply some of these results.

Definition A subset A of a topological space X is called a *retract* of X if there exists a continuous map $r : X \to A$ (called a *retraction*) such that $r(a) = a$ for any $a \in A$.

As we shall see shortly, it is a rather strong condition to require that a subset A be a retract of X. A simple example of a retract of a space is the "center circle" of a Möbius strip. (What is the retraction in this case?)

Now let $r : X \to A$ be a retraction, as in the above definition, and $i : A \to X$ the inclusion map. For any point $a \in A$, consider the induced homomorphisms

$$i_* : \pi(A, a) \to \pi(X, a),$$

$$r_* : \pi(X, a) \to \pi(A, a).$$

Because ri = identity map, we conclude that $r_* i_*$ = identity homomorphism of the group $\pi(A, a)$, by properties (*d*) and (*e*) given previously. From this we conclude that i_* *is a monomorphism* and r_* *is an epimorphism.* Moreover, the condition that $r_* i_*$ = identity imposes strong restrictions on the subgroup $i_* \pi(A, a)$ of $\pi(X, a)$.

We shall actually use this result later to prove that certain subspaces are not retracts.

Exercises

4.2 Show that a retract of a Hausdorff space must be a closed subset.

4.3 Prove that, if A is a retract of X, $r : X \to A$ is a retraction, $i : A \to X$ is the inclusion, and $i_* \pi(A)$ is a normal subgroup of $\pi(X)$, then $\pi(X)$ is the direct product of the subgroups image i_* and kernel r_* (see Section III.2 for the definition of direct product of groups).

4.4 Let A be a subspace of X, and let Y be a nonempty topological space. Prove that $A \times Y$ is a retract of $X \times Y$ if and only if A is a retract of X.

4.5 Prove that the relation "is a retract of" is transitive, i.e., if A is a retract of B and B is a retract of C, then A is a retract of C.

We now introduce the notion of *deformation retract.* The subspace A is a deformation retract of X if there exists a retraction $r : X \to A$ homotopic to the identity map $X \to X$. The precise definition is as follows:

Definition A subset A of X is a *deformation retract*[2] of X if there exists a retraction $r : X \to A$ and a homotopy $f : X \times I \to X$ such that

$$\left.\begin{aligned} f(x, 0) &= x \\ f(x, 1) &= r(x) \end{aligned}\right\} x \in X,$$

$$f(a, t) = a, \qquad a \in A \quad \text{and} \quad t \in I.$$

Theorem 4.2 *If A is a deformation retract of X, then the inclusion map $i : A \to X$ induces an isomorphism of $\pi(A, a)$ onto $\pi(X, a)$ for any $a \in A$.*

PROOF: As above, $r_* i_*$ is the identity map of $\pi(A, a)$. We will complete the proof by showing that $i_* r_*$ is the identity map of $\pi(X, a)$. This follows because ir is homotopic to the identity map $X \to X$ (relative to $\{a\}$); hence, Theorem 4.1 is applicable. Q.E.D.

We shall use this theorem in two different ways. On the one hand, we shall use it throughout the rest of this book to prove that two spaces have isomorphic fundamental groups. On the other hand, we can use it to prove that a subspace is not a deformation retract by proving the fundamental groups are not isomorphic. In particular, we shall be able to prove that certain retracts are not deformation retracts.

Definition A topological space X *is contractible to a point* if there exists a point $x_0 \in X$ such that $\{x_0\}$ is a deformation retract of X.

Definition A topological space X is *simply connected* if it is arcwise connected and $\pi(X, x) = \{1\}$ for some (and hence any) $x \in X$.

Corollary 4.3 *If X is contractible to a point, then X is simply connected.*

---

[2] Some authors define this term in a slightly weaker fashion.

Examples

4.1 A subset X of the plane or, more generally, of Euclidean n-space $\mathbf{R}^n$ is called *convex* if the line segment joining any two points of X lies entirely in X. We assert that *any convex subset X of $\mathbf{R}^n$ is contractible to a point.* To prove this, choose an arbitrary point $x_0 \in X$, and then define $f : X \times I \to X$ by the formula

$$f(x, t) = (1 - t)x + tx_0$$

for any $(x, t) \in X \times I$ [i.e., $f(x, t)$ is the point on the line segment joining x and x_0 which divides it in the ratio $(1 - t) : t$]. Then f is continuous, $f(x, 0) = x$, and $f(x, 1) = x_0$, as required. More generally, we may define a subset X of $\mathbf{R}^n$ to be *starlike with respect to the point* $x_0 \in X$ provided the line segment joining x and x_0 lies entirely in X for any $x \in X$. Then, the same proof suffices to show that, if X is starlike with respect to x_0, it is contractible to the point x_0.

4.2 We assert that the unit $(n - 1)$-sphere S^{n-1} is a deformation retract of $E^n - \{0\}$, the closed unit n-dimensional disc minus the origin. To prove this, define a map $f : X \times I \to X$, where

$$X = E^n - \{0\} = \{x \in \mathbf{R}^n : 0 < |x| \leqq 1\},$$

by the formula

$$f(x, t) = (1 - t)x + t \cdot \frac{x}{|x|}.$$

(The reader should draw a picture to show what happens here when $n = 2$ or $n = 3$.) Then f is continuous, $f(x, 0) = x$, $f(x, 1) = x/|x| \in S^{n-1}$, and, if $x \in S^{n-1}$, then $f(x, t) = x$ for all $t \in I$. In particular, for $n = 2$, we see that the boundary circle is a deformation retract of a punctured disc.

Exercises

4.6 Let x_0 be any point in the plane $\mathbf{R}^2$. Find a circle C in $\mathbf{R}^2$ which is a deformation retract of $\mathbf{R}^2 - \{x_0\}$. What is the n-dimensional analog of this fact?

4.7 Find a circle C which is a deformation retract of the Möbius strip.

4.8 Let T be a torus, and let X be the complement of a point in T. Find a subset of X which is homeomorphic to a figure "8" curve (i.e., the union of two circles with a single point in common) and which is a deformation retract of X.

4.9 Generalize Exercise 4.8 to arbitrary compact surfaces; i.e., let S be a compact surface and let X be the complement of a point in S. Find a subset A of X such that (a) A is homeomorphic to the union of a finite number of circles, and (b) A is a deformation retract of X. (HINT: Consider the representation of S as the space obtained by identifying in pairs the edges of a certain polygon.)

4.10 Let x and y be distinct points of a simply connected space X. Prove that there is a *unique* path class in X with initial point x and terminal point y.

4.11 Let X be a topological space, and for each positive integer n let X_n be an arcwise-connected subspace containing the base point $x_0 \in X$. Assume that the subspaces X_n are nested, i.e., $X_n \subset X_{n+1}$ for all n, that

$$X = \bigcup_{n=1}^{\infty} X_n,$$

and that for any compact subset A of X there exists an integer n such that $A \subset X_n$. (EXAMPLE: Each X_n is open.) Let $i_n : \pi(X_n) \to \pi(X)$ and $j_{mn} : \pi(X_m) \to \pi(X_n)$, $m < n$, denote homomorphisms induced by inclusion maps. Prove the following two statements: (a) For any $\alpha \in \pi(X)$, there exists an integer n and an element $\alpha' \in \pi(X_n)$ such that $i_n(\alpha') = \alpha$. (b) If $\beta \in \pi(X_m)$ and $i_m(\beta) = 1$, then there exists an integer $n \geqq m$ such that $j_{mn}(\beta) = 1$. [REMARK: These two statements imply that $\pi(X)$ is the *direct limit* of the sequence of groups $\pi(X_n)$ and homomorphisms j_{mn}. We shall see examples later on where the hypotheses of this exercise are valid.] If the homomorphisms $j_{n,n+1}$ are monomorphisms for all n, prove that each i_n is also a monomorphism, and that $\pi(X)$ is the union of the subgroups $i_n\pi(X_n)$.

5 The fundamental group of a circle is infinite cyclic

Let S^1 denote the unit circle in the Euclidean plane $\mathbf{R}^2$, $S^1 = \{(x, y) \in \mathbf{R}^2 : x^2 + y^2 = 1\}$ (or, equivalently, in the complex plane $\mathbf{C}$). Let $f : I \to S^1$ denote the closed path that goes around the circle exactly once, defined by

$$f(t) = (\cos 2\pi t, \sin 2\pi t), \qquad 0 \leqq t \leqq 1,$$

and denote the equivalence class of f by the symbol α.

Theorem 5.1 *The fundamental group $\pi(S^1, (1, 0))$ is an infinite cyclic group generated by the path class α.*

PROOF: Let $g : I \to S^1$, $g(0) = g(1) = (1, 0)$ be a closed path in S^1. We shall prove first that g belongs to the equivalence class α^m for some integer m (m may be positive, negative, or zero).

Let

$$U_1 = \{(x, y) \in S^1 : y > -\tfrac{1}{10}\},$$

$$U_2 = \{(x, y) \in S^1 : y < +\tfrac{1}{10}\}.$$

Then, U_1 and U_2 are connected open subsets of S^1, each of which is slightly larger than a semicircle, and $U_1 \cup U_2 = S^1$. Obviously U_1 and U_2 are each homeomorphic to an open interval of the real line, hence, each is contractible. In the case where $g(I) \subset U_1$ or $g(I) \subset U_2$, it is then clear that g is equivalent to the constant path, and hence belongs to the equivalence class of α^0. We put this case aside, and assume from now on that $g(I) \not\subset U_1$ and $g(I) \not\subset U_2$.

We next assert that it is possible to divide the unit interval into subintervals $[0, t_1], [t_1, t_2], \ldots, [t_{n-1}, 1]$, where $0 = t_0 < t_1 < \cdots < t_{n-1} < t_n = 1$, such that the following conditions hold:

(a) $g([t_i, t_{i+1}]) \subset U_1$ or
$g([t_i, t_{i+1}]) \subset U_2$ for $0 \leqq i < n$.

(b) $g([t_{i-1}, t_i])$ and $g([t_i, t_{i+1}])$
are not both contained in the *same* open set U_j, $j = 1$ or 2.

This assertion may be proved as follows. $\{g^{-1}(U_1), g^{-1}(U_2)\}$ is an open covering of the compact metric space I; let ε be a Lebesgue number[3] of this covering.

Divide the unit interval in any way whatsoever into subintervals of length $< \varepsilon$. With this subdivision, condition (a) will hold; however, condition (b) may not hold. If two consecutive subintervals are mapped by g into the same set U_j, then amalgamate these two subintervals into a single subinterval by omitting the common end point. Continue this process of amalgamation until condition (b) holds.

Let β denote the equivalence class of the path g, and let β_i denote the equivalence class of $g \mid [t_{i-1}, t_i]$ for $1 \leqq i \leqq n$. Then, obviously, β is a product,

$$\beta = \beta_1 \cdot \beta_2 \cdot \cdots \cdot \beta_n.$$

Each β_i is a path in U_1 or U_2. Because of condition (b), it is clear that $g(t_i) \in U_1 \cap U_2$. $U_1 \cap U_2$ has two components, one of which contains the point $(1, 0)$, and the other of which contains the point $(-1, 0)$. For each index i, $0 < i < n$, choose a path class γ_i in $U_1 \cap U_2$ with initial point $g(t_i)$ and terminal point $(1, 0)$ or $(-1, 0)$, depending on which component of $U_1 \cap U_2$ contains $g(t_i)$. Let

$$\begin{aligned} \delta_1 &= \beta_1\gamma_1, \\ \delta_i &= \gamma_{i-1}^{-1}\beta_i\gamma_i \quad \text{for} \quad 1 < i < n, \\ \delta_n &= \gamma_{n-1}^{-1}\beta_n. \end{aligned}$$

Then, it is clear that

$$\beta = \delta_1\delta_2 \cdots \delta_n \tag{2.5-1}$$

[3] We say ε is a *Lebesgue number* of a covering of a metric space X if the following condition holds: Any subset of X of diameter $< \varepsilon$ is contained in some set of the covering. It is a theorem that any open covering of a compact metric space has a Lebesgue number. The reader may either prove this as an exercise or look up the proof in a textbook on general topology.

where each δ_i is a path class in U_1 or U_2, having its initial and terminal points in the set $\{(1, 0), (-1, 0)\}$. For any index i, if δ_i is a closed path class, then $\delta_i = 1$, because U_1 and U_2 are simply connected. We may therefore assume that any such δ_i has been dropped from formula (2.5-1), and, changing notation if necessary, that $\delta_1, \delta_2, \ldots,$ and δ_n are not closed paths.

Because U_1 is simply connected, there is a unique path class η_1 in U_1 with initial point $(1, 0)$ and terminal point $(-1, 0)$ (see Exercise 4.10). Also, η_1^{-1} is the unique path class in U_1 with initial point $(-1, 0)$ and terminal point $(1, 0)$. Analogously, we denote by η_2 the unique path class in U_2 with initial point $(-1, 0)$ and terminal point $(1, 0)$. Note that $\eta_1\eta_2 = \alpha$.

Thus, we see that, for each index i,

$$\delta_i = \eta_1^{\pm 1} \quad \text{or} \quad \delta_i = \eta_2^{\pm 1}.$$

In view of condition (b) above, if $\delta_i = \eta_1^{\pm 1}$, then $\delta_{i+1} = \eta_2^{\pm 1}$, while if $\delta_i = \eta_2^{\pm 1}$, then $\delta_{i+1} = \eta_1^{\pm 1}$. Therefore only the following possibilities remain:

$$\beta = 1,$$

$$\beta = \eta_1\eta_2\eta_1\eta_2 \cdots \eta_1\eta_2,$$

or

$$\beta = \eta_2^{-1}\eta_1^{-1}\eta_2^{-1}\eta_1^{-1} \cdots \eta_2^{-1}\eta_1^{-1}.$$

In the second case $\beta = \alpha^m$ for some $m > 0$, whereas in the third case $\beta = \alpha^m$ for some integer $m < 0$. Thus, we have $\beta = \alpha^m$ in all cases.

From this it follows that $\pi(S^1)$ is a cyclic group. However, this argument gives no hint as to the order of $\pi(S^1)$. To prove that $\pi(S^1)$ is not a finite group, we now introduce the concept of the *degree* of a closed path in $\pi(S^1)$. The degree is an integer which, roughly speaking, tells how many times the path is wrapped around the circle.

To define the degree of a path, it will be convenient to regard S^1 as the unit circle in the complex plane $\mathbf{C}$,

$$S^1 = \{z \in \mathbf{C} : |z| = 1\}.$$

Because the product or quotient of any two complex numbers of absolute value 1 again has absolute value 1, S^1 is a group under multiplication. If $z \in S^1$, we denote by $a(z)$ the *angle* of z, i.e., the angle in radians from the positive real axis to the segment with end points 0 and z (see Figure 2.4). Thus, for any $z \in S^1$, $a(z)$ is a real number; however, it is not uniquely determined. If θ is a value of $a(z)$, then $\theta + 2k\pi$ is an equally good value for any integer k. We may also state this as follows. If

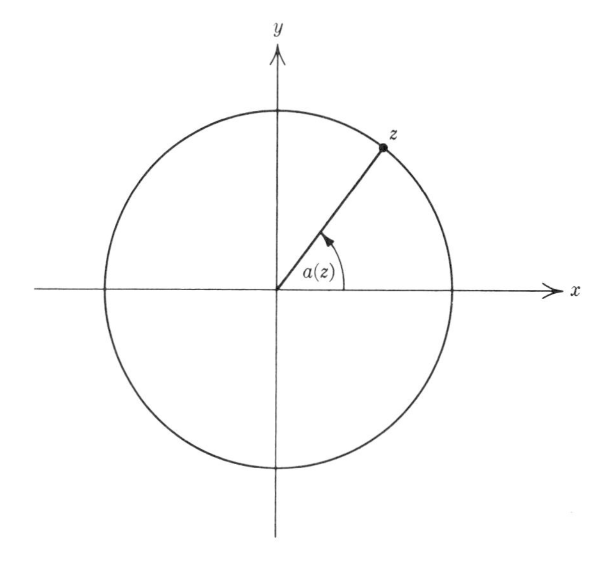

FIGURE 2.4 Definition of $a(z)$.

$z = e^{i\theta} = \cos\theta + i\sin\theta$, then θ is a determination of $a(z)$. Note that if θ_1 and θ_2 are values for $a(z_1)$ and $a(z_2)$, then $\theta_1 + \theta_2$ is a value of $a(z_1z_2)$, and $\theta_1 - \theta_2$ is a value of $a(z_1/z_2)$.

Let $h : I \to S^1$ be a closed path with $h(0) = h(1) = 1$. Choose a subdivision

$$0 = t_0 < t_1 < \cdots < t_n = 1$$

of the unit interval such that the following property holds: If t and t' both belong to the same subinterval $[t_{i-1}, t_i]$, then

$$|h(t') - h(t)| < 1. \tag{2.5-2}$$

We can prove the existence of such a subdivision by using the fact that h is uniformly continuous, or by the existence of a Lebesgue number of a certain covering of I. For each index i, $1 \leqq i \leqq n$, let θ_i be the unique determination of the angle of $h(t_i)/h(t_{i-1})$ which satisfies the inequality

$$-\frac{\pi}{2} < \theta_i < +\frac{\pi}{2}.$$

That θ_i can be so chosen follows from the inequality $|h(t_i) - h(t_{i-1})| < 1$, in view of (2.5-2). We now define

$$\text{degree of } h = \frac{1}{2\pi}\sum_{i=1}^{n} \theta_i.$$

It is clear that the degree of h is an integer, since

$$\sum_{i=1}^{n} \theta_i$$

is a determination of the angle of the complex number

$$\prod_{i=1}^{n} \frac{h(t_i)}{h(t_{i-1})} = \frac{h(t_n)}{h(t_0)} = \frac{1}{1} = 1.$$

To justify this definition, we must show that it is independent of the choice of the subdivision of the interval I. Since any two subdivisions of I into subintervals have a common refinement, it suffices to investigate what happens when we refine the given subdivision. Any refinement of a given subdivision may be obtained by a series of successive refinements such that only one new point of subdivision is added at each stage. Therefore we only need to investigate the case where we subdivide the subinterval $[t_{i-1}, t_i]$ by a point s, such that $t_{i-1} < s < t_i$. We then have to replace θ_i by $\theta_i' + \theta_i''$, where

$$\theta_i' = a\left(\frac{h(s)}{h(t_{i-1})}\right),$$

$$\theta_i'' = a\left(\frac{h(t_i)}{h(s)}\right),$$

and $|\theta_i'| < \pi/2$, $|\theta_i''| < \pi/2$. It is now clear that both θ_i and $\theta_i' + \theta_i''$ are determinations of the angle of the complex number $h(t_i)/h(t_{i-1})$; hence, they must differ by an integral multiple of 2π. But, because all are less than $\pi/2$ in absolute value, this is only possible if $\theta_i = \theta_i' + \theta_i''$. Thus, the definition of the degree of h is independent of the choice of subdivision.

Next, we will prove that, if $h \sim g$, then degree of h = degree of g. Since $h \sim g$, there exists a continuous map $F : I \times I \to S^1$ such that

$$\begin{aligned} F(t, 0) &= h(t), \\ F(t, 1) &= g(t), \\ F(0, s) &= F(1, s) = 1. \end{aligned} \tag{2.5-3}$$

We may now choose subdivisions $t_0 = 0 < t_1 < \cdots < t_n = 1$ and $s_0 = 0 < s_1 < \cdots < s_m = 1$ of the unit interval such that F maps each of the rectangles $[t_{i-1}, t_i] \times [s_{j-1}, s_j]$ into a subset of S^1 having diameter < 1; i.e., if (t, s) and (t', s') belong to $[t_{i-1}, t_i] \times [s_{j-1}, s_j]$, then

$$|F(t, s) - F(t', s')| < 1. \tag{2.5-4}$$

Now, let

$$\theta_i' = a\left(\frac{F(t_i, s_{j-1})}{F(t_{i-1}, s_{j-1})}\right), \qquad |\theta_i'| < \frac{\pi}{2},$$

$$\theta_i'' = a\left(\frac{F(t_i, s_j)}{F(t_{i-1}, s_j)}\right), \qquad |\theta_i''| < \frac{\pi}{2}.$$

We wish to prove that

$$\sum_{i=1}^{n} \theta_i' = \sum_{i=1}^{n} \theta_i''.$$

If we apply this argument for $j = 1, 2, \ldots, m$ in succession, it will then follow that degree of h = degree of g. For this purpose, let

$$\varphi_i = a\left(\frac{F(t_i, s_j)}{F(t_i, s_{j-1})}\right), \qquad |\varphi_i| < \frac{\pi}{2}$$

for $i = 0, 1, \ldots, n$. We can choose φ_i to satisfy the inequality $|\varphi_i| < \pi/2$ because condition (2.5-4) holds. $\theta_i'' - \theta_i'$ and $\varphi_i - \varphi_{i-1}$ are both determinations of the angle of the complex number

$$\frac{F(t_i, s_j)F(t_{i-1}, s_{j-1})}{F(t_{i-1}, s_j)F(t_i, s_{j-1})},$$

and hence they differ by an integral multiple of 2π. But in view of the restrictions on the absolute value of the numbers involved, we conclude that

$$\theta_i'' - \theta_i' = \varphi_i - \varphi_{i-1}.$$

Now sum from $i = 1$ to $i = n$:

$$\Sigma\theta_i'' - \Sigma\theta_i' = \Sigma(\varphi_i - \varphi_{i-1}) = \varphi_n - \varphi_0.$$

But, from condition (2.5-3), $\varphi_0 = \varphi_n = 0$. Therefore, $\Sigma\theta_i'' = \Sigma\theta_i'$ as desired.

Thus, we can assign to any element $\beta \in \pi(S^1)$ a unique integer, the degree of β. We leave it to the reader to verify by direct computation that, for any integer m, the map $h_m : I \to S^1$ defined by

$$h_m(t) = \cos 2m\pi t + i \sin 2\, m\pi t$$

has degree m. Thus, the group $\pi(S^1)$ is of infinite order, and hence is infinite cyclic. Q.E.D.

Remark: The basic idea involved in the above discussion of the degree of an element of $\pi(S^1)$ will be refined and generalized in the discussion of the fundamental group of a covering space in Chapter V.

As a corollary of Theorem 5.1, we see that the fundamental group of any space with a circle as deformation retract is infinite cyclic. Examples of such spaces are the Möbius strip, a punctured disc, the punctured plane, a region in the plane bounded by two concentric circles, etc. (see the exercises in the preceding section).

Exercises

5.1 Prove from the definitions that the mapping $\pi(S^1) \to \mathbf{Z}$ defined by $\beta \to$ (degree of β) is a homomorphism of $\pi(S^1)$ into the additive group of integers.

5.2 Give a direct proof (along the general lines of the first part of the proof of Theorem 5.1) that if $\beta \in \pi(S^1)$ and the degree of β is 0, then $\beta = 1$. (NOTE: This is an alternative way of proving Theorem 5.1.)

5.3 Let $\{U_i\}$ be an open covering of the space X having the following properties: (a) There exists a point x_0 such that $x_0 \in U_i$ for all i. (b) Each U_i is simply connected. (c) If $i \neq j$, then $U_i \cap U_j$ is arcwise connected. Prove that X is simply connected. [HINT: To prove any loop $f : I \to X$ based at x_0 is trivial, first consider the open covering $\{f^{-1}(U_i)\}$ of the compact metric space I, and make use of the Lebesgue number of this covering.]
Remark: The two most important cases of this exercise are the following: (1) A covering by two open sets, and (2) the sets U_i are linearly ordered by inclusion. The student should restate the exercise for these two special cases.

5.4 Use the result of Exercise 5.3, remark (1), to prove that the unit 2-sphere S^2 or, more generally, the n-sphere S^n, $n \geq 2$, is simply connected.

5.5 Prove that $\mathbf{R}^2$ and $\mathbf{R}^n$ are not homeomorphic if $n \neq 2$. (HINT: Consider the complement of a point in $\mathbf{R}^2$ or $\mathbf{R}^n$.)

5.6 Prove that any homeomorphism of the closed disc E^2 onto itself maps S^1 onto S^1 and U^2 onto U^2.

6 Application: The Brouwer fixed-point theorem in dimension 2

One of the best known theorems of topology is the following fixed-point theorem of L. E. J. Brouwer. Let E^n denote the closed unit ball in Euclidean n-space $\mathbf{R}^n$:

$$E^n = \{x \in \mathbf{R}^n : |x| \leqq 1\}.$$

Theorem 6.1 *Any continuous map f of E^n into itself has at least one fixed point, i.e., a point x such that $f(x) = x$.*

We shall only prove this theorem for $n \leqq 2$. Before going into the proof, it seems worthwhile to indicate why there should be interest in fixed-point theorems, such as this one.

Suppose we have a system of n equations in n unknowns:

$$\begin{aligned} g_1(x_1, \ldots, x_n) &= 0 \\ g_2(x_1, \ldots, x_n) &= 0 \\ &\cdot \\ &\cdot \\ &\cdot \\ g_n(x_1, \ldots, x_n) &= 0. \end{aligned} \tag{2.6-1}$$

Here the g_i's are assumed to be continuous real-valued functions of the real variables $x_1, \ldots, x_n$. It is often an important problem to be able to decide whether or not such a system of equations has a solution. We can transform this problem into a fixed-point problem as follows. Let

$$h_i(x_1, \ldots, x_n) = g_i(x_1, \ldots, x_n) + x_i$$

for $i = 1, 2, \ldots, n$. Then, for any point $x = (x_1, \ldots, x_n)$, we define

$$h(x) = (h_1(x), \ldots, h_n(x)).$$

Then, h is a continuous function mapping a certain subset of Euclidean n-space (depending on the domain of definition of the functions $g_1, \ldots, g_n$) into Euclidean n-space. If we can find a subset X of Euclidean n-space homeomorphic to E^n, such that h is defined in X and $h(X) \subset X$, then we can conclude by Brouwer's theorem that the function h has a fixed point in the set X; but any fixed point of the function h is readily seen to be a common solution of equations (2.6-1).

Brouwer's theorem has been extended from the subset E^n of Euclidean space to apply to certain subsets of function spaces. The resulting theorem can then be used to prove existence theorems for ordinary and partial differential equations; in fact, this is one of the most powerful methods of proving existence theorems for certain types of nonlinear equations.

PROOF OF THEOREM 6.1: For $n \leqq 2$: First we prove that, for any integer $n > 0$, the existence of a continuous map $f : E^n \to E^n$, which has no fixed points, implies that the $(n-1)$-sphere $S^{n-1} = \{x \in \mathbf{R}^n : |x| = 1\}$ is a retract of E^n. We do this by the following simple geometric construction. For any point $x \in E^n$, let $r(x)$ denote the point of intersection of S^{n-1} and the ray starting at the point $f(x)$ and going through the point x. Figure 2.5 shows the situation for the case where $n = 2$. Using vector notation, we can easily write a formula for $r(x)$ in terms of $f(x)$. From this formula, we see that r is a continuous map of E^n into S^{n-1}. If $x \in S^{n-1}$, it is clear that $r(x) = x$. Therefore, r is the desired retraction.

If we could prove that S^{n-1} is not a retract of E^n, then we would have a contradiction. For $n = 1$, this is clear, because E^1 is connected,

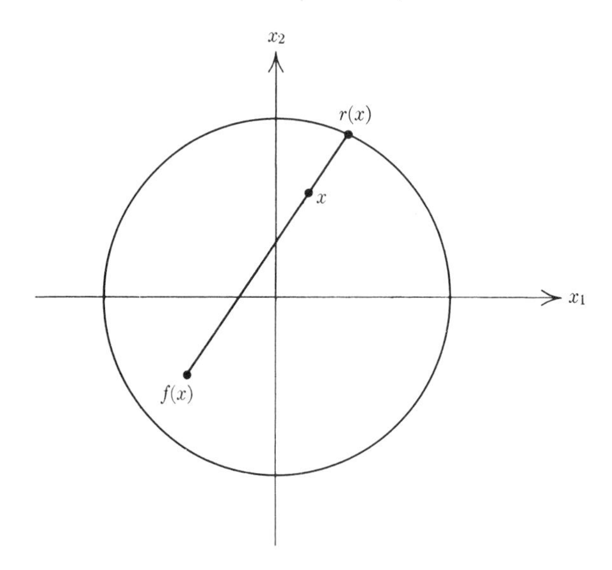

FIGURE 2.5 Proof of the Brouwer Fixed-Point Theorem.

but S^0 is disconnected. For $n = 2$, we invoke what we have learned about the fundamental groups of retracts. Because $\pi(S^1)$ is infinite cyclic, whereas $\pi(E^2)$ is a trivial group, it easily follows that S^1 is not a retract of E^2 (see the discussion of retracts in Section 4). Q.E.D.

7 The fundamental group of a product space

In this section, we shall prove that the fundamental group of the product of two spaces is naturally isomorphic to the direct product of their fundamental groups; in symbols,

$$\pi(X \times Y) \approx \pi(X) \times \pi(Y).$$

(For a review of the definition of the direct product of groups, see Section III.2.)

Let X, Y, and A be topological spaces. If $f : A \to X \times Y$ is any map, let us denote the coordinates of $f(a)$ by $(f_1(a), f_2(a))$ for any point $a \in A$. Then, f_1 and f_2 are maps of A into X and Y, respectively, and it is well known f is continuous if and only if both f_1 and f_2 are continuous. This is a basic property of the product topology. Thus, a natural one-to-one correspondence exists between continuous maps $f : A \to X \times Y$ and pairs of continuous maps $f_1 : A \to X$, $f_2 : A \to Y$. If we denote by

$p : X \times Y \to X$ and $q : X \times Y \to Y$ the projection of the product space onto its two factors, then $f_1 = pf$ and $f_2 = qf$.

Let us apply these considerations to the case where $A = I$, the unit interval. We see that there is a natural one-to-one correspondence between paths $f : I \to X \times Y$ in the product space and pairs of paths $f_1 : I \to X$, $f_2 : I \to Y$ in the factors. Note that $f_1 = pf$ and $f_2 = qf$ as before. This natural correspondence has the following obvious but important properties:

(a) If $f, g : I \to X \times Y$ are paths with the same initial and terminal points, then $f \sim g$ if and only if $f_1 \sim g_1$ and $f_2 \sim g_2$ (here $g_1 = pg$ and $g_2 = qg$).

(b) Let $f, g : I \to X \times Y$ be paths such that the terminal point of f is the initial point of g, and let $h = f \cdot g$. Then $h_1 = f_1 \cdot g_1$ and $h_2 = f_2 \cdot g_2$, where $h_1 = ph$ and $h_2 = qh$.

We can summarize these two statements by stating that the natural correspondence $f \leftrightarrow (f_1, f_2)$ is compatible with the equivalence relation and product we have defined between paths. We leave the verification of these statements to the reader.

Now let us apply these considerations to the study of the fundamental group of the product space, $\pi(X \times Y, (x, y))$. Let $p_* : \pi(X \times Y, (x, y)) \to \pi(X, x)$ and $q_* : \pi(X \times Y, (x, y)) \to \pi(Y, y)$ denote the homomorphisms induced by the projections p and q. From property (a), we see that the correspondence $\alpha \to (p_*\alpha, q_*\alpha)$ establishes a one-to-one correspondence between the sets $\pi(X \times Y, (x, y))$ and $\pi(X, x) \times \pi(Y, y)$. Moreover, it follows from property (b) that this correspondence preserves products, i.e., it is an isomorphism of groups. We summarize these results as follows:

Theorem 7.1 *The fundamental group of the product space, $\pi(X \times Y, (x, y))$, is naturally isomorphic to the direct product of fundamental groups, $\pi(X, x) \times \pi(Y, y)$. The isomorphism is defined by assigning to any element $\alpha \in \pi(X \times Y, (x, y))$ the ordered pair $(p_*\alpha, q_*\alpha)$, where $p : X \times Y \to X$ and $q : X \times Y \to Y$ denote the projections of the product space onto its factors.*

Obviously, this theorem can be extended to the product of any finite number of spaces.

Exercises

7.1 Describe the structure of the fundamental group of a torus.

7.2 Prove that the subset $S^1 \times \{x_0\}$ is a retract of $S^1 \times S^1$, but that it is not a deformation retract of $S^1 \times S^1$ for any point $x_0 \in S^1$.

7.3 Generalize Theorem 7.1 to obtain a description of the fundamental group of the product of an infinite collection of topological spaces.

7.4 Let $i : X \to X \times Y$ and $j : Y \to X \times Y$ be maps defined by $i(x) = (x, y_0)$ and $j(y) = (x_0, y)$, where $x_0 \in X$ and $y_0 \in Y$ are base points which are chosen once for all. Prove that the mapping of $\pi(X, x_0) \times \pi(Y, y_0)$ into $\pi(X \times Y, (x_0, y_0))$ defined by $(\beta, \gamma) \to (i_*\beta) \cdot (j_*\gamma)$ is an isomorphism of the first group onto the second. (HINT: Prove it is the inverse of the isomorphism described in Theorem 7.1.) Deduce as a corollary that the elements $i_*\beta$ and $j_*\gamma$ commute, i.e., $(i_*\beta)(j_*\gamma) = (j_*\gamma)(i_*\beta)$.

7.5 Assume that G is a topological space, $\mu : G \times G \to G$ is a continuous map, and $e \in G$ is such that the following conditions hold: For any $x \in G$, $\mu(x, e) = \mu(e, x) = x$. [An important example: G is a topological group, e is the identity element, and $\mu(x, y) =$ the product of x and y for any elements $x, y \in G$.] Let $i : G \to G \times G$ and $j : G \to G \times G$ be defined as in Exercise 7.4: $i(x) = (x, e)$ and $j(x) = (e, x)$ for any $x \in G$. Prove that, for any elements $\beta, \gamma \in \pi(G, e)$, $\mu_*[(i_*\beta)(j_*\gamma)] = \beta \cdot \gamma$. [HINT: Consider first the case where β or $\gamma = 1$.] Deduce as a corollary that $\pi(G, e)$ is an abelian group.

7.6 Let G, e, and μ be as in Exercise 7.5. Assume in addition that there exists a continuous map $c : G \to G$ such that $\mu(x, c(x)) = \mu(c(x), x) = e$ for any $x \in G$. [An important example: G is a topological group and $c(x) = x^{-1}$ for any $x \in G$.] Prove that, for any element $\beta \in \pi(G, e)$, $c_*(\beta) = \beta^{-1}$.

8 Homotopy type and homotopy equivalence of spaces

Before we can prove the next theorem, we need to develop some preliminary material about the topology of certain subsets of the plane. A topological space will be called a *closed disc* if it is homeomorphic to the set

$$E^2 = \{(x, y) \in \mathbf{R}^2 : x^2 + y^2 \leqq 1\};$$

it will be called an *open disc* if it is homeomorphic to the set

$$U^2 = \{(x, y) \in \mathbf{R}^2 : x^2 + y^2 < 1\}.$$

The *boundary* of a closed disc is the subset that corresponds to the circle S^1 under a homeomorphism of the disc onto E^2; it can be proved that this subset is independent of the choice of the homeomorphism (see exercise 5.6).

We shall now consider some elementary properties of discs.

(a) Any compact, convex subset E of the plane with nonempty interior is a closed disc.

PROOF: We can set up a homeomorphism between E and E^2 as follows. Choose a point x_0 belonging to the interior of E. Consider any ray in the plane starting at the point x_0; the intersection of this ray with E must be a closed interval having x_0 as one end point. Map this interval linearly onto the unit interval on the parallel ray through the origin. If we do this for each ray through x_0, we obtain a one-to-one correspondence between the points of E and E^2 which can be proved to be continuous in both directions.

(b) Let E_1 and E_2 be closed discs with boundaries B_1 and B_2, respectively. Then, any continuous map $f : B_1 \to B_2$ can be extended to a continuous map $F : E_1 \to E_2$. If f is a homeomorphism, then we can choose F to be a homeomorphism also.

PROOF: In view of the definition of a closed disc, it suffices to prove this statement in the case where $E_1 = E_2 = E^2$ and $B_1 = B_2 = S^1$. We leave this proof to the reader.

(c) Let E_1 be a closed disc. Let E_2 denote the quotient space of E_1 obtained by identifying a closed segment of the boundary of E_1 to a point. Then, this quotient space E_2 is again a closed disc.

PROOF: In view of property (b), it suffices to prove this assertion for the case of a particular closed disc and a particular segment on the boundary of that disc. We are at liberty to choose the particular disc and segment in any convenient way. We choose E_1 to be the trapezoid $ABDE$ in the xy plane shown in Figure 2.6, and E_2 to be the triangle ABC. We shall define a map $f : E_1 \to E_2$ such that the segment DE of the boundary of E_1 is mapped onto the vertex C of E_2, but otherwise f is one-to-one.

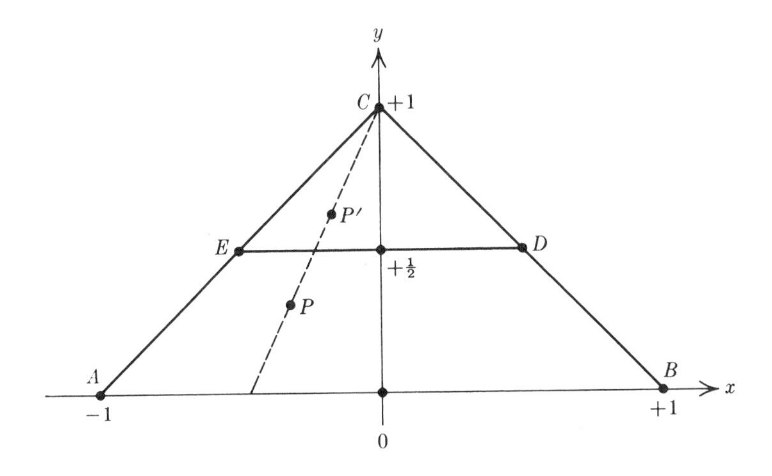

FIGURE 2.6 Proof of Theorem (8.3).

Then, we shall complete the proof by showing that E_2 has the quotient topology determined by f.

We define f by the condition that, for any point $P \in E_1$, the points P, $P' = f(P)$ and $C = (0, 1)$ will lie on a straight line, and the y coordinate of P' will be twice that of P. If (x, y) are the coordinates of P and (x', y') are the coordinates of P', then we find that

$$\left\{\begin{aligned} x' &= x\left(\frac{2y-1}{y-1}\right) \\ y' &= 2y \end{aligned}\right\} \qquad 0 \leqq y \leqq \tfrac{1}{2},$$

or

$$\left\{\begin{aligned} x &= x'\left(\frac{y'-2}{2y'-2}\right) \\ y &= \tfrac{1}{2}y' \end{aligned}\right\} \qquad 0 \leqq y' < 1.$$

The first pair of formulas shows that f is continuous, whereas the second pair of formulas shows that f is one-to-one except on the segment DE; obviously the segment DE is all mapped into the point C. Because E_1 is compact and E_2 is Hausdorff, f is a closed map, and hence E_2 has the quotient topology. Q.E.D.

We are now ready to state and prove a key lemma. Let D denote a closed disc, let B denote its boundary (which is a circle), and let $g : I \to B$ denote a continuous map which wraps the interval exactly once around the circle; i.e., $g(0) = g(1) = d_0 \in B$, and g maps the open interval $(0, 1)$ homeomorphically onto $B - \{d_0\}$. Let X be a topological space.

Lemma 8.1 *A continuous map $f : B \to X$ can be extended to a map $D \to X$ if and only if the closed loop $fg : I \to X$ is equivalent to the constant loop at the base point $f(d_0)$.*

PROOF: First assume that $f : B \to X$ can be extended to a continuous map $F : D \to X$. Consider the unit square $\{(x, y) \in \mathbf{R}^2 : 0 \leqq x \leqq 1$ and $0 \leqq y \leqq 1\}$. Define a continuous map h of the boundary of this square into B as follows:

$$\begin{aligned} h(x, 0) &= g(x), \qquad 0 \leqq x \leqq 1, \\ h(x, 1) &= h(0, y) = h(1, y) = d_0 \end{aligned} \tag{2.8-1}$$

for $x \in I$ or $y \in I$. By property (b), we can extend h to a continuous map H of the unit square. Then, the existence of the composite map FH proves the loop fg is equivalent to the constant path.

Next, assume the loop fg is equivalent to the constant path. By definition, this means there exists a continuous map G of the unit square into X such that

$$G(x, 0) = f(g(x)),$$
$$G(x, 1) = G(0, y) = G(1, y) = f(d_0).$$

Because G maps the top and the two sides of this square into the single point $f(d_0)$, it is clear that G induces a continuous map of the quotient space of the square (obtained by identifying the top and two sides of the square to a single point) into X. By property (c), this quotient space is a closed disc, which we may take to be D, and the natural map of the boundary of the square onto the quotient space may be taken to be the map h in equations (2.8-1). The induced map of the disc D into X is clearly an extension of f, as desired. Q.E.D.

In applying this lemma, it is convenient to use the following "abuse of language": We shall say that the map $f : B \to X$ "*represents*" the equivalence class of the loop fg.

To state the next theorem, let $\varphi_0, \varphi_1 : X \to Y$ be continuous maps, and let $\varphi : X \times I \to Y$ be a homotopy between φ_0 and φ_1; i.e., $\varphi(x, 0) = \varphi_0(x)$ and $\varphi(x, 1) = \varphi_1(x)$. Choose a base point $x_0 \in X$. Then, φ_0 and φ_1 induce homomorphisms

$$\varphi_{0*} : \pi(X, x_0) \to \pi(Y, \varphi_0(x_0)),$$
$$\varphi_{1*} : \pi(X, x_0) \to \pi(Y, \varphi_1(x_0)).$$

Let γ denote the homotopy class of the path $t \to \varphi(x_0, t)$, $0 \leqq t \leqq 1$, in Y. This defines an isomorphism $u : \pi(Y, \varphi_0(x_0)) \to \pi(Y, \varphi_1(x_0))$ by the formula

$$u(\alpha) = \gamma^{-1}\alpha\gamma, \qquad \alpha \in \pi(Y, \varphi_0(x_0)).$$

Theorem 8.2 *Under the above hypotheses, the following diagram is commutative:*

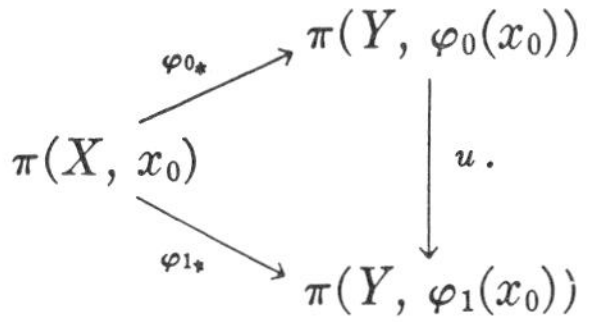

$$\begin{array}{ccc} & \overset{\varphi_{0*}}{\nearrow} & \pi(Y, \varphi_0(x_0)) \\ \pi(X, x_0) & & \downarrow u \\ & \underset{\varphi_{1*}}{\searrow} & \pi(Y, \varphi_1(x_0)) \end{array}$$

This theorem is the natural and full generalization of Theorem 4.1.

PROOF: Let $\alpha \in \pi(X, x_0)$; we must prove that

$$\varphi_{1*}(\alpha) = \gamma^{-1}(\varphi_{0*}\alpha)\gamma.$$

Choose a closed path $f : I \to X$ representing the path α. Consider the map

$$g : I \times I \to Y$$

defined by

$$g(x, y) = \varphi(f(x), y).$$

Then, for $x, y \in I$, we have

$$g(x, 0) = \varphi_0(f(x)),$$

$$g(x, 1) = \varphi_1(f(x)),$$

$$g(0, y) = g(1, y) = \varphi(x_0, y).$$

Hence, the map g of the bottom of the square represents $\varphi_{0*}(\alpha)$, on the top of the square it represents $\varphi_{1*}(\alpha)$, and on the two sides of the square it represents γ. If we read around the boundary of the square, the map represents $(\varphi_{0*}\alpha)\gamma(\varphi_{1*}\alpha)^{-1}\gamma^{-1}$. Now apply Lemma 8.1 to conclude that

$$(\varphi_{0*}\alpha)\gamma(\varphi_{1*}\alpha)^{-1}\gamma^{-1} = 1.$$

From this the desired equation follows [multiply on the right by $\gamma(\varphi_{1*}\alpha)$ and then on the left by γ^{-1}]. Q.E.D.

Definition Two spaces X and Y are of the *same homotopy type* if there exist continuous maps (called *homotopy equivalences*) $f : X \to Y$, $g : Y \to X$ such that $gf \simeq$ identity: $X \to X$ and $fg \simeq$ identity: $Y \to Y$.

Obviously, two homeomorphic spaces are of the same homotopy type, but the converse is not true.

Exercise

8.1 Prove that, if A is a deformation retract of X, then the inclusion $i : A \to X$ is a homotopy equivalence. (Actually, one of the conditions in the definition of a deformation retract given in Section 4 is superfluous here; omission of this condition leads to the notion of a "deformation retract in the weak sense." For spaces which are sufficiently "nice," it can be proved that the two notions agree.)

Theorem 8.3 *If $f : X \to Y$ is a homotopy equivalence, then f_* : $\pi(X, x) \to \pi(Y, f(x))$ is an isomorphism for any $x \in X$.*

PROOF: Because $gf \simeq$ identity : $X \to X$, we obtain the following diagram (which is commutative by Theorem 8.2):

$$\begin{array}{ccc} \pi(X, x) & \xrightarrow{f_*} & \pi(Y, f(x)) \\ & \searrow^{u} & \downarrow g_* \\ & & \pi(X, gf(x)) \end{array}.$$

Here u is an isomorphism induced by a certain path from x to $gf(x)$. Therefore, we conclude f_* is a monomorphism, and g_* is an epimorphism.

If we apply the same argument to the homotopy $fg \simeq$ identity : $Y \to Y$, we obtain the following commutative diagram:

$$\begin{array}{ccc} \pi(Y, f(x)) & & \\ {\scriptstyle g_*}\downarrow & \searrow^{v} & \\ \pi(X, gf(x)) & \longrightarrow & \pi(Y, fgf(x)) \end{array}$$

Therefore, we conclude g_* is a monomorphism. Because g_* is both an epimorphism and a monomorphism, it is an isomorphism. Because

$$g_*f_* = u$$

and both g_* and u are isomorphisms, we conclude that f_* is also an isomorphism. Q.E.D.

This theorem will be used as an aid in the determination of the fundamental group of certain spaces, and as a method of proving that certain spaces are not of the same homotopy type (and hence are not homeomorphic).

Exercise

8.2 Assume that G, μ, and e satisfy the hypotheses of Exercise 7.5. Use Lemma 8.1 to prove directly that for any elements $\alpha, \beta \in \pi(G, e)$, $\alpha\beta\alpha^{-1}\beta^{-1} = 1$. (HINT: Choose D to be a square, and choose a map of B into G which represents $\alpha\beta\alpha^{-1}\beta^{-1}$. Use the existence of μ to define the required extension.) Deduce that $\pi(G, e)$ is abelian.

NOTES

The fundamental group was introduced by the great French mathematician Henri Poincaré in 1895 ("Analysis Situs." *J. Ecole Polytechn.*, *1*, 1895, pp. 1–121). The notion of two spaces being of the same homotopy type was introduced by Witold Hurewicz in a series of four papers, in 1935–36, which appeared in the *Proceedings of the Koninklijke Nederlandse Akademie van Wetenschapen.* In these papers, Hurewicz also introduced higher dimensional analogs of the fundamental group, called *homotopy groups.* These ideas of Hurewicz have played a significant role in algebraic topology since 1935.

The reader who is interested in the proof of existence theorems in analysis by the use of fixed-point theorems is referred to the following book by Jane Cronin: Mathematical Surveys. No. 11, *Fixed Points and Topological Degree in Non-linear Analysis.* Providence: American Mathematical Society, 1964.

REFERENCES

1. Crowell, R. H., and R. H. Fox. *Introduction to Knot Theory*. Boston: Ginn, 1963, Chapters II and V.
2. Hilton, P., and S. Wylie. *Homology Theory: An Introduction to Algebraic Topology*. Cambridge: The University Press, 1960, Chapter VI.
3. Pontrjagin, L. *Topological Groups*. Princeton, N.J.: Princeton University Press, 1939, Section 46.
4. Seifert, H., and W. Threlfall. *A Textbook of Topology*. New York: Academic Press, 1980. Chapter 7.

CHAPTER THREE

Free Groups and Free Products of Groups

1 Introduction

In the preceding chapters we have introduced the fundamental group of a space and actually determined its structure in some of the simplest cases. In more complicated cases we need a larger vocabulary and a greater knowledge of group theory to describe its structure and actually to make use of its properties. The object of this chapter is to supply this need. We first discuss the case of abelian groups, because this case is simpler and more closely related to the student's previous experience. Then we discuss the general case of not necessarily abelian groups. Here the results are entirely analogous to the abelian case, but the possibilities are more varied and less intuitive.

The three main group theoretic concepts introduced in this chapter are the following: free group, free product of groups, and presentation of a group by generators and relations. These concepts will be used throughout the rest of this book. The definition of a free group or a free product of groups involves a mathematical concept of wide application, the so-called "universal mapping problem," which is also a basic concept in Chapter IV.

2 The weak product of abelian groups

Possibly the student is already familiar with the concept of *product*, or *direct product*, or *cartesian product*, of two groups; in any case the definition is very simple, and we repeat it here. Let G_1 and G_2 be groups. Their *product*, denoted by $G_1 \times G_2$, is the set of all ordered pairs (g_1, g_2), $g_1 \in G_1$, $g_2 \in G_2$, with multiplication defined componentwise according to the following rule:

$$(g_1, g_2)(g_1', g_2') = (g_1g_1', g_2g_2').$$

The verification that $G_1 \times G_2$ is actually a group is a simple routine matter.

For any positive integer n, we may define in a similar way the product of n groups, $G_1, \ldots, G_n$; it is denoted by $G_1 \times G_2 \times \cdots \times G_n$ or

$$\prod_{i=1}^{n} G_i.$$

In fact, we may even define in this way the product of an infinite sequence of groups $G_1, G_2, G_3, \ldots$, denoted by

$$\prod_{i=1}^{\infty} G_i.$$

In each case, the underlying set is the cartesian product of the underlying sets of the groups involved, and the multiplication is defined componentwise. At this stage, the reader will probably recall that in set theory the cartesian product of *any* (nonempty) collection of sets is well defined; we need not confine ourselves to the case of a countable collection of sets. In a similar way, we can define the product of *any* (nonempty) collection of groups $\{G_i : i \in I\}$, where I denotes some index set, countable or not (here I does not denote the unit interval). First we form the set-theoretic product, and then we define multiplication componentwise: For any elements

$$g, g' \in \prod_{i \in I} G_i,$$

and any index $i \in I$, the ith component of the product gg' is given by the formula

$$(gg')_i = (g_i)(g'_i).$$

In words, the ith component of the product is the product of the ith components of the factors.

Let $\{G_i : i \in I\}$ be any collection of groups, and let

$$G = \prod_{i \in I} G_i$$

be their product.

Definition The *weak product*[1] of the collection $\{G_i : i \in I\}$ is the subgroup of their product G consisting of all elements $g \in G$ such that g_i is the identity element of G_i for all except a finite number of indices i.

[1] When each group G_i is abelian and the group operation is addition, it is customary to call the weak product the "direct sum." In this definition, we do not require that any two groups in the collection $\{G_i\}$ be nonisomorphic. In fact, it may even occur that all of the groups of the collection are isomorphic to some given group.

Obviously, if $\{G_i : i \in I\}$ is a finite collection of groups, then the product and weak product are the same.

If G denotes either the product or weak product of the collection $\{G_i : i \in I\}$, then, for each index $i \in I$, there is a *natural monomorphism* $\varphi_i : G_i \to G$ defined by the following rule: For any element $x \in G_i$ and any index $j \in I$,

$$(\varphi_i x)_j = \begin{cases} x & \text{if } j = i, \\ 1 & \text{if } j \neq i. \end{cases}$$

In the case where each G_i is an *abelian* group, the following theorem gives an important characterization of their weak product G and the monomorphisms φ_i.

Theorem 2.1 *If $\{G_i : i \in I\}$ is a collection of abelian groups and G is their weak product, then for any abelian group A and any collection of homomorphisms*

$$\psi_i : G_i \to A, \qquad i \in I,$$

there exists a unique homomorphism $f : G \to A$ such that for any $i \in I$ the following diagram is commutative:

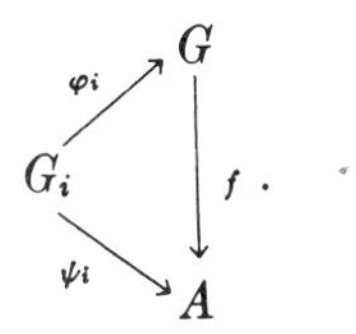

PROOF: Given the ψ_i's, define f by the following rule: For any $x \in G$, $f(x)$ will be the product of the elements $\psi_i(x_i)$ for all $i \in I$. Because $x_i = 1$ for all except a finite number of indices i, this product is really a finite product; and because all the groups involved are abelian, the order of multiplication is immaterial. Thus, $f(x)$ is well defined, and it is readily verified that f is a homomorphism, which renders the given diagram commutative. It is easy to see that f is the unique homomorphism having this property. Q.E.D.

Our next proposition states that this theorem actually characterizes the weak product of abelian groups.

Proposition 2.2 *Let $\{G_i\}$, G, and $\varphi_i : G_i \to G$ be as in Theorem 2.1; let G' be any abelian group and let $\varphi_i' : G_i \to G'$ be any collection of homomorphisms such that the conclusion of Theorem 2.1 holds with G' and φ_i' sub-*

stituted for G *and* φ_i*, respectively. Then, there exists a unique isomorphism* $h : G \to G'$ *such that the following diagram is commutative for any* $i \in I$*:*

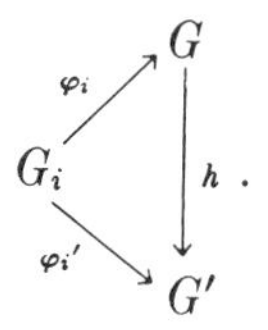

PROOF: The existence of a *homomorphism* $h : G \to G'$ making the required diagram commutative is assured by Theorem 2.1. Because Theorem 2.1 also applies to G' and the φ_i' (by hypothesis), there exists a unique homomorphism $k : G' \to G$ such that the following diagram is commutative for any index $i \in I$:

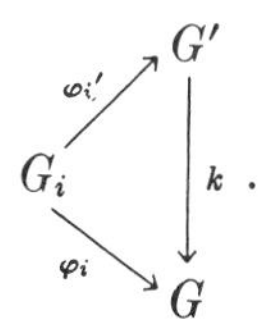

From these facts, we readily conclude that the following two diagrams are commutative for any $i \in I$:

However, these two diagrams would also be commutative if we replaced kh by the identity map $G \to G$ in the first, and hk by the identity map $G' \to G'$ in the second. We now invoke the uniqueness statement in the conclusion of Theorem 2.1 to conclude that kh and hk are both identity maps. Hence, h and k are inverse isomorphisms of each other. Q.E.D.

The student should reflect on the significance of the characterization of the weak product given by Theorem 2.1. We may consider any other abelian group A with definite homomorphisms $\psi_i : G_i \to A$ as a candidate for some kind of a "product" of the abelian groups G_i; then this theorem asserts that the weak product G is the "freest" among all such candidates, in the sense that there exists a homomorphism of G into A commuting with φ_i and ψ_i for all i. Here we use the word "freest" in the sense of "fewest possible relations imposed," and the general philosophy is that if certain relations hold for the group G, they also hold for any homomorphic image of G; of course, additional relations may hold for the homomorphic image. This same philosophy also holds for other kinds of algebraic objects, such as rings, etc.

As we shall see, the argument used to prove Proposition 2.2 applies almost verbatim to many other cases.

Since the weak product G of a collection $\{G_i\}$ of abelian groups is completely characterized by the properties of the monomorphisms $\varphi_i : G_i \to G$ stated in Theorem 2.1, we could just as well ignore the fact that G is a subgroup of the product

$$\prod_{i \in I} G_i$$

and focus our attention instead on the group G and the homomorphisms φ_i. Furthermore, because each φ_i is a monomorphism, we can identify G_i with its image in G under φ_i, and consider φ_i as an inclusion map, if this is convenient. In this case, we say that G is the weak product of the subgroups G_i, it being understood that each φ_i is an inclusion map.

3 Free abelian groups

We recall that, if S is a subset of a group G, then S is said to *generate* G in case every element of G can be written as a product of positive and negative powers of elements of S. (An equivalent condition is the following: S is not contained in any proper subgroup of G.) For example, if G is a cyclic group of order n,

$$G = \{x, x^2, x^3, \ldots, x^n = 1\},$$

then the set $S = \{x\}$ generates G.

If the set S generates the group G, certain products of elements of S may be the identity element of G. For example,

(a) If $x \in S$, then $xx^{-1} = 1$.
(b) If G is a cyclic group of order n generated by $\{x\}$, then $x^n = 1$.

Any such product of elements of S that is equal to the identity is often called a *relation* between the elements of the generating set S. Roughly speaking, we may distinguish between two types of relations between generators: *trivial relations*, as in example (a), which are a direct consequence of the axioms for a group and thus hold no matter what the choice of G and S, and *nontrivial relations*, such as example (b), which are not a consequence of the axioms for a group, but depend on the particular choice of G and S.

These notions lead naturally to the following definition: Let S be a set of generators for the group G. We say that G is *freely generated by* S or a *free group* on S in case there are no nontrivial relations between the elements of S. For example, if G is an infinite cyclic group consisting of all positive and negative powers of the element x, then G is a free group on the set $S = \{x\}$.

These notions also lead to the idea that we can completely prescribe a group by listing the elements of a generating set S and listing the nontrivial relations between them.

The ideas described in the preceding paragraphs have been current among group theorists for a long time. Unfortunately, when stated as above, these ideas are lacking in mathematical precision. For example, what precisely is a nontrivial relation? It cannot be an element of G, because considered as elements of G, all relations give the identity. Also, under what conditions are two relations to be considered the same? For example, in a cyclic group of order n, are the relations

$$x^n = 1$$

$$x^{n+1}x^{-1} = 1$$

to be considered the same or different?

We should emphasize that it was not an easy matter for mathematicians to find an entirely satisfactory and precise way of treating these questions. Fortunately, such a treatment has been found in recent years. This treatment has the advantage that it applies not only to groups, but also to other algebraic structures such as rings, and even to many situations in other branches of mathematics. As so often happens in mathematics, the method of definition finally chosen seems rather roundabout and nonobvious.[2] This method of definition depends on the following rather simple observations:

(1) Let S be a set of generators for G, and let $f : G \to G'$ be an epimorphism; i.e., G' is a homomorphic image of G. Then, the set $f(S)$ is a set of generators for G'. Moreover, *any relation which holds between the elements of* S *also holds between the elements of* $f(S)$. Thus, the group G' satisfies at least as many relations as or more relations than G.

(2) Let S be a set of generators for G, and let $f : G \to G'$ be an arbitrary homomorphism. Then, f is completely determined by its restriction to the set S. However, we do not assert that any map $g : S \to G'$ can be extended to a homomorphism $f : G \to G'$ (the student should give a

[2] An analogous situation occurs in the problem of precisely defining limits in the calculus. The $\varepsilon - \delta$ technique which is standard today seems rather far removed from our intuitive notion of a variable quantity approaching a limit.

counter-example). The intuitive reason for this is clear: Given a map $g : S \to G'$ there may be nontrivial relations between the elements of S which do *not* hold between the elements of $g(S)$.

We shall now give a precise definition of a free *abelian* group on a given set S; in Section 5 we shall discuss the case of general (i.e., not necessarily abelian) groups. The case of abelian groups is discussed first because it is simpler.

Definition Let S be an arbitrary set. A *free abelian group* on the set S is an abelian group F together with a function $\varphi : S \to F$ such that the following condition holds: For any abelian group A and any function $\psi : S \to A$, there exists a unique homomorphism $f : F \to A$ such that the following diagram is commutative:

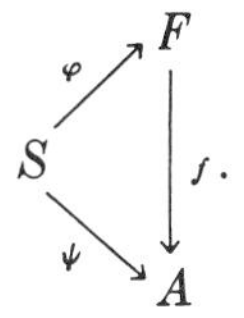

First, we show that this definition does indeed characterize free abelian groups on a given set S.

Proposition 3.1 *Let F and F' be free abelian groups on the set S with respect to the functions $\varphi : S \to F$ and $\varphi' : S \to F'$, respectively. Then, there exists a unique isomorphism $h : F \to F'$ such that the following diagram is commutative:*

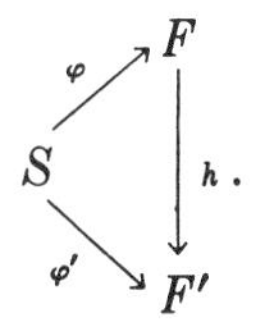

PROOF: The proof is completely analogous to that of Proposition 2.2, and may be left to the reader.

Let us emphasize that all we have done so far is make a definition; given the set S, it is not at all clear that there exists a free abelian group F on the set S. Moreover, even if F exists, it is conceivable that the map φ need not be one-to-one, or that F may not be generated by the subset $\varphi(S)$ in the sense of the definition at the beginning of this section. We shall clarify all these points by actually proving the existence of F and elucidating its structure completely.

Exercise

3.1 Prove directly from the definition that $\varphi(S)$ generates F. [HINT: Assume not; consider the subgroup F' generated by $\varphi(S)$.]

As a first step, we consider the following situation. Assume that $\{S_i : i \in I\}$ is a family of nonempty subsets of S, which are pairwise disjoint and such that

$$S = \bigcup_{i \in I} S_i.$$

For each index $i \in I$, let F_i be a free abelian group on the set S_i with respect to a function $\varphi_i : S_i \to F_i$. Let F denote the weak product of the groups F_i for all $i \in I$, and let $\eta_i : F_i \to F$ denote the natural monomorphism. Since the S_i are pairwise disjoint, we can define a function $\varphi : S \to F$ by the rule

$$\varphi \mid S_i = \eta_i \varphi_i.$$

Proposition 3.2 *Under the above hypotheses, F is a free abelian group on the set S with respect to the function $\varphi : S \to F$.*

Roughly speaking, this proposition means that the weak product of any collection of free abelian groups is a free abelian group.

PROOF: Let A be an abelian group and let $\psi : S \to A$ be a function. We have to prove the existence of a unique homomorphism $f : F \to A$ such that $\psi = f\varphi$. For each index i, let $\psi_i : S_i \to A$ denote the restriction of ψ to the subset S_i. Because F_i is a free abelian group on the set S_i, there exists a unique homomorphism $f_i : F_i \to A$ such that the following diagram is commutative:

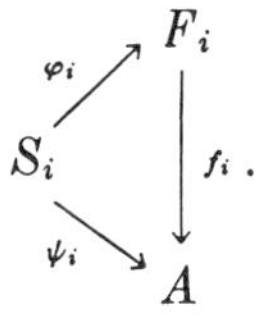

$$\begin{array}{ccc} & & F_i \\ & \overset{\varphi_i}{\nearrow} & \\ S_i & & \downarrow f_i \\ & \underset{\psi_i}{\searrow} & \\ & & A \end{array} \qquad (3.3\text{-}1)$$

We now invoke the fundamental property of the weak product of groups contained in Theorem 2.1 to conclude that there exists a unique homomorphism $f : F \to A$ such that the following diagram is commutative for any index i:

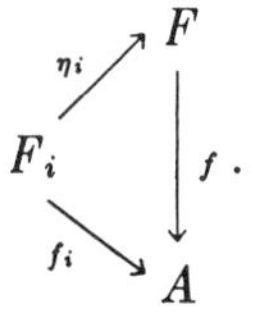

$$\begin{array}{ccc} & & F \\ & \overset{\eta_i}{\nearrow} & \\ F_i & & \downarrow f \\ & \underset{f_i}{\searrow} & \\ & & A \end{array} \qquad (3.3\text{-}2)$$

We can put these two commutative diagrams together into a single diagram as follows:

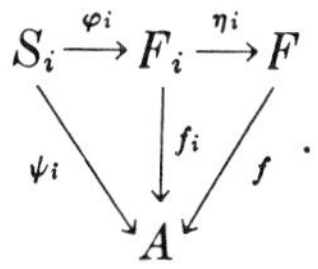

$$\begin{array}{ccccc} S_i & \xrightarrow{\varphi_i} & F_i & \xrightarrow{\eta_i} & F \\ & {\scriptstyle \psi_i}\searrow & \downarrow {\scriptstyle f_i} & \swarrow {\scriptstyle f} & \\ & & A & & \end{array} \quad . \qquad (3.3\text{-}3)$$

Because $\varphi \mid S_i = \eta_i\varphi_i$, we conclude that the following diagram is commutative for each index i.

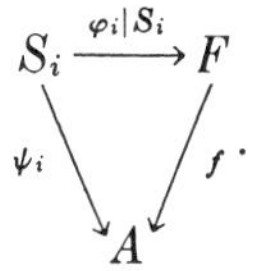

$$\begin{array}{ccc} S_i & \xrightarrow{\varphi_i|S_i} & F \\ {\scriptstyle \psi_i}\searrow & & \swarrow {\scriptstyle f} \\ & A & \end{array} \quad . \qquad (3.3\text{-}4)$$

Finally, because $\psi_i = \psi \mid S_i$ for each i and $S = \cup S_i$, we conclude that $\psi = f\varphi$, as required.

To prove uniqueness, let f be any homomorphism $F \to A$ having the required property. Define $f_i : F_i \to A$ by $f_i = f\eta_i$. With this definition, it follows that diagram (3.3-1) is commutative for each index i; for,

$$\begin{aligned} f_i\varphi_i &= f\eta_i\varphi_i = f(\varphi \mid S_i) = (\psi \mid S_i) \\ &= \psi_i. \end{aligned}$$

Because F_i is the free abelian group on S_i (with respect to φ_i), it follows that each f_i is unique. Then because (3.3-2) is commutative for each i, and F is the weak product of the F_i, it follows that f is unique. Q.E.D.

We now apply this theorem as follows: Suppose that

$$S = \{x_i : i \in I\}.$$

For each index i, let S_i denote the subset $\{x_i\}$ having only one element, and let F_i be an infinite cyclic group consisting of all positive and negative powers of the element x_i:

$$F_i = \{x_i^n : n \in \mathbf{Z}\}.$$

Let $\varphi_i : S_i \to F_i$ denote the inclusion map, i.e., $\varphi_i(x_i) = x_i^1$. It is clear that F_i is a free abelian group on the set S_i. Therefore, all the hypotheses of Proposition 3.2 are satisfied. Thus, we conclude that a free abelian group on any set S is a weak product of a collection of infinite cyclic groups, with the cardinal number of the collection equal to that of S.

Because F is the weak product of the F_i, any element $g \in F$ is of the following form: For any index i, the ith component $g_i = x_i^{n_i}$ where each

$n_i \in \mathbf{Z}$ and $n_i = 0$ for all but a finite number of indices i. Moreover, the function φ is defined by the following rule: For any index $j \in I$,

$$(\varphi x_i)_j = \begin{cases} x_i^1 & \text{if } i = j \\ x_j^0 & \text{if } i \neq j. \end{cases}$$

From this formula, it is clear that φ is a one-to-one map.

As φ is a one-to-one map, if we wish, we can identify each $x_i \in S$ with its image $\varphi(x_i) \in F$. Then S becomes a subset of F, and it is clear that we can express each element $g \neq 1$ of F uniquely in the following form:

$$g = x_{i_1}^{n_1} x_{i_2}^{n_2} \cdots x_{i_k}^{n_k} \tag{3.3-5}$$

where the indices $i_1, i_2, \ldots, i_k$ are all distinct, and $n_1, n_2, \ldots, n_k$ are nonzero integers. This expression for the element g is unique except for the order of the factors. Moreover, each such product of the x_i's represents a unique element $g \neq 1$ of F. From this it is clear that F is generated by the subset $S = \varphi(S)$.

This identification of S and $\varphi(S)$ is quite customary in the discussion of free abelian groups. When this is done, $\varphi : S \to F$ becomes an inclusion map, and often it is not even mentioned in the discussion.

An alternative approach to the topic of free abelian groups would be to *define* an abelian group F to be free on the subset $\{x_i : i \in I\} \subset F$ if every element $g \neq 1$ of F admits an expression of the form (3.3-5), which is unique up to order of the factors. Actually, this procedure would be somewhat quicker and easier than the one we have chosen. However, it would suffer from the disadvantage that it could not be generalized to non-abelian groups and other situations which will actually be our main concern.

One reason for the importance of free abelian groups is the following proposition.

Proposition 3.3 *Any abelian group is the homomorphic image of a free abelian group; i.e., given any abelian group A, there exists a free abelian group F and an epimorphism $f : F \to A$.*

PROOF: The proof is very simple. Let $S \subset A$ be a set of generators for A (e.g., we could take $S = A$), and let F be a free group on the set S with respect to a function $\varphi : S \to F$. Let $\psi : S \to A$ denote the inclusion map. By definition, there exists a homomorphism $f : F \to A$ such that $f\varphi = \psi$. It is clear that f must be an epimorphism, since S was chosen to be a set of generators for A. Q.E.D.

This proposition enables us to attach a precise meaning to the notion "nontrivial relation between the generators S," mentioned earlier. Let

A, S, F, and f have the meaning just described; then we define any element $r \neq 1$ of kernel f to be a nontrivial relation between the set of generators S. If $\{r_i : i \in I\}$ is any collection of such relations, and r is an element of the subgroup of F generated by the r_i's, then the relation r is said to be a *consequence* of the relations r_i. This implies that r can be expressed as a product of the r_i's and their inverses. If the collection $\{r_i : i \in I\}$ generates the kernel of f, then the group A is completely determined up to isomorphism by the set of generators S and the set of relations $\{r_i : i \in I\}$; for, A is isomorphic to the quotient group of F modulo the subgroup generated by the r_i's.

It is clear that, if S and S' are sets having the same cardinal number, and F and F' are free abelian groups on S and S', respectively, then F and F' are isomorphic. We shall now show that the converse of this statement is true, at least for the case of finite sets. For this purpose, we make the following definition. If G is any group, and n is any positive integer, then G^n denotes the subgroup of G generated by the set

$$\{g^n : g \in G\}.$$

If the group G is abelian, then the set $\{g^n : g \in G\}$ is actually already a subgroup.

Lemma 3.4 *Let F be a free abelian group on a set consisting of k elements. Then, the quotient group F/F^n is a finite group of order n^k.*

PROOF: We leave the proof to the reader; it is not difficult if one makes use of the explicit structure of free abelian groups described above.

Corollary 3.5 *Let S and S' be finite sets whose cardinals are not equal, and let F and F' be free abelian groups on S and S', respectively. Then, F and F' are nonisomorphic.*

PROOF: The proof is by contradiction. Any isomorphism between F and F' would induce an isomorphism between the quotient groups F/F^n and F'/F'^n, which is impossible by the lemma.

Exercise

3.2 Prove that the statement of this corollary is still true if S is a finite set and S' is an infinite set.

Let F be a free abelian group on a set S. The cardinal number of the set S is called the *rank* of F. We have proved that *two free abelian groups are isomorphic if and only if they have the same rank,* at least in the case where one of them has finite rank.

We shall conclude this section on abelian groups with a brief discussion of the structure of finitely generated abelian groups. Let A be an abelian group; the set of all elements of A which have finite order is readily seen to be a subgroup, called the *torsion subgroup of* A. When the torsion subgroup consists of the element 1 alone, A is called a *torsion-free* abelian group. On the other hand, if every element of A has finite order, then A is called a torsion group. If we denote the torsion subgroup by T, then the quotient group A/T is obviously torsion free. It is clear that, if A and A' are isomorphic, then so are their torsion subgroups, T and T', and their torsion-free quotient groups, A/T and A'/T'. However, the converse is not true in general; we cannot conclude that A is isomorphic to A' if $T \approx T'$ and $A/T \approx A'/T'$. However, for abelian groups which are generated by a finite subset we have the following theorem which describes their structure completely:

Theorem 3.6 (*a*) *Let A be a finitely generated abelian group and let T be its torsion subgroup. Then, T and A/T are also finitely generated, and A is isomorphic to the direct product $T \times A/T$. Hence, the structure of A is completely determined by its torsion subgroup T and its torsion-free quotient group A/T.* (*b*) *Every finitely generated torsion-free abelian group is a free abelian group of finite rank.* (*c*) *Every finitely generated torsion abelian group T is isomorphic to a product $C_1 \times C_2 \times \cdots \times C_n$, where each C_i is a finite cyclic group of order ε_i such that ε_i is a divisor of ε_{i+1} for $i = 1, 2, \ldots, n-1$. Moreover, the integers $\varepsilon_1, \varepsilon_2, \ldots, \varepsilon_n$ are uniquely determined by the torsion group T and they completely determine its structure.*

The numbers $\varepsilon_1, \ldots, \varepsilon_n$ are called the *torsion coefficients* of T, or more generally, if T is the torsion subgroup of A, they are called the torsion coefficients of A. Similarly, the rank of the free group A/T is called the rank of A. With this terminology, we can summarize Theorem 3.6 by stating that the rank and torsion coefficients are a complete set of invariants of a finitely generated abelian group. Theorem 3.6 asserts that every finitely generated abelian group is a direct product of cyclic groups, but it also asserts much more. Note that a finitely generated torsion group is actually of finite order.

A word of explanation about the various isomorphisms mentioned in Theorem 3.6 seems in order here. These isomorphisms are not *natural*, or uniquely determined in any way. In each case, there are usually many different choices for the isomorphism in question and one choice is as good as another.

Theorem 3.7 *Let F be a free abelian group on a set S, and let F' be a subgroup of F. Then, F' is a free abelian group on a certain set S', and the cardinal of S' is less than or equal to that of S.*

Although the proofs of Theorems 3.6 and 3.7 are not difficult, we shall not give them here, because they properly belong in the study of linear algebra and modules over a principal ideal domain.

Exercises

3.3 Give an example of a torsion-free abelian group which is not free.

3.4 Let A be an abelian group which is a direct product of two cyclic groups of orders 12 and 18, respectively. What are the torsion coefficients of A? (Note that the torsion coefficients are required to satisfy a divisibility condition.)

3.5 Give an example to show that in Theorem 3.7 the subset $S \subset F$ and the subgroup $F' \subset F$ may be disjoint, even in the case where the cardinals of S and S' are equal.

4 Free products of groups

The free product of a collection of groups is the exact analog for arbitrary (i.e., not necessarily abelian) groups of the weak product for abelian groups. (It should be emphasized that any groups considered in this section may be either abelian or non-abelian, unless the contrary is explicitly stated.)

Definition Let $\{G_i : i \in I\}$ be a collection of groups, and assume there is given for each index i a homomorphism φ_i of G_i into a fixed group G. We say that G is the *free product* of the groups G_i (with respect to the homomorphisms φ_i) if and only if the following condition holds: For any group H and any homomorphisms

$$\psi_i : G_i \to H, \qquad i \in I,$$

there exists a unique homomorphism $f : G \to H$ such that for any $i \in I$, the following diagram is commutative:

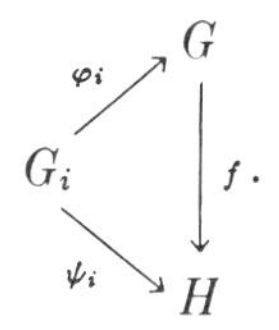

First, we have the following uniqueness proposition about free products:

Proposition 4.1 *Assume that G and G' are free products of a collection $\{G_i : i \in I\}$ of groups (with respect to homomorphisms $\varphi_i : G_i \to G$ and*

$\varphi_i' : G_i \to G'$, respectively). Then, there exists a unique isomorphism $h : G \to G'$ such that the following diagram is commutative for any $i \in I$:

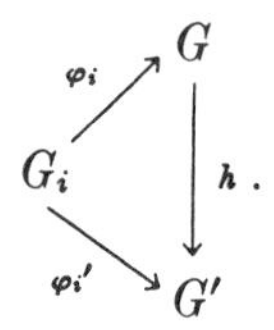

PROOF: The proof is almost word for word that of Proposition 2.2.

Although we have defined free products of groups and proved their uniqueness, it still remains to prove that they always exist. We shall also show that each of the homomorphisms φ_i occurring in the definition is a monomorphism, that the free product is generated by the union of the images $\varphi_i(G_i)$, and get more detailed insight into the algebraic structure of a free product.

Theorem 4.2 *Given any collection $\{G_i : i \in I\}$ of groups, their free product exists.*

PROOF: We define a *word* in the G_i's to be a finite sequence $(x_1, x_2, \ldots, x_n)$ where each x_k belongs to one of the groups G_i, *any two successive terms in the sequence belong to different groups, and no term is the identity element of any G_i.* The integer n is the *length* of the word. We also include the empty word, i.e., the unique word of length 0. Let W denote the set of all such words.

For each index i, we now define left operations of the group G_i on the set W (see Appendix B). Let $g \in G_i$ and $(x_1, \ldots, x_n) \in W$; we must define $g(x_1, \ldots, x_n)$.

Case 1: $x_1 \notin G_i$. Then, if $g \neq 1$,

$$g(x_1, \ldots, x_n) = (g, x_1, \ldots, x_n).$$

We shall also define the action of g on the empty word by a similar formula, i.e., $g(\ ) = (g)$. If $g = 1$, then,

$$g(x_1, \ldots, x_n) = (x_1, \ldots, x_n).$$

Case 2: $x_1 \in G_i$. Then,

$$g(x_1, \ldots, x_n) = \begin{cases} (gx_1, x_2, \ldots, x_n) & \text{if } gx_1 \neq 1, \\ (x_2, \ldots, x_n) & \text{if } gx_1 = 1. \end{cases}$$

[When $gx_1 = 1$ and $n = 1$, it is understood, of course, that $g(x_1)$ is the empty word.]

We must now verify that the requirements for left operations of G_i on W are actually satisfied; i.e., for any word w,

$$1w = w,$$
$$(gg')w = g(g'w).$$

This verification is a trivial checking of various cases.

It is clear that each of the groups G_i acts effectively. Thus, each element g of G_i may be considered as a permutation of the set W, and G_i may be considered as a subgroup of the group of all permutations of W (see Appendix B). Let G denote the subgroup of the group of all permutations of W which is generated by the union of the G_i's. Then, G contains each G_i as a subgroup; we let

$$\varphi_i : G_i \to G$$

denote the inclusion map.

Any element of G may be expressed as a finite product of elements from the various G_i's. If two consecutive factors in this product come from the same G_i, it is clear that they may be replaced by a single factor. Thus, any element $g \neq 1$ of G may be expressed as a finite product of elements from the G_i's in *reduced form*, i.e., so no two consecutive factors belong to the same group, and so no factor is the identity element. We now assert that *the expression of any element* $g \neq 1$ *of* g *in reduced form is unique:* If

$$g = g_1 g_2 \cdots g_m = h_1 h_2 \cdots h_n$$

with both products in reduced form, then $m = n$ and $g_i = h_i$ for $1 \leqq i \leqq m$. To see this, consider the effect of the permutations $g_1 g_2 \cdots g_m$ and $h_1 h_2 \cdots h_n$ on the empty word; the results are the words $(g_1, g_2, \ldots, g_m)$ and $(h_1, h_2, \ldots, h_n)$, respectively. Because these two words must be equal, the conclusion follows.

It is clear how to form the inverse of an element of G written in reduced form, and how to form the product of two such elements.

It is now an easy matter to verify that G is actually the free product of the G_i's with respect to the φ_i's. For, let H be any group and let $\psi_i : G_i \to H$, $i \in I$, be any collection of homomorphisms. Define a function $f : G \to H$ as follows. Express any given $g \neq 1$ in reduced form,

$$g = g_1 g_2 \cdots g_m, \qquad g_k \in G_{i_k}, \qquad 1 \leqq k \leqq m$$

and then set

$$f(g) = (\psi_{i_1} g_1)(\psi_{i_2} g_2) \cdots (\psi_{i_m} g_m).$$

We also set $f(1) = 1$, of course. It is clear that f is a homomorphism, and that f makes the required diagrams commutative. It is also clear that f is the only homomorphism that makes these diagrams commutative.

Q.E.D.

Because the homomorphisms $\varphi_i : G_i \to G$ are monomorphisms, it is customary to identify each group G_i with its image under φ_i, and to regard it as a subgroup of the free product G. Then, φ_i becomes an inclusion map, and it is not usually necessary to mention it explicitly.

The two most important facts to remember from the proof of Theorem 4.2 are the following:

(a) Any element $g \neq 1$ of the free product can be expressed uniquely as a product in reduced form of elements from the groups G_i.
(b) The rules for multiplying two such products in reduced form (or for forming their inverses) are the obvious and natural ones.

These facts give one great insight into the structure of a free product of groups.

Example

4.1 Let G_1 and G_2 be cyclic groups of order 2, $G_1 = \{1, x_1\}$ and $G_2 = \{1, x_2\}$. Then, any element $g \neq 1$ of their free product can be written uniquely as a product of x_1 and x_2, with the factors x_1 and x_2 alternating. For example, the following are such elements:

$$x_1,\ x_1x_2,\ x_1x_2x_1,\ x_1x_2x_1x_2,\ \text{etc.},$$

or

$$x_2,\ x_2x_1,\ x_2x_1x_2,\ x_2x_1x_2x_1,\ \text{etc.}$$

Note that the elements x_1x_2 and x_2x_1 are both of infinite order, and they are different. Note also the great difference between the direct product or weak product of G_1 and G_2 and their free product in this case. The direct product is an abelian group of order 4, whereas the free product is a non-abelian group with elements of infinite order.

Notation: We denote the free product of groups $G_1, G_2, \ldots, G_n$ by $G_1 * G_2 * \cdots * G_n$ or

$$\prod_{1 \leq i \leq n}^{*} G_i.$$

The free product of the family of groups $\{G_i : i \in I\}$ is denoted by

$$\prod_{i \in I}^{*} G_i.$$

Exercises

4.1 Let $\{G_i : i \in I\}$ be a collection containing more than one group, each of which has more than one element. Prove that their free product is non-abelian, contains elements of infinite order, and that its center consists of the identity element alone.

4.2 For each index i, let G_i' be a subgroup of G_i (proper or improper). Prove that the free product of the collection $\{G_i' : i \in I\}$ may be considered as a subgroup of the free product of the G_i.

4.3 Let $\{G_i : i \in I\}$ and $\{G_i' : i \in I\}$ be two families of groups indexed by the same set I. Assume that for each index $i \in I$ there is given a homomorphism $f_i : G_i \to G_i'$. Prove that there exists a unique homomorphism $f : G \to G'$ of the free product of the first family of groups into the free product of the second family such that the following diagram is commutative for each index i:

$$\begin{array}{ccc} G_i & \xrightarrow{\varphi_i} & G \\ {\scriptstyle f_i}\downarrow & & \downarrow{\scriptstyle f} \\ G_i' & \xrightarrow[\varphi_i']{} & G' \end{array} .$$

Show that if each f_i is a monomorphism (respectively, epimorphism) then f is a monomorphism (respectively, epimorphism).

4.4 Prove that if an element x of the free product $G * H$ has finite order, then x is an element of G or H, or is conjugate to an element of G or H. (HINT: Express x as a word in reduced form; then make the proof by induction on the length of the word.) Deduce that if G and H are cyclic groups of orders m and n respectively, where $m > 1$ and $n > 1$, then the maximum order of any element of $G * H$ is $\max(m, n)$.

4.5 Let $\{G_i : i \in I\}$ be a collection of abelian groups, and let G be their free product with respect to homomorphisms $\varphi_i : G_i \to G$. Let $G' = G/[G, G]$ be the quotient of G by its commutator[3] subgroup and let $\varphi_i' : G_i \to G'$ be the composition of φ_i with the natural homomorphism $G \to G'$. Prove that G' is a weak product of the groups $\{G_i\}$ with respect to the homomorphisms φ_i' (i.e., the conclusion of Proposition 2.1 holds).

4.6 Let G, H, G', and H' be cyclic groups of orders m, n, m', and n', respectively. If $G * H$ is isomorphic to $G' * H'$, then $m = m'$ and $n = n'$ or else $m = n'$ and $n = m'$. (HINT: Apply Exercise 4.5 to $G * H$ and $G' * H'$; thus we see that, if we "abelianize" $G * H$ and $G' * H'$, we obtain finite abelian groups of orders mn and $m'n'$, respectively. Now apply Exercise 4.4.)

4.7 Let H and H' be conjugate subgroups of G. Prove that if f is any homomorphism of G into some other group such that $f(H) = 1$, then $f(H') = 1$ also.

4.8 Let G be the free product of the family of groups $\{G_i : i \in I\}$, where it is assumed that $G_i \neq \{1\}$ for any index i. Prove that, for any two distinct indices i and $i' \in I$, the subgroups G_i and $G_{i'}$ of G are not conjugate. (HINT: Apply Exercise 4.7. Use Exercise 4.3 to construct a homomorphism f of G into another free product with the required properties.)

4.9 Let $G = G_1 * G_2$, and let N be the least normal subgroup of G which contains G_1. Prove that G/N is isomorphic to G_2. (HINT: Use Exercise 4.3. Let

[3] This terminology and notation is explained in the following section just before the statement of Proposition 5.3.

$G_1' = \{1\}$, $G_2' = G_2$, $f_1 : G_1 \to G_1'$ be the trivial homomorphism, and let $f_2 : G_2 \to G_2'$ be the identity map. Prove that N is the kernel of the induced homomorphism $f : G \to G'$.)

4.10 Let G admit two different decompositions as a free product:

$$G = G_0 * (\prod_{i \in I}{}^* G_i) = H_0 * (\prod_{i \in I}{}^* H_i)$$

with the same index set I. Assume that, for each index $i \in I$, G_i and H_i are conjugate subgroups of G. Prove that G_0 and H_0 are isomorphic. (HINT: The method of proof is similar to that of Exercise 4.9.)

5 Free groups

As the reader may have guessed, the definition of a free group is entirely analogous to that of a free abelian group.

Definition Let S be an arbitrary set. A *free group on the set* S (or a *free group generated by* S) is a group F together with a function $\varphi : S \to F$ such that the following condition holds: For any group H and any function $\psi : S \to H$, there exists a unique homomorphism $f : F \to H$ such that the following diagram is commutative:

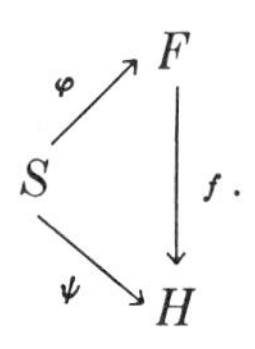

Exactly as in the previous cases we have encountered, this definition completely characterizes a free group. To be precise:

Proposition 5.1 *Let F and F' be free groups on the set S with respect to functions $\varphi : S \to F$ and $\varphi' : S \to F'$, respectively. Then, there exists a unique isomorphism $h : F \to F'$ such that the following diagram is commutative:*

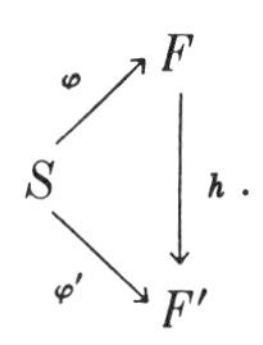

It still remains to prove that, given any set S, there exists a free group on the set S, and to establish its principal properties. We shall do this by exactly the same method as that used for the case of free abelian groups.

Assume, then, that

$$S = \bigcup_{i \in I} S_i,$$

where the subsets S_i are disjoint and nonempty. For each index i, let F_i be a free group on the set S_i with respect to a function $\varphi_i : S_i \to F_i$. Let F denote the free product of the groups F_i with respect to homomorphisms $\eta_i : F_i \to F$ (recall that we have proved that each η_i is actually a monomorphism!). Because the subsets S_i are pairwise disjoint, we can define a function $\varphi : S \to F$ by the rule

$$\varphi \mid S_i = \eta_i \varphi_i.$$

Proposition 5.2 *Under the above hypotheses, F is the free group on the set S with respect to the function $\varphi : S \to F$.*

The proof of this proposition is the same as that of Proposition 3.2 except for obvious modifications. Hence, it is not necessary to go through these details again. This proposition may be restated as follows: The free product of any collection of free groups is a free group.

We shall now apply this proposition to prove the existence of free groups exactly as we applied Proposition 3.2 to prove the existence of free abelian groups. The details are as follows: Let $S = \{x_i : i \in I\}$ be an arbitrary nonempty set, and, for each index i, let $S_i = \{x_i\}$. Let f_i denote an infinite cyclic group generated by x_i,

$$F_i = \{x_i^n : n \in \mathbf{Z}\},$$

and let $\varphi : S_i \to F_i$ denote the inclusion map. Then, F_i is readily seen to be a free group on the set S_i with respect to the map φ_i (as we shall see later, this case, where S has only one element, is the only one where the free group on a set S and the free abelian group on S are the same). The hypotheses of Proposition 5.2 are all satisfied; we conclude that F is a free group on the set S with respect to the function $\varphi : S \to F$. Note that F is a free product of infinite cyclic groups. From what we have learned about free products, we see that *every element $g \neq 1$ of the free group F can be expressed uniquely in the form*

$$g = x_1^{n_1} x_2^{n_2} \cdots x_k^{n_k},$$

where $x_1, x_2, \ldots, x_k$ are elements of S such that any two successive elements are different, and $n_1, n_2, \ldots, n_k$ are nonzero integers, positive or negative.

Such an expression for g is called a *reduced word* in the elements of S. To avoid exceptions, we say that the identity 1 is represented by the *empty word*. The rules for forming inverses and products of reduced words are the obvious ones.

From these facts, it is clear that the function $\varphi : S \to F$ is one-to-one, and that F is actually generated by the subset $\varphi(S)$ in the sense defined earlier.

In many cases it is convenient to take S to be a subset of F and φ to be the inclusion map. If this is the case, we may as well omit any mention of φ.

Exercises

5.1 Prove that a free group on a nonempty set S is abelian if and only if S has exactly one element.

5.2 Prove that the center of a free group on a set having more than one element consists of the identity element alone.

5.3 Let g and h be two elements of a free group on a set S having more than one element. Give a necessary and sufficient condition for g and h to be conjugate in terms of their expressions as reduced words. (HINT: Consider cyclic permutations of the factors of a reduced word.)

We shall conclude this section by considering the relation between free groups and free abelian groups. Recall that, if x and y are any two elements of a group G, the notation $[x, y]$ denotes the element $xyx^{-1}y^{-1} \in G$, and it is called the *commutator* of x and y (in the given order). The notation $[G, G]$ denotes the subgroup of G generated by all commutators; it is called the *commutator* subgroup and is readily verified to be a normal subgroup. The quotient group $G/[G, G]$ is abelian. Conversely, if N is any normal subgroup of G such that G/N is abelian, then $N \supset [G, G]$.

Proposition 5.3 *Let F be a free group on the set S with respect to a function $\varphi : S \to F$, and let $\pi : F \to F/[F, F]$ denote the natural projection of F onto the quotient group. Then, $F/[F, F]$ is a free abelian group on S with respect to the function $\pi\varphi : S \to F/[F, F]$.*

The proof is a nice exercise in the use of the definitions and the facts stated in the preceding paragraph.

Corollary 5.4 *If F and F' are free groups on finite sets S and S', then F and F' are isomorphic if and only if S and S' have the same cardinal number.*

PROOF: Any isomorphism of F onto F' would induce an isomorphism of the quotient groups, $F/[F, F]$ and $F'/[F', F']$. We now reach a contradiction by using the preceding proposition and Corollary 3.5. This proves the "only if" part of the corollary. The proof of the "if" part is trivial.

Exercise

5.4 Prove that this corollary is still true if S is a finite set and S' is an arbitrary set.

If F is a free group on a set S, the cardinal number of S is called the *rank* of F. Corollary 5.4 shows that the rank is an invariant of the group at least in the case of free groups of finite rank. It can also be proved that the rank of a free group is an invariant even in the case where it is an infinite cardinal. The proof is more of an exercise in the arithmetic of cardinal numbers than in group theory, and we shall not give it here.

If F is a free group on the set S with respect to the function $\varphi : S \to F$, because φ is one-to-one it is usually convenient to consider S as a subset of F and φ as an inclusion map, as we mentioned above. With this convention, S is called a *basis* for F. In other words, a basis for F is any subset S of F such that F is a free group on S with respect to the inclusion map $S \to F$. A free group has many different bases.

Several important theorems about free groups will be proved in Chapters VI and VII.

6 The presentation of groups by generators and relations

We begin with a result that is the analog for arbitrary groups of Proposition 3.3.

Proposition 6.1 *Any group is the homomorphic image of a free group. To be precise, if S is any set of generators for the group G, and F is a free group on S, then the inclusion map $S \to G$ determines a unique epimorphism of F onto G.*

The proof is the same as that of Proposition 3.3. This proposition enables us to give a mathematically precise meaning to the term "nontrivial relation between generators" by a method analogous to that used in the case of abelian groups. There is one slight difference between the abelian case and the present case because, in the case of abelian groups, any subgroup can be the kernel of a homomorphism, whereas in the case of non-abelian groups, only a *normal* subgroup can be a kernel. For this reason we shall give a complete discussion of this case.

Let S be a set of generators for the group G let F be a free group on the set S with respect to a map $\varphi : S \to F$, let $\psi : S \to G$ be the inclusion map, and let $f : F \to G$ be the unique homomorphism such that $f\varphi = \psi$. Any element $r \neq 1$ of the kernel of F is (by definition) a *relation* between the generators of S for the group G. In view of what we have proved, r can be expressed uniquely as a reduced word in the elements of S. Because every element of S is also an element of G, this reduced word can also be considered as a product in G; however, in G, this product reduces to the identity element. Thus, by this device of introducing the free group F on the set S, we have given the relation r a "place to live," to use a figure of speech. If $\{r_j\}$ is any collection of relations, then any other relation r is said to be a *consequence* of the relations r_j if and only if r is contained in the least *normal* subgroup of F which contains the relation r_j. In the case where every relation is a consequence of the set of relations $\{r_j\}$, the kernel of f is completely determined by the set $\{r_j\}$; it is the intersection of all *normal* subgroups of F which contain the set $\{r_j\}$. In this case, the group G is completely determined up to isomorphism by the set of generators S and the set of relations $\{r_j\}$, because it is isomorphic to the quotient of F modulo the least normal subgroup containing the set $\{r_j\}$. Such a set of relations is called a *complete* set of relations.

Definition A *presentation* of a group G is a pair $(S, \{r_j\})$ consisting of a set of generators for G and a complete set of relations between these generators. The presentation is said to be *finite* in case both S and $\{r_j\}$ are finite sets, and the group G is said to be *finitely presented* in case it has at least one finite presentation.

Let us emphasize that any group admits many different presentations, which may look quite different. Conversely, given two presentations $(S, \{r_j\})$ and $(S', \{r'_k\})$, it is often nearly impossible to determine whether or not the two groups thus defined are isomorphic.

Examples

6.1 A cyclic group of order n admits a presentation with one generator x and one relation x^n.

6.2 We shall prove later that the fundamental group of the Klein Bottle admits the following two different presentations (among others):

(a) Two generators a and b and one relation $baba^{-1}$.
(b) Two generators a and c and one relation a^2c^2.

The relationship between the two presentations in this case is fairly simple: $c = ba^{-1}$ or $b = ca$. To be precise, let $F(a, b)$ and $F(a, c)$ denote free groups on the sets $\{a, b\}$ and $\{a, c\}$, respectively. Define homomorphisms $f : F(a, b) \to F(a, c)$ and $g : F(a, c) \to F(a, b)$ by the following conditions:

$$f(a) = a, \qquad f(b) = ca,$$

$$g(a) = a, \qquad g(c) = ba^{-1}.$$

It follows directly from the definition of a free group that these equations define unique homomorphisms. We compute that:

$$g[f(a)] = a, \qquad g[f(b)] = b,$$

$$f[g(a)] = a, \qquad f[g(c)] = c.$$

Therefore, gf is the identity map of $F(a, b)$, and fg is the identity map of $F(a, c)$. Hence, f and g are isomorphisms which are the inverse of each other. Next, we check that

$$a^2c^2 = c^{-1}[f(baba^{-1})]c,$$

$$baba^{-1} = (ba^{-1})[g(a^2c^2)](ba^{-1})^{-1}.$$

Therefore, the normal subgroup of $F(a, b)$, generated by $baba^{-1}$, and the normal subgroup of $F(a, c)$, generated by a^2c^2, correspond under the isomorphisms f and g. Hence, f and g induce isomorphisms of the corresponding quotient groups.

Note that the essence of the above argument is contained in the following two simple calculations:

(a) If $b = ca$, then $baba^{-1} = ca^2c$ and $a^2c^2 = c^{-1}[baba^{-1}]c$.
(b) If $c = ba^{-1}$, then $a^2c^2 = a^2ba^{-1}ba^{-1}$ and $baba^{-1} = (ba^{-1})(a^2c^2)(ba^{-1})^{-1}$.

6.3 Consider the following two group presentations:

(a) Two generators a and b and one relation a^3b^{-2}.
(b) Two generators x and y and one relation $xyxy^{-1}x^{-1}y^{-1}$.

We assert that these are presentations of isomorphic groups. The relationship between the two different pairs of generators is given by the following system of equations:

$$a = xy, \qquad b = xyx,$$
$$x = a^{-1}b, \qquad y = b^{-1}a^2.$$

We leave it to the reader to work out the details. We shall see in Section IV.6 that this is a presentation of the fundamental group of the complement of a certain knotted circle in Euclidean 3-space.

In dealing with groups presented by means of generators and relations, it is often convenient to take a more informal approach. To illustrate what we mean, consider the first presentation in Example 6.3. The group G under consideration is the quotient of a free group F on two generators a and b by the least normal subgroup containing the element a^3b^{-2}. Let us denote the image of the generators a and b in the group G by the same symbols. Then, $a^3b^{-2} = 1$ in G, or $a^3 = b^2$. When computing with elements of G (which are products of powers of a and b) we can use the equation $a^3 = b^2$ in whatever way is convenient.

Exercise

6.1 Suppose we are given presentations of two groups G_1 and G_2 by means of generators and relations. Show how to obtain from this a presentation of the direct product $G_1 \times G_2$, the free product $G_1 * G_2$, and the commutator quotient group $G_1/[G_1, G_1]$.

7 Universal mapping problems

In the preceding sections of this chapter we have defined and studied the following types of algebraic objects: weak products of abelian groups, free abelian groups, free products of groups, and free groups. In each of these cases, the algebraic object in question was actually a system consisting of two things with a mapping between them, e.g., $\varphi : S \to G$. This system consisting of two things and a mapping between them was characterized by a certain triangular diagram, e.g.,

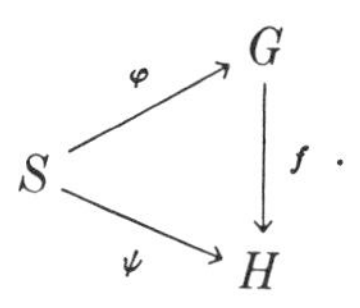

As the reader will recall, the object H and the map ψ in this diagram could be chosen in a fairly arbitrary manner, subject only to minor restrictions. It was then required that there exist a unique map f making the diagram commutative.

This method of characterizing the system $\varphi : S \to G$ is usually referred to by the statement that $\varphi : S \to G$ (or for brevity, G) is the solution of a "universal mapping problem." We shall see another important example of such a universal mapping problem in the next chapter. Defining or characterizing mathematical objects as the solution to a universal mapping problem has become very common in recent years. For example, one of the most prominent contemporary algebraists (C. Chevalley) has written a textbook on algebra [10] that has universal mapping problems as one of its main themes.

If a mathematical object is defined or characterized as being the solution to a universal mapping problem, it follows easily (by the method used to prove Proposition 2.2) that this object is unique up to an isomorphism. In fact, the isomorphism is even uniquely determined! However, the *existence* of an object satisfying a given universal mapping problem is another question. The reader will note that in the four cases discussed in this chapter, at least three different constructions were given to prove the existence of a solution. However, in each case, the existence proof carried with it a bonus, in that it gave great insight into the actual structure of the desired mathematical object.

There exists a rather general method for proving the existence of solutions of universal mapping problems (see [9] and [11]). However, this general method gives absolutely no insight into the mathematical structure of the solution. It is a pure existence proof.

We now give two more examples of the characterization of mathematical objects as solutions of universal mapping problems. The examples are given for illustrative purposes only and will not be used in any of the succeeding chapters.

Examples

7.1 *Free commutative ring with a unit.* Let $\mathbf{Z}[x_1, x_2, \ldots, x_n]$ denote, as usual, the ring of all polynomials with integral coefficients in the "variables" or "indeterminates" $x_1, x_2, \ldots, x_n$. Each nonzero element of this ring can be expressed uniquely as a finite linear combination with integral coefficients of the monomials $x_1^{k_1}x_2^{k_2} \ldots x_n^{k_n}$, where $k_1, k_2, \ldots, k_n$ are non-negative integers. This ring may be considered to be the free commutative ring with unit generated by the set $S = \{x_1, \ldots, x_n\}$. We make this assertion precise, as follows: Let $\varphi : S \to \mathbf{Z}[x_1, \ldots, x_n]$ denote the inclusion map. Then, for any commutative ring R (with unit) and any function $\psi : S \to R$, there exists a unique ring homomorphism $f : \mathbf{Z}[x_1, \ldots, x_n] \to R$ [with $f(1) = 1$] such that the following diagram is commutative:

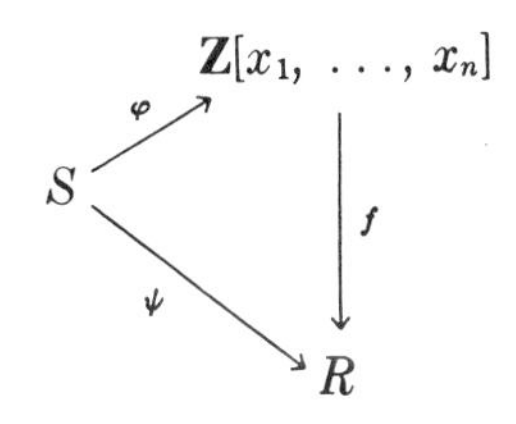

7.2 *The Stone-Čech Compactification.* For any Tychonoff space X, there is defined a certain compact Hausdorff space $\beta(X)$ which contains X as an everywhere dense subset; it is called the Stone-Čech Compactification of X. Let $\varphi : X \to \beta(X)$ denote the inclusion map. Then, we have the following characterization: For any compact Hausdorff space Y and any continuous map $\psi : X \to Y$, there exists a unique continuous map $f : \beta(X) \to Y$ such that the following diagram is commutative:

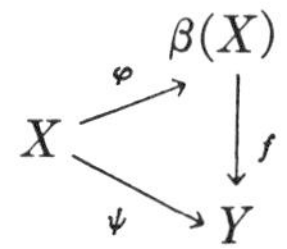

For a more complete discussion see J. L. Kelley, *General Topology.* Princeton, N.J.: Van Nostrand, 1955, pp. 152–153.

For a precise, axiomatic treatment of universal mapping problems and further examples, see references [9] and [11].

NOTES

Definition of free groups, free products, etc.

The concepts of free abelian group, free group, free product of groups, etc., are rather old. The main difference between a modern treatment of the subject and one of the older treatments is the method of defining these algebraic objects. Formerly, they were defined in terms of what are now considered some of their characteristic properties. For example, a free group on set S was defined to be the collection of all equivalence classes of "words" formed from the elements of S. From a strictly logical point of view, there can be no objection to this procedure. However, from a conceptual point of view, it has the disadvantage that the definition of each type of free object requires new insight and ingenuity, and may be a difficult problem. The idea of defining free objects as solutions to universal mapping problems, which gradually evolved during the time of World War II and immediately thereafter, seems to be one of the important unifying ideas in modern mathematics.

The elegant proof given in the text for the existence of free products of groups (Theorem 4.2), which is simpler than the older proofs, is due to B. L. Van der Waerden (*Amer. J. Math.*, *70*, 1948, pp. 527–528). In a more recent paper (*Proc. Kon. Ned. Akad. Weten.* (series A), *69*, 1966, pp. 78–83), Van der Waerden has pointed out how the basic idea of the procedure used for the proof of Theorem 4.2 is applicable to prove the existence of solutions to universal mapping problems in many other algebraic situations.

Different levels of abstraction in mathematics

The first time the student encounters the material in this chapter, it may seem rather foreign to him. The probable reason is that it is on a higher level of abstraction than any of his previous studies in mathematics. To make this point clearer, we shall try to describe briefly the different levels of abstraction that seem to occur naturally in mathematics.

The lowest level of abstraction is the level of most high school and beginning undergraduate mathematics courses. This level is characterized by a concern with a few very explicit mathematical objects, e.g., the integers, rational numbers, real numbers, the complex numbers, the Euclidean plane, etc. The next level of abstraction occurs when certain properties common to several different concrete mathematical objects are isolated and studied for their own sake. This leads to the study of such abstract and general mathematical systems as groups, rings, fields, vector spaces, topological spaces, etc. Ordinarily the mathematics student makes the transition to this level of abstraction some time in his undergraduate career.

The material of this chapter provides an introduction to the next higher level of abstraction. As was pointed out in Example 4.1, the weak direct product of two abelian groups, G_1 and G_2, and their free product $G_1 * G_2$, are quite different types of groups. Yet there is a strong analogy between the weak direct product of abelian groups and the free product of arbitrary groups. To perceive this analogy, it is necessary to consider the category of all abelian groups and the category of all (i.e., not necessarily abelian) groups, respectively. This is characteristic of this next level of abstraction: the simultaneous consideration of all mathematical systems (e.g., groups, rings, or topological spaces) of a certain kind, and the study of the properties of such a collection of mathematical systems.

The history of mathematics in the last two hundred years or so has been characterized by the considerations of mathematical systems on ever higher levels of abstraction. Presumably this trend will continue in the future. It should be emphasized strongly, however, that this movement is not a case of abstraction for the sake of abstraction itself. Rather, it has been forced on mathematicians for various reasons, such as bringing out the analogies between seemingly quite different phenomena.

Presentations of groups by generators and relations

Let us emphasize that the specification of a group by means of generators and relations is very unsatisfactory in many respects, because some of the most natural problems that arise in connection with group presentations are very difficult or impossible. For a further discussion of this point, see the texts by Kurosh [3], Chap. X, or Rotman [5], Chap. 12.

REFERENCES

Group theory

1. Crowell, R. H., and R. H. Fox. *Introduction to Knot Theory.* Boston: Ginn and Co., 1963. (Reprinted by Springer-Verlag, New York, 1977.) Chapters III and IV.
2. Hall, M. *The Theory of Groups.* New York: Macmillan, 1959. Chapters 7 and 17.

3. Kurosh, A. G. *The Theory of Groups.* Trans. and ed. by K. A. Hirsch. 2 vols. New York: Chelsea, 1955–56. Chapters IX and X.
4. Reidemeister, K. *Einführung in die kombinatorische Topologie.* Braunschweig: Friedr. Vieweg & Sohn, 1932. Chapter II.
5. Rotman, J. J. *The Theory of Groups.* Boston: Allyn and Bacon, 1965. Chapter 11.
6. Schenkman, E. *Group Theory.* Princeton, N.J.: Van Nostrand, 1965. Chapter V.
7. Scott, W. R. *Group Theory.* Englewood Cliffs, N.J.: Prentice-Hall, 1964. Chapter 8.
8. Specht, W. *Gruppentheorie* (Die Grundlehren der Mathematischen Wissenschaften, Band LXXXII). Berlin-Göttingen-Heidelberg: Springer-Verlag, 1956. Chapters 2.1 and 2.2.

Universal mapping problems

9. Bourbaki, N. *Théorie des Ensembles.* Paris: Hermann et Cie., 1970. Chapter IV, Section 3.
10. Chevalley, C. *Fundamental Concepts of Algebra.* New York: Academic Press, 1956.
11. Samuel, P. "On Universal Mappings and Free Topological Groups." *Bull. Amer. Math. Soc., 54*, 1948, pp. 591–598.

CHAPTER FOUR

Seifert and Van Kampen Theorem on the Fundamental Group of the Union of Two Spaces. Applications

1 Introduction

So far we have actually determined the structure of the fundamental group of only a very few spaces (e.g., contractible spaces, the circle). To be able to apply the fundamental group to a wider variety of problems, we must know methods for determining its structure for more spaces. In this chapter, we shall develop rather general means for doing this.

Assume that we wish to determine the fundamental group of an arcwise-connected space X, which is the union of two subspaces U and V, each of which is arcwise connected, and whose fundamental group is known. Choose a base point $x_0 \in U \cap V$; it seems plausible to expect that there should be relations between the groups $\pi(U, x_0)$, $\pi(V, x_0)$, and $\pi(X, x_0)$. The main theorem of this chapter (discovered independently by H. Seifert and E. Van Kampen) asserts that, if U and V are both open sets, and it is assumed that their intersection $U \cap V$ is also arcwise connected, then $\pi(X, x_0)$ is *completely* determined by the following diagram of groups and homomorphisms:

$$\begin{array}{ccc} & & \pi(U) \\ & \overset{\varphi_1}{\nearrow} & \\ \pi(U \cap V) & & \\ & \underset{\varphi_2}{\searrow} & \\ & & \pi(V) \end{array} \qquad (4.1\text{-}1)$$

Here φ_1 and φ_2 are induced by inclusion maps. The way in which $\pi(X, x_0)$ is determined by this diagram can be roughly described as follows. The above diagram can be completed by forming the following commutative diagram:

$$\begin{array}{ccccc} & & \pi(U) & & \\ & \overset{\varphi_1}{\nearrow} & & \overset{\psi_1}{\searrow} & \\ \pi(U \cap V) & & \longrightarrow & & \pi(X). \\ & \underset{\varphi_2}{\searrow} & & \underset{\psi_2}{\nearrow} & \\ & & \pi(V) & & \end{array} \qquad (4.1\text{-}2)$$

Here all arrows denote homomorphisms induced by inclusion maps, and the base point x_0 is systematically omitted. Then, the Seifert-Van Kampen theorem asserts that $\pi(X)$ is the *freest possible* group we can use to complete diagram (4.1-1) to a commutative diagram like (4.1-2). As usual, the phrase "freest possible" is made precise by the consideration of a certain universal mapping problem.

Actually, we shall state and prove a more general version of the theorem, in that we allow X to be the union of any number of arcwise-connected open subsets rather than just two. This more general version is no more difficult to prove, and in some situations it is the only applicable version.

After proving the Seifert-Van Kampen theorem, we state several corollaries and then use these corollaries to determine the structure of the fundamental groups of the various compact surfaces and certain other spaces. In the final section of this chapter we show how these methods can be applied to distinguish between certain knots.

2 Statement and proof of the theorem of Seifert and Van Kampen

First, we give a precise statement of the theorem. Assume that U and V are arcwise-connected open subsets of X such that $X = U \cup V$ and $U \cap V$ is nonempty and arcwise connected. Choose a base point $x_0 \in U \cap V$ for all fundamental groups under consideration.

Theorem 2.1 *Let H be any group, and ρ_1, ρ_2, and ρ_3 any three homomorphisms such that the following diagram is commutative:*

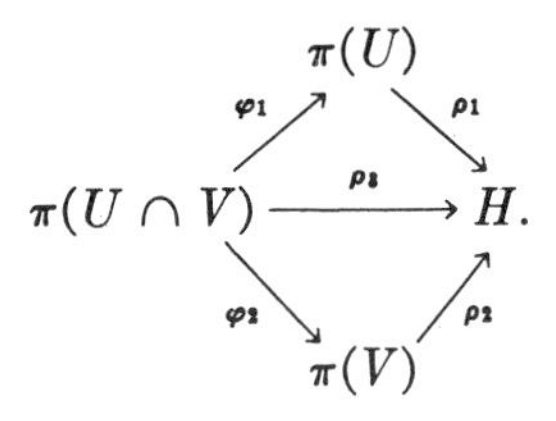

Then, there exists a unique homomorphism $\sigma : \pi(X) \to H$ such that the following three diagrams are commutative:

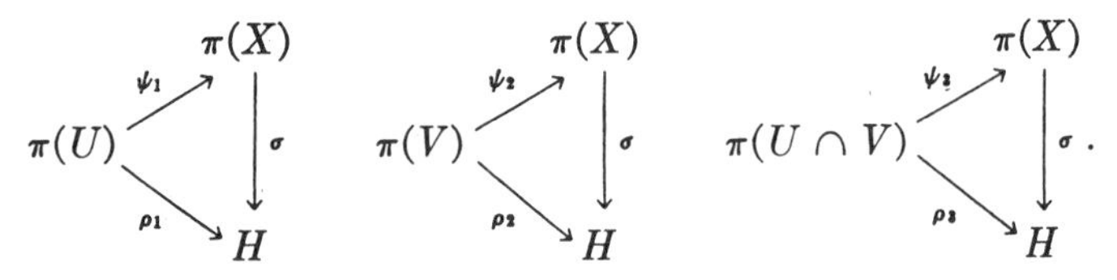

(Here the homomorphisms φ_i and ψ_i, $i = 1, 2\ 3$, are induced by inclusion maps.)

By the methods used in Chapter 3, we can prove that the group $\pi(X)$ is characterized up to isomorphism by this theorem. We leave the precise statement and proof of this fact to the reader.

We shall next state the more general version of the Seifert-Van Kampen theorem. The generalization consists in allowing a covering of the space X by any number of open sets instead of just by two open sets as in Theorem 2.1. Of course, the open sets must all be arcwise connected, and the intersection of any finite number of them must be arcwise connected and contain the base point. To be precise, we assume the following hypotheses:

(a) X is an arcwise-connected topological space and $x_0 \in X$.
(b) $\{U_\lambda : \lambda \in \Lambda\}$ is a covering of X by arcwise-connected open sets such that for all $\lambda \in \Lambda$, $x_0 \in U_\lambda$.
(c) For any two indices $\lambda_1, \lambda_2 \in \Lambda$ there exists an index $\lambda \in \Lambda$ such that $U_{\lambda_1} \cap U_{\lambda_2} = U_\lambda$ (we express this fact by saying that the family of sets $\{U_\lambda\}$ is "closed under finite intersections").

We now consider the fundamental groups of these various sets with base point x_0. For brevity, we omit the base point from the notation.

If $U_\lambda \subset U_\mu$, then the notation

$$\varphi_{\lambda\mu} : \pi(U_\lambda) \to \pi.(U_\mu)$$

denotes the homomorphism induced by the inclusion map. Similarly, for any index λ,

$$\psi_\lambda : \pi(U_\lambda) \to \pi(X)$$

is induced by the inclusion map $U_\lambda \to X$. Note that, if $U_\lambda \subset U_\mu$, the following diagram is commutative:

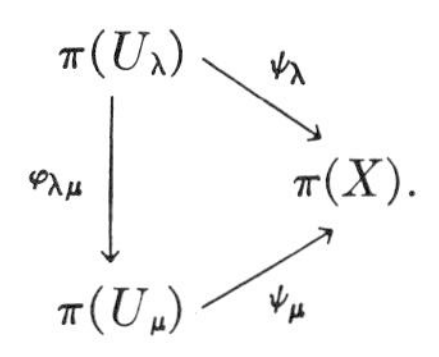

Theorem 2.2 *Under the above hypotheses the group $\pi(X)$ satisfies the following universal mapping condition: Let H be any group and let ρ_λ :*

$\pi(U_\lambda) \to H$ be any collection of homomorphisms defined for all $\lambda \in \Lambda$ such that if $U_\lambda \subset U_\mu$ the following diagram is commutative:

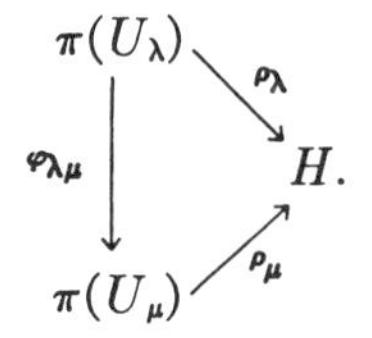

Then, there exists a unique homomorphism $\sigma : \pi(X) \to H$ such that for any $\lambda \in \Lambda$ the following diagram is commutative:

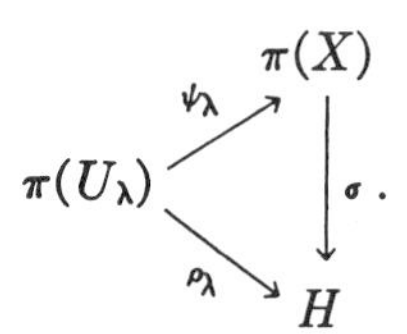

Moreover, this universal mapping condition characterizes $\pi(X)$ up to a unique isomorphism.

The proof of the last sentence of the theorem is a routine matter which may be left to the reader. We shall now give the proof of the rest of this theorem. Applications of this theorem are given in Sections 3–6.

Lemma 2.3 *The group $\pi(X)$ is generated by the union of the images $\psi_\lambda[\pi(U_\lambda)]$, $\lambda \in \Lambda$.*

PROOF: Let $\alpha \in \pi(X)$; choose a closed path $f : I \to X$ representing α. Choose an integer n so large that $1/n$ is less than the Lebesgue number of the open covering $\{f^{-1}(U_\lambda) : \lambda \in \Lambda\}$ of the compact metric space I. Subdivide the interval I into the closed subintervals $J_i = [i/n, (i+1)/n]$, $0 \leqq i \leqq n-1$. For each subinterval J_i, choose an index $\lambda_i \in \Lambda$ such that $f(J_i) \subset U_{\lambda_i}$. Choose a path g_i in $U_{\lambda_{i-1}} \cap U_{\lambda_i}$ joining the point x_0 to the point $f(i/n)$, $1 \leqq i \leqq n-1$. Let $f_i : I \to X$ denote the path represented by the composite function

$$I \xrightarrow{h_i} J_i \xrightarrow{f|J_i} X$$

where h_i is the unique orientation-preserving linear homeomorphism. Then $f_0 \cdot g_1^{-1}$, $g_1 \cdot f_1 \cdot g_2^{-1}$, $g_2 \cdot f_2 \cdot g_3^{-1}$, $\ldots$, $g_{n-2} \cdot f_{n-2} \cdot g_{n-1}^{-1}$, $g_{n-1} \cdot f_{n-1}$ are closed paths, each contained in a single open set U_λ, and their product in the order given is equivalent to f. Hence, we can write

$$\alpha = \alpha_0 \cdot \alpha_1 \cdot \alpha_2 \cdot \cdots \cdot \alpha_{n-1},$$

where

$$\alpha_i \in \psi_{\lambda_i}[\pi(U_{\lambda_i})], \qquad 0 \leqq i \leqq n-1.$$

This completes the proof of the lemma.

Remark: The hypotheses could be slightly weakened for the purposes of this lemma. Actually it is only required that $\{U_\lambda\}$ be an open covering by arcwise-connected subsets of X such that the intersection of any two sets be arcwise connected. It does not matter whether or not the intersection of three sets is arcwise connected.

PROOF OF THEOREM 2.2: Let H be any group and let $\rho_\lambda : \pi(U_\lambda) \to H$, $\lambda \in \Lambda$, be a set of homomorphisms satisfying the hypotheses of the theorem. We must demonstrate the existence of a unique homomorphism $\sigma : \pi(X) \to H$ such that the following diagram is commutative for any $\lambda \in \Lambda$:

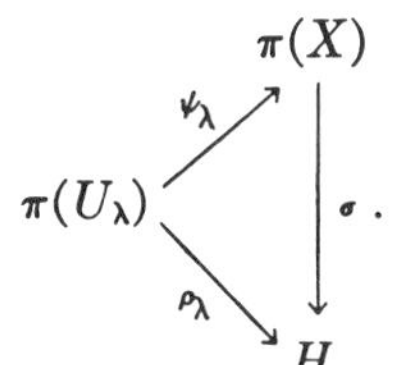

From the lemma just proved, it is clear that such a homomorphism σ, if it exists, must be unique, and must be defined according to the following rule. Let $\alpha \in \pi(X)$. Then, by Lemma 2.3, we have

$$\alpha = \psi_{\lambda_1}(\alpha_1) \cdot \psi_{\lambda_2}(\alpha_2) \cdot \ldots \cdot \psi_{\lambda_n}(\alpha_n), \tag{4.2-1}$$

where $\alpha_i \in \pi(U_{\lambda_i})$, $i = 1, 2, \ldots, n$. Hence, if the homomorphism σ exists, we must have

$$\sigma(\alpha) = \rho_{\lambda_1}(\alpha_1) \cdot \rho_{\lambda_2}(\alpha_2) \cdot \ldots \cdot \rho_{\lambda_n}(\alpha_n). \tag{4.2-2}$$

Our strategy will be to take equation (4.2-2) as a definition of σ. To justify this definition, we must show that it is independent of the choice of the representation of α in the form (4.2-1). Clearly, if it is independent of the form of the representation of α, then it is a homomorphism, and the desired commutativity relations must hold.

To prove that σ is independent of the representation of α in the form (4.2-1), it suffices to prove the following lemma:

Lemma 2.4 *Let $\beta_i \in \pi(U_{\lambda_i})$, $i = 1, \ldots, q$ be such that*

$$\psi_{\lambda_1}(\beta_1) \cdot \psi_{\lambda_2}(\beta_2) \cdot \ldots \cdot \psi_{\lambda_q}(\beta_q) = 1.$$

Then, the product

$$\rho_{\lambda_1}(\beta_1)\rho_{\lambda_2}(\beta_2) \cdots \rho_{\lambda_q}(\beta_q) = 1.$$

[It is suggested that the proof of this lemma be omitted on the first reading of this section.]

PROOF: Choose closed paths

$$f_i : \left[\frac{i-1}{q}, \frac{i}{q}\right] \to U_{\lambda_i}$$

representing β_i for $i = 1, 2, \ldots, q$. Then, the product

$$\prod_{i=1}^{q} \psi_{\lambda_i}(\beta_i)$$

is clearly represented by the closed path $f : [0, 1] \to X$ defined by

$$f \left| \left[\frac{i-1}{q}, \frac{i}{q} \right] = f_i, \qquad i = 1, 2, \ldots, q. \right.$$

By hypothesis, f is equivalent to the constant path. Hence, there exists a continuous map

$$F : I \times I \to X$$

such that, for any $s, t \in I$,

$$F(s, 0) = f(s),$$
$$F(s, 1) = F(0, t) = F(1, t) = x_0.$$

Let ε denote the Lebesgue number of the open covering $\{F^{-1}(U_\lambda) : \lambda \in \Lambda\}$ of the compact metric space $I \times I$ (we give $I \times I$ the metric it has as a subset of the Euclidean plane). We now subdivide the square $I \times I$ into smaller rectangles of diameter $< \varepsilon$ as follows. Choose numbers

$$s_0 = 0, \qquad s_1, s_2, \ldots, s_m = 1,$$
$$t_0 = 0, \qquad t_1, t_2, \ldots, t_n = 1,$$

such that the following three conditions hold: (a) $s_0 < s_1 < s_2 < \cdots < s_m$ and $t_0 < t_1 < t_2 < \cdots < t_n$. (b) The fractions $1/q, 2/q, \ldots, (q-1)/q$ are included among the numbers $s_1, s_2, \ldots, s_m$. (c) If we subdivide the unit square $I \times I$ into rectangles by the vertical and horizontal lines,

$$s = s_i, \qquad i = 0, 1, \ldots, m,$$
$$t = t_j, \qquad j = 0, 1, \ldots, n,$$

the length of the diagonal of each rectangle is less than ε. Clearly, such a subdivision is possible.

Before proceeding further with the proof, we must introduce a rather elaborate notation for the various vertices, edges, and rectangles of this subdivision as follows:

Vertices:

$$v_{ij} = (s_i, t_j), \qquad 0 \leqq i \leqq m, \qquad 0 \leqq j \leqq n.$$

Subintervals of $I = [0, 1]$:

$$J_i = [s_{i-1}, s_i], \qquad 1 \leqq i \leqq m,$$
$$K_j = [t_{j-1}, t_j], \qquad 1 \leqq j \leqq n.$$

Rectangles:

$$R_{ij} = J_i \times K_j, \qquad 1 \leqq i \leqq m, \qquad 1 \leqq j \leqq n.$$

Horizontal edges:

$$a_{ij} = J_i \times \{t_j\}, \qquad 1 \leqq i \leqq m, \qquad 0 \leqq j \leqq n.$$

Vertical edges:

$$b_{ij} = \{s_i\} \times K_j, \qquad 0 \leqq i \leqq m, \qquad 1 \leqq j \leqq n.$$

In Figure 4.1 we indicate how a typical rectangle of this subdivision and its vertices and edges are labeled. We also need the following notation for certain paths:

$$A_{ij} : J_i \to X, \qquad A_{ij}(s) = F(s, t_j), \qquad s \in J_i.$$

$$B_{ij} : K_j \to X, \qquad B_{ij}(t) = F(s_i, t), \qquad t \in K_j.$$

With a slight abuse of notation, we can write

$$A_{ij} = F \mid a_{ij},$$

$$B_{ij} = F \mid b_{ij}.$$

For each rectangle R_{ij}, choose an open set $U_{\lambda(i,j)}$ such that

$$F(R_{i,j}) \subset U_{\lambda(i,j)}.$$

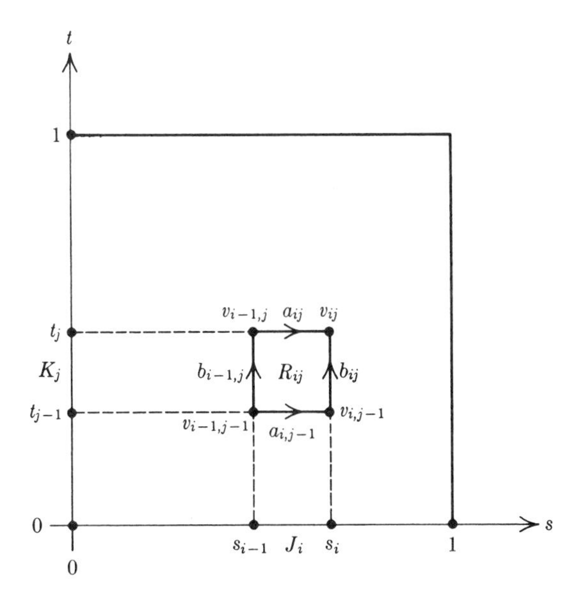

FIGURE 4.1 Notation used in the proof of Lemma 2.4.

Condition (c) on the subdivisions assures us that such a choice is possible. Each vertex v_{ij} is a vertex of 1, 2, or 4 of the rectangles R_{kl}; let $U_{\mu(i,j)}$ denote the intersection of the corresponding 1, 2, or 4 open sets $U_{\lambda(k,l)}$. Then, $U_{\mu(i,j)}$ is an open set of the given covering and

$$F(v_{ij}) \in U_{\mu(i,j)}.$$

Choose a path

$$g_{ij} : I \to U_{\mu(i,j)}$$

with initial point x_0 and terminal point $F(v_{ij})$; if $F(v_{ij}) = x_0$, we require that g_{ij} be the constant path.

Having introduced most of the necessary notation, we now interpolate a sublemma.

Sublemma *Let U_λ and U_μ be two sets of the given open covering of X and let*

$$h : I \to U_\lambda \cap U_\mu, \qquad h(0) = h(1) = x_0,$$

be a closed path. Let $\alpha \in \pi(U_\lambda, x_0)$ and $\beta \in \pi(U_\mu, x_0)$ denote the equivalence class of the loop h in the two different groups. Then, $\rho_\lambda(\alpha) = \rho_\mu(\beta)$.

PROOF OF SUBLEMMA: The set $U_\nu = U_\lambda \cap U_\mu$ also belongs to the covering by hypothesis, and h represents an element $\gamma \in \pi(U_\nu, x_0)$. Then, clearly,

$$\alpha = \varphi_{\nu\lambda}(\gamma),$$

$$\beta = \varphi_{\nu\mu}(\gamma).$$

Hence,

$$\rho_\lambda(\alpha) = \rho_\lambda\varphi_{\nu\lambda}(\gamma) = \rho_\nu(\gamma),$$

$$\rho_\mu(\beta) = \rho_\mu\varphi_{\nu\mu}(\gamma) = \rho_\nu(\gamma).$$

Q.E.D.

This sublemma enables us to adopt a certain sloppiness of notation without fear of ambiguity. We can denote the element $\rho_\lambda(\alpha) = \rho_\mu(\beta) \in H$ by the notation $\rho(h)$; we need not worry about whether we should take the equivalence class of h in the group $\pi(U_\lambda)$ or in the group $\pi(U_\mu)$.

With this convention, let

$$\alpha_{ij} = \rho[(g_{i-1,j}A_{ij})(g_{ij})^{-1}],$$

$$\beta_{ij} = \rho[(g_{i,j-1}B_{ij})(g_{ij})^{-1}].$$

[Here $(g_{ij})^{-1}$ denotes the path defined by $t \to g_{ij}(1 - t)$.] Note that α_{ij} and β_{ij} are both well-defined elements of H.

Next, we assert that, corresponding to each rectangle R_{ij}, there is a relation of the following form in the group H:

$$\alpha_{i,j-1}\beta_{ij} = \beta_{i-1,j}\alpha_{ij}. \tag{4.2-3}$$

To prove this, note first that we have the following equivalence between (nonclosed) paths in $U_{\lambda(i,j)}$:

$$A_{i,j-1}B_{ij} \sim B_{i-1,j}A_{ij}.$$

This equivalence is a consequence of Lemma II.8.1 applied to the mapping $F \mid R_{ij} : R_{ij} \to U_{\lambda(i,j)}$ and Exercise II.3.3. As a result, we have the following equivalence between closed paths in $U_{\lambda(i,j)}$:

$$g_{i-1,j-1}A_{i,j-1}(g_{i,j-1})^{-1}g_{i,j-1}B_{ij}(g_{ij})^{-1}$$
$$\sim g_{i-1,j-1}B_{i-1,j}(g_{i-1,j})^{-1}g_{i-1,j}A_{ij}(g_{ij})^{-1}. \tag{4.2-4}$$

If we now take the equivalence class in $\pi(U_{\lambda(i,j)})$ of both sides, and then apply the homomorphism $\rho_{\lambda(i,j)}$, we obtain equation (4.2-3). [NOTE: To be strictly correct, since multiplication of paths is not associative, parentheses should be inserted in (4.2-4). However, it does not matter how the parentheses are inserted.]

The next relation we need is

$$\prod_{i=1}^{m} \alpha_{i0} = \prod_{k=1}^{q} \rho_{\lambda_k}(\beta_k), \tag{4.2-5}$$

which is an easy consequence of requirement (b) that the points $1/q, 2/q, \ldots, (q-1)/q$ be included in the set $\{s_i : 0 < i < m\}$ together with the definitions and constructions we have made. Finally, we have the relations

$$\alpha_{in} = 1, \qquad 1 \leqq i \leqq m, \tag{4.2-6}$$

$$\beta_{0j} = \beta_{mj} = 1, \qquad 1 \leqq j \leqq n. \tag{4.2-7}$$

These relations result from the fact that

$$F(s, 1) = F(0, t) = F(1, t) = x_0$$

for any $s, t \in I$.

In view of relation (4.2-5), we must prove

$$\prod_{i=1}^{m} \alpha_{i0} = 1. \tag{4.2-8}$$

We shall now do this by using relations (4.2-3), (4.2-6), and (4.2-7). First, we show that

$$\prod_{i=1}^{m} \alpha_{i,j-1} = \prod_{i=1}^{m} \alpha_{i,j} \tag{4.2-9}$$

for any integer j, $1 \leqq j \leqq n$. Indeed, we have

$$\begin{aligned}
\alpha_{1,j-1}\alpha_{2,j-1} \cdots \alpha_{m,j-1} &= \alpha_{1,j-1}\alpha_{2,j-1} \cdots \alpha_{m,j-1}\beta_{m,j} && \text{by (4.2-7)},\\
&= \alpha_{1,j-1}\alpha_{2,j-1} \cdots \alpha_{m-1,j-1}\beta_{m-1,j}\alpha_{m,j} && \text{by (4.2-3)},\\
&= \alpha_{1,j-1}\alpha_{2,j-1} \cdots \beta_{m-2,j}\alpha_{m-1,j}\alpha_{m,j} && \text{by (4.2-3)},\\
&\quad\vdots && \quad\vdots\\
&= \beta_{0j}\alpha_{1,j}\alpha_{2,j} \cdots \alpha_{m-1,j}\alpha_{m,j} && \text{by (4.2-3)},\\
&= \alpha_{1,j}\alpha_{2,j} \cdots \alpha_{m-1,j}\alpha_{m,j} && \text{by (4.2-7)}.
\end{aligned}$$

In all, we must apply (4.2-3) m times. If we now apply (4.2-9) with $j = 1, 2, \ldots, n$ in succession, we obtain

$$\prod_{i=1}^{m} \alpha_{i0} = \prod_{i=1}^{m} \alpha_{in}.$$

But, by use of (4.2-6),

$$\prod_{i=1}^{m} \alpha_{in} = 1.$$

This completes the proof of (4.2-8), and hence of Lemma 2.4. Q.E.D.

3 First application of Theorem 2.1

Assume, as in the statement of Theorem 2.1, that X is the union of the open sets U and V and that U, V, and $U \cap V$ are all arcwise connected. Let φ_i and ψ_i have the meaning assigned to them in Section 2.

Theorem 3.1 *If $U \cap V$ is simply connected, then $\pi(X)$ is the free product of the groups $\pi(U)$ and $\pi(V)$ with respect to the homomorphisms $\psi_1 : \pi(U) \to \pi(X)$ and $\psi_2 : \pi(V) \to \pi(X)$.*

PROOF: This is a direct corollary of Theorem 2.1. If $\pi(U \cap V) = \{1\}$, then the diagram

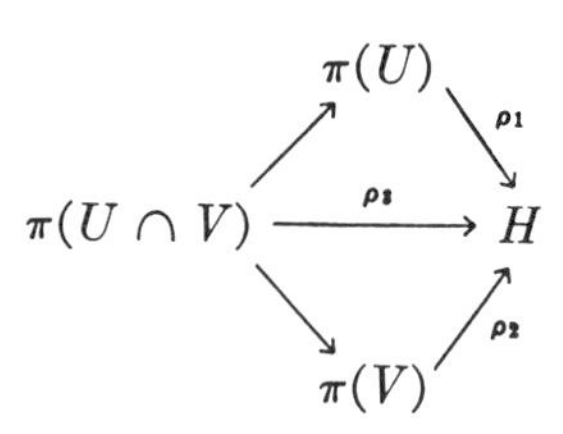

will be commutative for any choice of ρ_1 and ρ_2; hence, these two homomorphisms are completely arbitrary, whereas ρ_3 is uniquely determined. Similarly, the diagram

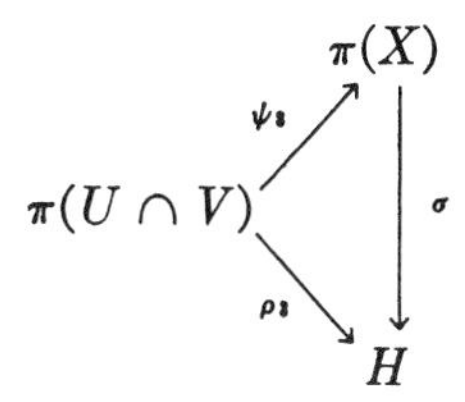

will be commutative for any choice of σ; requiring it to be commutative imposes no condition on σ. The remaining two conditions on σ in Theorem 2.1 are exactly those which occur in the definition of the free product of two groups. Q.E.D.

We now give some examples where this theorem is applicable. These examples will, in turn, be used later to study other examples.

Examples

3.1 Let X be a space such that $X = A \cup B$, $A \cap B = \{x_0\}$, and A and B are each homeomorphic to a circle S^1 (see Figure 4.2). X may be visualized as a curve shaped like a figure "8."

If A and B were open subsets of X, we could apply Theorem 3.1 with $U = A$ and $V = B$ to determine the structure of $\pi(X)$. Unfortunately, A and B are not open.

However, a slight modification of this strategy will work. Choose points $a \in A$ and $b \in B$ such that $a \neq x_0$ and $b \neq x_0$. Let $U = X - \{b\}$, and let $V = X - \{a\}$. U and V are each homeomorphic to a circle with two "whiskers." Then, it is clear that A and B are deformation retracts of U and V, respectively, and that $U \cap V = X - \{a, b\}$ is contractible, hence, simply connected. Thus, we conclude that $\pi(X)$ is the free product of the groups $\pi(U)$ and $\pi(V)$ or, equivalently, the free product of $\pi(A)$ and $\pi(B)$ [because $\pi(A) \approx \pi(U)$ and $\pi(B) \approx \pi(V)$]. Because A and B are circles, $\pi(A)$ and $\pi(B)$ are infinite cyclic groups. Therefore, $\pi(X)$ is the free product of two infinite cyclic groups; by

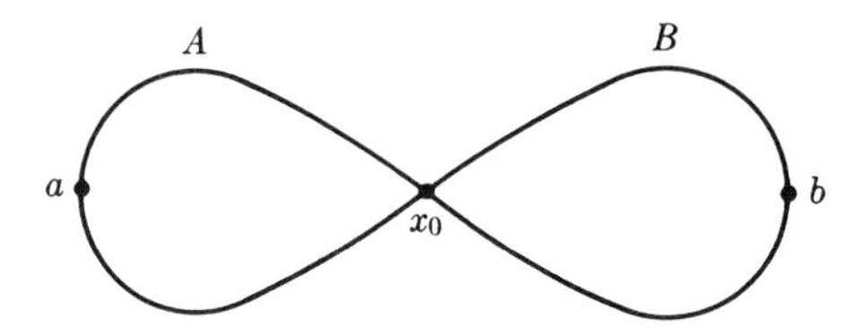

FIGURE 4.2 Example 3.1, a figure "8" curve.

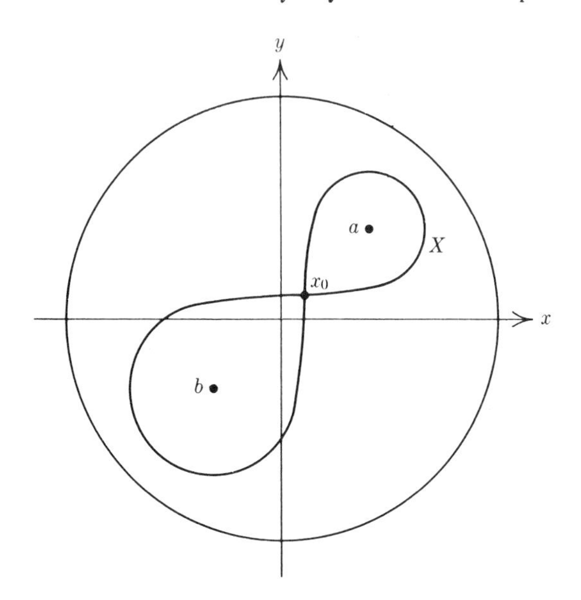

FIGURE 4.3 Example 3.2, a disc with two holes.

Proposition III.5.2, *$\pi(X)$ is a free group on two generators.* We can take as generators closed path classes α and β based at x_0, which go once around A and B, respectively.

3.2 Let E^2 be the closed unit disc in the plane, let a and b be distinct interior points of E^2, and let $Y = E^2 - \{a, b\}$. It is easily seen that we can find a subset $X \subset Y$, such that X is the union of two circles with a single point in common, as in Example 3.1, and X is a deformation retract of Y (see Figure 4.3). Therefore, $\pi(Y) \approx \pi(X)$, and $\pi(Y)$ is a free group on two generators. We can take as generators path classes α and β based at x_0 which go once around the "holes" a and b.

There is an experimental physical verification of this result that appeals to one's geometric intuition. Take a piece of plywood or some other strong, light material in the shape of a circular disc, and at the points a and b attach vertical pegs several inches long. Fasten both ends of a piece of string a few feet long to the plywood at the point x_0 with a thumbtack. Any element $\neq 1$ of the fundamental group of Y can be represented uniquely as a "reduced word" in α and β; and for any such reduced word, we can choose a representative path in Y and then lay out the string on the board to represent this path. We can then test experimentally whether or not this path is equivalent to the constant path by moving the string about on the board. Of course, it is not permissible to lift the string over the pegs while doing this.

3.3 The same argument applies if Y is an open disc minus two points, or the entire plane minus two points, or a sphere minus three points. It also applies if, instead of removing isolated points from a disc, we remove small circular discs, either open or closed.

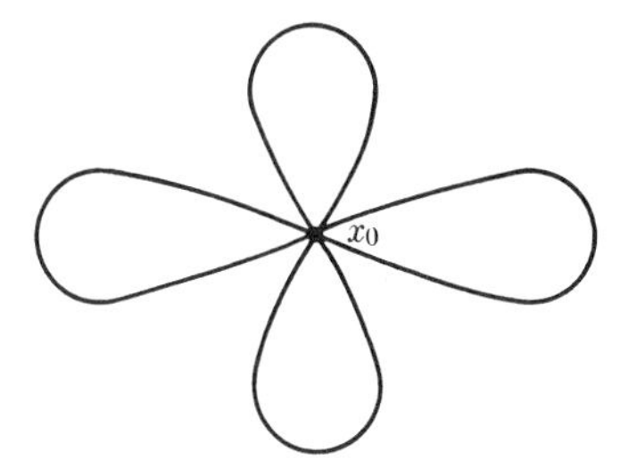

FIGURE 4.4 Example 3.4 for the case $n = 4$.

3.4 Let X be the union of n circles with a single point in common, $n > 2$; i.e.,

$$X = A_1 \cup A_2 \cup \cdots \cup A_n,$$

where each A_i is homeomorphic to S^1, and, if $i \neq j$, $A_i \cap A_j = \{x_0\}$. The space X can be pictured as an "n-leafed rose" in the plane (see Figure 4.4 for the case where $n = 4$). We will prove by induction on n that *$\pi(X)$ is a free group on n generators, $\alpha_1, \alpha_2, \cdots, \alpha_n$, where α_i is represented by a path that goes around the circle A_i once.* We have already proved this in the case where $n = 2$. To make this induction, we apply Theorem 3.1 as follows. Choose a point $a_i \in A_i$ such that $a_i \neq x_0$. Let

$$U = X - \{a_n\},$$

$$V = X - \{a_1, a_2, \ldots, a_{n-1}\}.$$

Then, U and V are open sets, $A_1 \cup \cdots \cup A_{n-1}$ is a deformation retract of U, A_n is a deformation retract of V, and $U \cap V$ is contractible. Thus, using Theorem 3.1, we can conclude $\pi(X, x_0)$ is the free product of $\pi(U)$ and $\pi(V)$ or, equivalently, of $\pi(A_1 \cup \cdots \cup A_{n-1})$ and $\pi(A_n)$. Proposition III.5.2 can now be applied to complete the proof of the inductive step.

3.5 We can use the result just proved to discuss the following example: Let Y be a space obtained by removing n points from a disc (open or closed) or from the entire plane. By the same type of argument as that used in Example 3.2, we conclude $\pi(Y)$ is a free group on n generators, $\alpha_1, \alpha_2, \cdots, \alpha_n$. Roughly speaking, α_i is represented by a closed path which goes around the ith hole once.

We leave it to the reader to discuss a physical model illustrating that $\pi(Y)$ is a free group on n generators, as was done for Example 3.2.

These examples illustrate an important point: To actually apply the Seifert-Van Kampen theorem, it is usually necessary to use the properties of deformation retracts. For future reference, we now give a formal statement of the principle involved here. Let $\{U_\lambda : \lambda \in \Lambda\}$ be an open covering of the space X such that all the hypotheses of Theorem 2.2 are satisfied. Assume that $\{A_\lambda : \lambda \in \Lambda\}$ is a family of subsets of the space X such that the following conditions hold:

(a) For all $\lambda \in \Lambda$, $x_0 \in A_\lambda$.
(b) For each $\lambda \in \Lambda$, $A_\lambda \subset U_\lambda$ and the inclusion map induces an isomorphism $\pi(A_\lambda, x_0) \approx \pi(U_\lambda, x_0)$.
(c) If $U_\lambda \subset U_\mu$, then $A_\lambda \subset A_\mu$.

Let $\psi'_\lambda : \pi(A_\lambda) \to \pi(X)$ and $\varphi'_{\lambda\mu} : \pi(A_\lambda) \to \pi(A_\mu)$ denote homomorphisms induced by inclusion maps.

Lemma 3.2 *Under the above hypotheses, Theorem 2.2 remains true if for each λ we replace $\pi(U_\lambda)$ by $\pi(A_\lambda)$ and ψ_λ by ψ'_λ, and for each pair (λ, μ) such that $U_\lambda \subset U_\mu$ we replace $\varphi_{\lambda\mu}$ by $\varphi'_{\lambda\mu}$.*

PROOF: The proof is obvious.

The most common case in practice is that in which each A_λ is a closed subset of X which is a deformation retreat of U_λ.

We leave it to the reader to state explicitly the special cases of Lemma 3.2 which correspond to Theorems 2.1 and 3.1.

Exercises

3.1 Prove the following generalization of Theorem 3.1. Let $\{W\} \cup \{V_i : i \in I\}$ be a covering of X by open arcwise-connected sets having the following properties: (a) W is a *proper* subset of V_i for all $i \in I$. (b) For any two distinct indices $i, j \in I$, $V_i \cap V_j = W$. (c) W is simply connected. (d) $x_0 \in W$. Using Theorem 2.2, prove that $\pi(X, x_0)$ is the free product of the groups $\pi(V_i, x_0)$ [with respect to the homomorphisms $\psi_i : \pi(V_i) \to \pi(X)$ induced by inclusion maps].

3.2 Let

$$X = \bigcup_{i \in I} A_i,$$

where each A_i is homeomorphic to S^1, be such that, for any two distinct indices $i, j \in I$, $A_i \cap A_j = \{x_0\}$, and the topology on X satisfies the Hausdorff separation axiom and the following condition: A subset B of X is closed (open) if and only if $B \cap A_i$ is a closed (open) subset of A_i for all $i \in I$. For each index i, let α_i be a generator of the infinite cyclic group $\pi(A_i, x_0)$. Use the result of Exercise 3.1 to prove that $\pi(X, x_0)$ is a free group on the set $\{\alpha_i : i \in I\}$.

3.3 Give an example of a compact Hausdorff space

$$X = \bigcup_{i=1}^{\infty} A_i,$$

where each A_i is homeomorphic to S^1, $A_i \cap A_j = \{x_0\}$ for $i \neq j$, and yet X does not satisfy the condition of the previous exercise. (SUGGESTION: there exists a subset of the Euclidean plane having the required properties.) Is $\pi(X, x_0)$ a free group on the set $\{\alpha_i\}$, as in Exercise 3.2?

3.4 Let Y be the complement of the following subset of the plane $\mathbf{R}^2$:

$$\{(x, 0) \in \mathbf{R}^2 : x \text{ is an integer}\}.$$

Prove that $\pi(Y)$ is a free group on a countable set of generators.

3.5 Let X be a Hausdorff space such that $X = A \cup B$, where A and B are each homeomorphic to a torus, and $A \cap B = \{x_0\}$. What is the structure of $\pi(X, x_0)$?

3.6 Let M_1 and M_2 be disjoint, connected n-manifolds. Prove that the following method of constructing the connected sum $M_1 \# M_2$ is equivalent to the definition given in Section I.4 in the case where $n = 2$. Choose points $m_i \in M_i$, and open neighborhoods U_i of m_i such that there exist homeomorphisms h_i of U_i onto $\mathbf{R}^n$ with $h_i(m_i) = 0$, $i = 1, 2$. Define $M_1 \# M_2$ to be the quotient space of $(M_1 - \{m_1\}) \cup (M_2 - \{m_2\})$ obtained by identifying points $x_1 \in U_1 - \{m_1\}$ and $x_2 \in U_2 - \{m_2\}$ if and only if

$$h_1(x_1) = \frac{h_2(x_2)}{|h_2(x_2)|^2}.$$

3.7 If M_1 and M_2 are connected n-manifolds, $n > 2$, prove that $\pi(M_1 \# M_2)$ is the free product of $\pi(M_1)$ and $\pi(M_2)$.

4 Second application of Theorem 2.1

Once again we assume the hypotheses and notation of Theorem 2.1: U, V, and $U \cap V$ are arcwise-connected open subsets of X, $X = U \cup V$, and $x_0 \in U \cap V$.

Theorem 4.1 *Assume that V is simply connected. Then, $\psi_1 : \pi(U) \to \pi(X)$ is an epimorphism, and its kernel is the smallest normal subgroup of $\pi(U)$ containing the image $\varphi_1[\pi(U \cap V)]$.*

Note that this theorem completely specifies the structure of $\pi(X)$: It is isomorphic to the quotient group of $\pi(U)$ modulo the stated normal subgroup.

PROOF: Consider the following commutative diagram:

$$\begin{array}{ccccc} & & \pi(U) & & \\ & \overset{\varphi_1}{\nearrow} & & \overset{\psi_1}{\searrow} & \\ \pi(U \cap V) & & \xrightarrow{\quad \psi_3 \quad} & & \pi(X). \\ & \underset{\varphi_2}{\searrow} & & \underset{\psi_2}{\nearrow} & \\ & & \pi(V) & & \end{array}$$

Because $\pi(V) = \{1\}$, it readily follows that ψ_3 is a trivial homomorphism and that image φ_1 is contained in kernel ψ_1. It is also clear that ψ_1 is an epimorphism; this follows from Lemma 2.3, or we could prove it directly from Theorem 2.1.

Thus, the only thing remaining is to prove that the kernel of ψ_1 is the *smallest* normal subgroup of $\pi(U)$ containing image φ_1 (conceivably, it could be a larger normal subgroup containing image φ_1). For this purpose, take $H = \pi(U)/N$, where N is the smallest normal subgroup of $\pi(U)$ containing image φ_1, and let $\rho_1 : \pi(U) \to H$ be the natural map of $\pi(U)$ onto its quotient group. Let $\rho_2 : \pi(V) \to H$ and $\rho_3 : \pi(U \cap V) \to H$ be trivial homomorphisms. Then, the hypotheses of Theorem 2.1 are satisfied. Hence, we conclude that there exists a homomorphism $\sigma : \pi(X) \to H$ such that the following diagram is commutative:

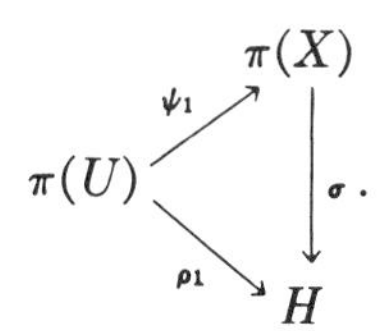

From this, it follows that

$$\text{kernel } \psi_1 \subset \text{kernel } \rho_1 = N.$$

Because we already know that

$$N \subset \text{kernel } \psi_1,$$

we can conclude that

$$\text{kernel } \psi_1 = N$$

as required. Q.E.D.

In the next section we combine this theorem with our preceding results to determine the structure of the fundamental groups of the various compact, connected 2-manifolds.

Exercise

4.1 Assuming the hypotheses and using the notation of Theorem 2.1, prove the following assertions:

(a) If φ_2 is an isomorphism onto, then so is ψ_1.

(b) If both φ_1 and φ_2 are epimorphisms, then ψ_3 is also an epimorphism, and its kernel is the smallest normal subgroup of $\pi(U \cap V)$ which contains both the kernel of φ_1 and the kernel of φ_2.

(c) If $\pi(U \cap V)$ is a cyclic group with generator α, then $\pi(X)$ is isomorphic to the quotient group of the free product of $\pi(U)$ and $\pi(V)$ by the least normal subgroup containing $(\varphi_1\alpha)(\varphi_2\alpha)^{-1}$.

(d) $\pi(X)$ is isomorphic to the quotient group of the free product $\pi(U) * \pi(V)$

by the smallest normal subgroup containing

$$\{(\varphi_1\alpha)(\varphi_2\alpha)^{-1} : \alpha \in \pi(U \cap V)\}.$$

(e) Assume that you are given presentations for the groups $\pi(U)$ and $\pi(V)$, also a set of generators for $\pi(U \cap V)$. Show how to obtain a presentation for $\pi(X)$ from this data and the knowledge of the homomorphisms φ_1 and φ_2. Prove that, if $\pi(U)$ and $\pi(V)$ have finite presentations, and $\pi(U \cap V)$ is finitely generated, then $\pi(X)$ has a finite presentation.

(f) If φ_2 is an epimorphism, then so is ψ_1. Describe the kernel of ψ_1 in this case.

(g) If there exists a homomorphism $r : \pi(V) \to \pi(U \cap V)$ such that $r\varphi_2$ is the identity, then there exists a homomorphism $s : \pi(X) \to \pi(U)$ such that $s\psi_1$ is the identity, and $\varphi_1 r = s\psi_2$.

5 Structure of the fundamental group of a compact surface

We shall show by examples how Theorem 4.1 can be used to determine the structure of the fundamental group of the various compact, connected 2-manifolds.

Examples

5.1 *The torus, T.* Because $T = S^1 \times S^1$, we already know by Theorem II.7.1 that

$$\pi(T) \approx \pi(S^1) \times \pi(S^1)$$

is the product of two infinite cyclic groups, i.e., a free abelian group on two generators. However, we shall derive this result from Theorem 4.1. This simple case serves as a good introduction to the rest of the examples.

Represent the torus as the space obtained by identifying the opposite faces of a square, as shown in Figure 4.5. Under the identification the sides a and b each

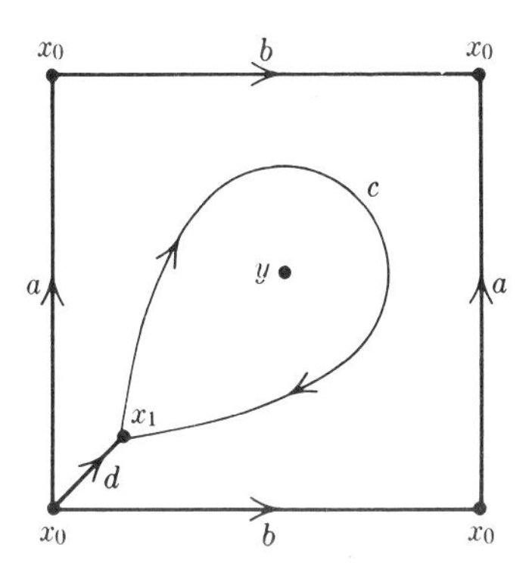

FIGURE 4.5 Determination of the fundamental group of a torus.

become circles which intersect in the point x_0. Let y be the center point of the square, and let $U = T - \{y\}$. Let V be the image of the interior of the square under the identification. Then, U and V are open subsets, U, V, and $U \cap V$ are arcwise connected, and V is simply connected (V is homeomorphic to an open disc). Thus, we can apply Theorem 4.1. We conclude that

$$\psi_1 : \pi(U, x_1) \to \pi(T, x_1)$$

is an epimorphism, and its kernel is the smallest normal subgroup containing the image of the homomorphism

$$\varphi_1 : \pi(U \cap V, x_1) \to \pi(U, x_1).$$

Because the boundary of a square is a deformation retract of the whole square minus a point, it is clear that the union of the two circles a and b is a deformation retract of U. Therefore, $\pi(U, x_1)$ is a free group on two generators. To be more precise, $\pi(U, x_0)$ is a free group on two generators α and β, where α and β are represented by the circles a and b, respectively. Hence, $\pi(U, x_1)$ is a free group on the two generators

$$\alpha' = \delta^{-1}\alpha\delta,$$

$$\beta' = \delta^{-1}\beta\delta,$$

where δ is the equivalence class of a path d from x_0 to x_1 (see Figure 4.5). It is also clear that $U \cap V$ has the homotopy type of a circle. Therefore, $\pi(U \cap V, x_1)$ is an infinite cyclic group generated by γ, the equivalence class of a closed path c which circles around the point y once. It is also clear from Figure 4.5 that

$$\varphi_1(\gamma) = \alpha'\beta'\alpha'^{-1}\beta'^{-1}.$$

Hence, $\pi(T, x_1)$ is isomorphic to the free group on the generators α' and β' modulo the normal subgroup generated by the element $\alpha'\beta'\alpha'^{-1}\beta'^{-1}$. Changing to the base point x_0, we see that $\pi(T, x_0)$ is isomorphic to the free group on the generators α and β modulo the normal subgroup generated by $\alpha\beta\alpha^{-1}\beta^{-1}$.

This means exactly that we have a presentation of the group $\pi(T)$ (see Section III.6). In this case, we can readily determine the structure of $\pi(T)$ from this presentation. On the one hand, it follows that the generators α and β of $\pi(T)$ commute; from this it follows that $\pi(T)$ is a commutative group, and therefore the least normal subgroup of the free group on α and β containing $\alpha\beta\alpha^{-1}\beta^{-1}$ contains the commutator subgroup. On the other hand, it is obvious that this normal subgroup is contained in the commutator subgroup. Therefore, the two subgroups are equal. Hence, by Proposition III.5.3, $\pi(T)$ is a free abelian group on the generators α and β.

5.2 *The real projective plane, P_2.* We shall prove that $\pi(P_2)$ is cyclic of order 2 by using Theorem 4.1. We consider P_2 the space obtained by identifying the opposite sides of a 2-sided polygon, as shown in Figure 4.6. Under the identification, the edge a becomes a circle. Let y be the center point of the polygon,

$U = P_2 - \{y\},$

V = image of the interior of the polygon under the identification.

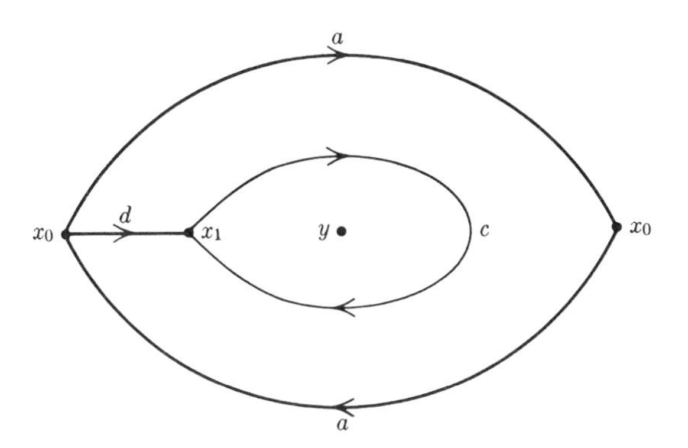

FIGURE 4.6 Determination of the fundamental group of a projective plane.

Then, the conditions for the application of Theorem 4.1 hold. In this case the circle a is a deformation retract of U; therefore, $\pi(U, x_0)$ is an infinite cyclic group generated by an element α represented by the closed path a. Also, $\pi(U, x_1)$ is an infinite cycle group generated by $\alpha' = \delta^{-1}\alpha\delta$, where δ has the same meaning as in Example 5.1. Finally, $\pi(U \cap V, x_1)$ is an infinite cyclic group with generator γ represented by a closed path c which goes around the point y once. It is clear that

$$\varphi_1(\gamma) = \alpha'^2.$$

Therefore, $\pi(P_2, x_1)$ is the quotient of an infinite cyclic group generated by α' modulo the subgroup generated by α'^2 ; equivalently, $\pi(P_2, x_0)$ is the quotient of an infinite cyclic group generated by α modulo the subgroup generated by α^2. Thus, $\pi(P_2)$ is a cyclic group of order 2.

5.3 *The connected sum of n tori.* Here the method is completely analogous to the two preceding examples, but the final result is new and more complicated. We can represent M, the sum of n tori, as a $4n$-gon with the sides identified in pairs, as shown in Figure 4.7. Under the identification, the edges a_1, b_1, a_2, b_2, . . ., a_n, b_n become circles on M, and any two of these circles intersect only in the base point x_0. As before, let $U = M - \{y\}$, the complement of the center point y, and let V be the image of the interior of the polygon under the identification; V is an open disc in M. The union of the $2n$ circles a_1, b_1, . . ., a_n, b_n is a deformation retract of U; therefore, $\pi(U, x_0)$ is a free group on the $2n$ generators α_1, β_1, α_2, β_2, . . ., α_n, β_n, where α_i is represented by the circle a_i, and β_i is represented by the circle b_i. As before, $\pi(U \cap V, x_1)$ is an infinite cyclic group with generator γ represented by the circle c, and

$$\varphi_1(\gamma) = \prod_{i=1}^{n} [\alpha_i', \beta_i'],$$

where $[\alpha_i', \beta_i']$ denotes the commutator $\alpha_i'\beta_i'\alpha_i'^{-1}\beta_i'^{-1}$, and

$$\alpha_i' = \delta^{-1}\alpha_i\delta,$$
$$\beta_i' = \delta^{-1}\beta_i\delta.$$

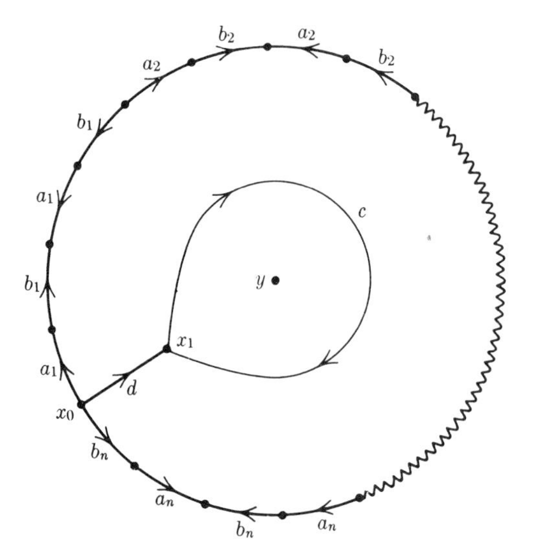

FIGURE 4.7 Determination of the fundamental group of an orientable surface of genus n.

As a result, $\pi(M, x_0)$ is the quotient of the free group on the generators $\alpha_1, \beta_1, \ldots, \alpha_n, \beta_n$ modulo the normal subgroup generated by the element

$$\prod_{i=1}^{n} [\alpha_i, \beta_i];$$

i.e., $\pi(M, x_0)$ has a presentation consisting of the set of generators $\{\alpha_1, \beta_1, \ldots, \alpha_n, \beta_n\}$ and the single relation

$$\prod_{i=1}^{n} [\alpha_i, \beta_i].$$

In the case where $n > 1$, there is no simple, invariant description of this group. It is readily seen however that if we "abelianize" $\pi(M, x_0)$ (i.e., if we take its quotient modulo its commutator subgroup), we obtain a free abelian group on $2n$ generators. This is a consequence of the single relation's obviously being contained in the commutator subgroup of the free group on the generators $\alpha_1, \beta_1, \ldots, \alpha_n, \beta_n$. From this it follows that, if $m \neq n$, the connected sum of m tori and the connected sum of n tori have nonisomorphic fundamental groups. Therefore, they are not of the same homotopy type. This is a stronger result than that proved in Chapter I, where it was shown that these spaces were not homeomorphic (assuming the proof that the Euler characteristic is a topological invariant).

5.4 *The connected sum of n projective planes.* The connected sum M of n projective planes can be obtained by identifying in pairs the sides of a $2n$-gon, as shown in Figure 4.8. By carrying out exactly the same procedure as before,

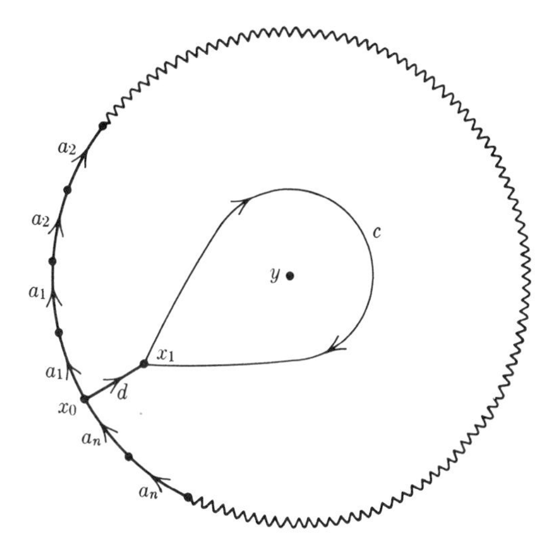

FIGURE 4.8 Determination of the fundamental group of a nonorientable surface of genus n (first method).

we find that the fundamental group $\pi(M, x_0)$ has a presentation consisting of the set of generators

$$\{\alpha_1, \alpha_2, \ldots, \alpha_n\},$$

where α_i is represented by the circle a_i, and one relation

$$\alpha_1^2\alpha_2^2 \ldots \alpha_n^2.$$

For $n > 1$, this is again a group with no simple invariant description. If we abelianize, we obtain an abelian group which also has a presentation consisting of n generators and one relation. The reader who is familiar with the theory of finitely generated abelian groups can easily determine the rank and torsion coefficients of this group by reducing a certain integer matrix to canonical form. We shall do this by a more geometric procedure.

Using Theorem I.7.2, we see that M, a nonorientable surface of genus n, has the following alternative representation:

(a) For n odd, M is homeomorphic to the connected sum of an orientable surface of genus $\frac{1}{2}(n - 1)$ and a projective plane.
(b) For n even, M is homeomorphic to the connected sum of an orientable surface of genus $\frac{1}{2}(n - 2)$ and a Klein Bottle.

This leads to the representation M as the space obtained by identifying the edges of $2n$-gon in pairs as shown in Figure 4.9(a) and (b). In case (a), we see that $\pi(M, x_0)$ has a presentation with generators

$$\{\alpha_1, \beta_1, \ldots, \alpha_k, \beta_k, \varepsilon\}$$

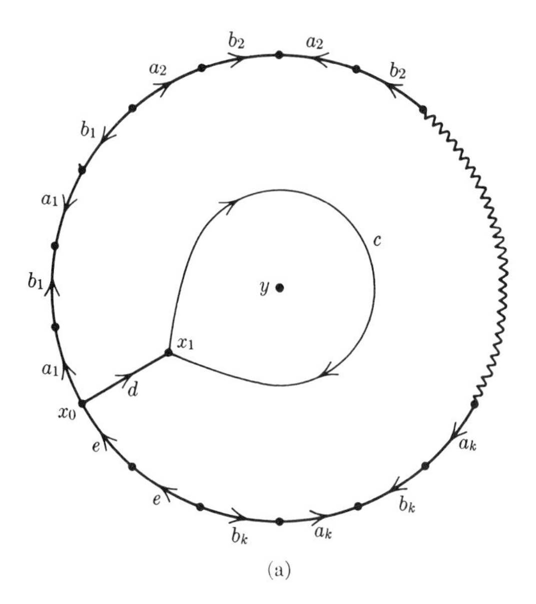

(a)

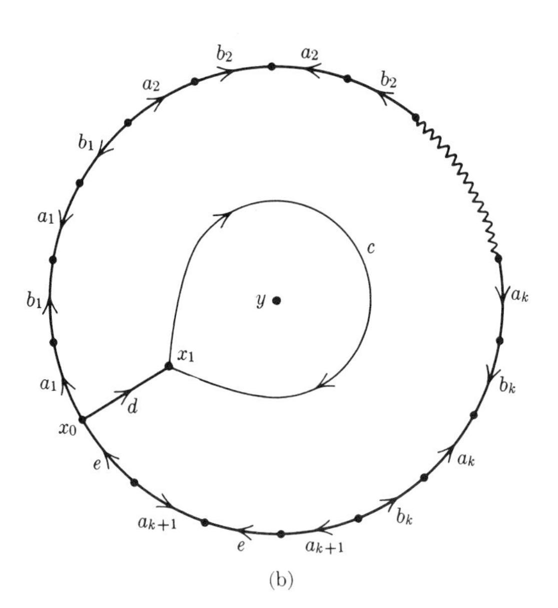

(b)

FIGURE 4.9 Determination of the fundamental group of a nonorientable surface of genus n (second method). (a) n odd, $k = \frac{1}{2}(n - 1)$. (b) n even, $k = \frac{1}{2}(n - 2)$.

and one relation

$$[\alpha_1, \beta_1][\alpha_2, \beta_2] \cdots [\alpha_k, \beta_k]\varepsilon^2;$$

whereas in case (b) there is a presentation of $\pi(M, x_0)$ with generators

$$\{\alpha_1, \beta_1, \ldots, \alpha_k, \beta_k, \alpha_{k+1}, \varepsilon\}$$

and the one relation

$$[\alpha_1, \beta_1][\alpha_2, \beta_2] \cdots [\alpha_k, \beta_k]\alpha_{k+1}\varepsilon\alpha_{k+1}^{-1}\varepsilon.$$

Using this presentation, we can easily determine the structure of the abelianized group,

$$\frac{\pi(M)}{[\pi(M), \pi(M)]}.$$

In case (a) it is the direct product of a free abelian group on the $2k$ generators $\{\alpha_1, \beta_1, \ldots, \alpha_k, \beta_k\}$ and a cyclic group of order 2 (generated by ε); i.e., it is an abelian group of rank $2k = n - 1$ and with one torsion coefficient of order 2. In case (b) it is the direct product of a free abelian group on the $2k + 1$ generators $\{\alpha_1, \beta_1, \ldots, \alpha_k, \beta_k, \alpha_{k+1}\}$ and a cyclic group of order 2 (generated by ε); i.e., it is an abelian group of rank $2k + 1 = n - 1$ with one torsion coefficient of order 2.

We can summarize our results on the abelianized fundamental groups as follows:

Proposition 5.1 *If M is the connected sum of n tori, then the abelianized fundamental group, $\pi(M)/[\pi(M), \pi(M)]$ is a free abelian group of rank $2n$. If M is the connected sum of n projective planes, then the abelianized fundamental group is of rank $n - 1$, and has one torsion coefficient, which is of order 2.*

From this result we see that a compact, connected orientable manifold is never of the same homotopy type as a compact, connected nonorientable manifold, because the abelianized fundamental group of a nonorientable manifold always contains an element of order 2, whereas in the orientable case, every element is of infinite order. It also follows that, if $m \neq n$, then the connected sum of m projective planes and of n projective planes are not of the same homotopy type.

These results are a slight improvement on those of Chapter I, obtained by using the Euler characteristic.

Exercises

5.1 Using the results of Sections I.10–I.12, prove that the fundamental group of a compact, connected bordered surface is a free group. Express the rank of the fundamental group in terms of the Euler characteristic and number of boundary components in both the orientable and nonorientable cases.

5.2 Show how to obtain geometrically the two different presentations of the fundamental group of a Klein Bottle mentioned as an example in Section III.6.

5.3 Consider the presentation of the fundamental group of the Klein Bottle with two generators, a and b, and one relation, $baba^{-1}$. Prove that the subgroup generated by b is a normal subgroup, and that the quotient group is infinite cyclic. Prove also that the subgroup generated by a is infinite cyclic.

5.4 The fact that the connected sum of three projective planes is homeomorphic to the connected sum of a torus and a projective plane gives rise to two different presentations of the fundamental group (as in Problem 5.2). Prove algebraically that these presentations represent isomorphic groups.

5.5 For any integer $n > 2$, show how to construct a space whose fundamental group is cyclic of order n.

5.6 Prove that the fundamental group of a compact nonorientable surface of genus n has a presentation consisting of n generators, $\alpha_1, \ldots, \alpha_n$, and one relation, $\alpha_1\alpha_2 \ldots \alpha_n\alpha_1^{-1}\alpha_2^{-1} \ldots \alpha_{n-1}^{-1}\alpha_n$ (see Exercise I.8.8).

5.7 Prove that the fundamental group of a compact, orientable surface of genus n has a presentation consisting of $2n$ generators, $\alpha_1, \alpha_2, \ldots, \alpha_{2n}$, and one relation, $\alpha_1\alpha_2 \ldots \alpha_{2n}\alpha_1^{-1}\alpha_2^{-1} \ldots \alpha_{2n}^{-1}$ (see Exercise I.8.9).

6 Application to knot theory

A *knot* is, by definition, a simple closed curve in Euclidean 3-space. It is a mathematical abstraction of our intuitive idea of a knot tied in a piece of string; the two ends of the string are to be thought of as spliced together so that the knot can not become untied.

It is also necessary to define when two knots are to be thought of as equivalent or nonequivalent. Here it would be highly desirable to frame the definition so that it corresponds to the usual notion of two knots in two different pieces of string being the same. Of several alternative ways of doing this, the following definition is now universally accepted (as the result of many years of experience) as being the most suitable.

Definition Two knots K_1 and K_2 contained in $\mathbf{R}^3$ *are equivalent* if there exists an orientation-preserving homeomorphism $h : \mathbf{R}^3 \to \mathbf{R}^3$ such that $h(K_1) = K_2$.

Obviously, if K_1 and K_2 are equivalent according to this definition, then h maps $\mathbf{R}^3 - K_1$ homeomorphically onto $\mathbf{R}^3 - K_2$. Therefore, $\mathbf{R}^3 - K_1$ and $\mathbf{R}^3 - K_2$ have isomorphic fundamental groups. Thus, given two knots K_1 and K_2 in $\mathbf{R}^3$, if we can prove that the groups $\pi(\mathbf{R}^3 - K_1)$ and $\pi(\mathbf{R}^3 - K_2)$ are nonisomorphic, then we know the knots K_1 and K_2 are nonequivalent. This is the most common method of distinguishing between knots. The fundamental group $\pi(\mathbf{R}^3 - K)$ is called the *group of the knot K*.

We shall show how it is possible to use the Seifert-Van Kampen theorem to determine a presentation of the group of certain knots, and then discuss the problem of proving that these groups are nonisomorphic.

In certain cases, it will be convenient to think of the knots we shall consider as being imbedded in the 3-sphere S^3,

$$S^3 = \{x \in \mathbf{R}^4 : |x| = 1\}$$

rather than being imbedded in $\mathbf{R}^3$. This makes little difference, because S^3 is homeomorphic to the Alexandroff 1-point compactification of $\mathbf{R}^3$; this can be proved by stereographic projection (see Newman, M. H. A. *Elements of the Topology of Plane Sets of Points.* Cambridge: The University Press, 1951, pp. 64–65).

Exercise

6.1 If K is a knot in $\mathbf{R}^3$ and we regard S^3 as the 1-point compactification of $\mathbf{R}^3$, prove that the fundamental groups $\pi(\mathbf{R}^3 - K)$ and $\pi(S^3 - K)$ are isomorphic. (HINT: Use Theorem 4.1.)

We shall consider a class of knots called *torus knots* because they are contained in a torus imbedded in $\mathbf{R}^3$ in the standard way (i.e., the torus is obtained by rotating a circle about a line in its plane). Recall that a torus may be considered as the space obtained by identifying the opposite edges of the unit square,

$$\{(x, y) \in \mathbf{R}^2 : 0 \leqq x \leqq 1, \qquad 0 \leqq y \leqq 1\},$$

or, alternatively, as the space obtained from the entire plane $\mathbf{R}^2$ by identifying two points (x, y) and (x', y') if and only if $x - x'$ and $y - y'$ are both integers. Let $p : \mathbf{R}^2 \to T$ be the identification map. Let L be a line through the origin in $\mathbf{R}^2$ with slope m/n, where $1 < m < n$, and m and n are relatively prime integers. It is readily seen that the image

$$K = p(L)$$

is a simple closed curve on the torus T; it spirals around the torus m times while going around it n times the other way. If we now assume that T is imbedded in $\mathbf{R}^3$ in the standard way, then

$$K \subset T \subset \mathbf{R}^3,$$

and K is a knot in $\mathbf{R}^3$ called a *torus knot of type* (m, n). Such knots will be our main object of study.

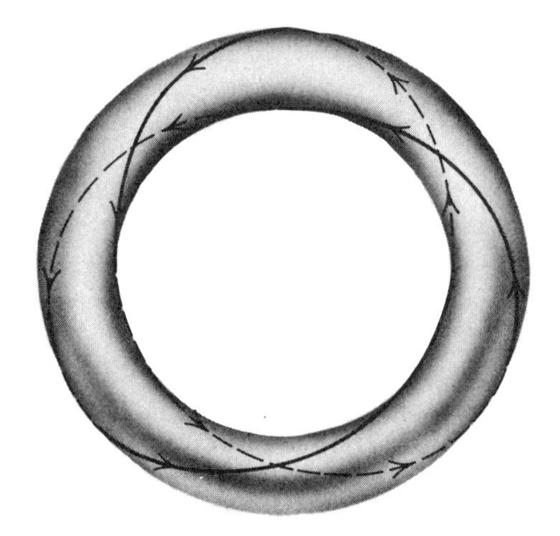

FIGURE 4.10 Torus knot of type (2, 3).

We shall also consider *unknotted circles* in $\mathbf{R}^3$, i.e., any knot equivalent to an ordinary Euclidean circle in a plane in $\mathbf{R}^3$.

To begin, we obtain a presentation of the group of a torus knot of type (m, n) and of the group of an unknotted circle. The first step is to obtain a certain decomposition of the 3-sphere S^3 into two pieces, which is necessary for the use of the Seifert-Van Kampen theorem. Let

$$A = \{(x_1, x_2, x_3, x_4) \in S^3 : x_1^2 + x_2^2 \leqq x_3^2 + x_4^2\},$$

$$B = \{(x_1, x_2, x_3, x_4) \in S^3 : x_1^2 + x_2^2 \geqq x_3^2 + x_4^2\}.$$

It is clear that A and B are closed subsets of S^3, that $A \cup B = S^3$, and that

$$A \cap B = \{(x_1, x_2, x_3, x_4) \in S^3 : x_1^2 + x_2^2 = \tfrac{1}{2} \text{ and } x_3^2 + x_4^2 = \tfrac{1}{2}\}$$

From this it is clear that $A \cap B$ is a torus; in fact, it is the Cartesian product of the circle $x_1^2 + x_2^2 = \frac{1}{2}$ [in the (x_1, x_2) plane] and the circle $x_3^2 + x_4^2 = \frac{1}{2}$ [in the (x_3, x_4) plane].

We now assert that A and B are each solid tori (i.e., homeomorphic to the product of a disc and a circle). We shall prove this by exhibiting a homeomorphism. Let

$$D = \{(x_1, x_2) \in \mathbf{R}^2 : x_1^2 + x_2^2 \leqq \tfrac{1}{2}\}$$

$$S^1 = \{(x_3, x_4) \in \mathbf{R}^2 : x_3^2 + x_4^2 = \tfrac{1}{2}\}$$

be a closed disc and a circle, each of radius $\frac{1}{2}\sqrt{2}$. Define a map

$$f : D \times S^1 \to A$$

by the formula

$$f(x_1, x_2, x_3, x_4)$$
$$= (x_1, x_2, \sqrt{2}\, x_3[1 - (x_1^2 + x_2^2)]^{1/2}, \sqrt{2}\, x_4[1 - (x_1^2 + x_2^2)]^{1/2}).$$

This function is obviously continuous. We leave it to the reader to verify that it is one-to-one and onto, and hence a homeomorphism. A similar proof applies to the set B. It is also clear from this that the torus $A \cap B$ is the common boundary of the two solid tori A and B.

We leave it to the reader to verify that, under stereographic projection, the torus $A \cap B$ corresponds to a torus imbedded in $\mathbf{R}^3$ in the standard way.

First, we consider the group of an unknotted circle K in S^3. We can take as our unknotted circle the "center line" of the solid torus A:

$$K = \{(x_1, x_2, x_3, x_4) \in A : x_1 = x_2 = 0\}.$$

Then, K is the unit circle in the (x_3, x_4) plane. Clearly, the boundary of A is a deformation retract of $A - K$; therefore B is a deformation retract of $S^3 - K$. It is also clear that the center line of B,

$$\{(x_1, x_2, x_3, x_4) \in B : x_3 = x_4 = 0\},$$

is a deformation retract of B. Therefore, the center line of B is a deformation retract of $S^3 - K$. Hence, $S^3 - K$ has the homotopy type of a circle, and the group of K is infinite cyclic. Thus, we have proved

Proposition 6.1 *The group of an unknotted circle in* $\mathbf{R}^3$ *is infinite cyclic.*

Next, we consider a torus knot K of type (m, n) in S^3. We can consider K a subset of the torus $A \cap B \subset S^3$. It would be convenient to apply the Seifert-Van Kampen theorem to determine the fundamental group of $S^3 - K$ by using the fact that

$$S^3 - K = (A - K) \cup (B - K).$$

Then, $A - K$, $B - K$, and $(A - K) \cap (B - K)$ are all arcwise connected, but unfortunately $A - K$ and $B - K$ are not open subsets of $S^3 - K$. The way around this difficulty is clear: We enlarge A and B slightly to obtain open sets with the same homotopy type as A and B.

To be precise, choose a number $\varepsilon > 0$ small enough so that, if N denotes a tubular neighborhood of K of radius ε, then $S^3 - N$ is a deformation retract of $S^3 - K$. It is clear that this will be the case provided ε

is sufficiently small; the precise meaning of the phrase "sufficiently small" depends on the integers m and n. Then, let U and V be the $\frac{1}{2}\varepsilon$ neighborhoods of A and B, respectively. It is clear that U and V are each homeomorphic to the product of an open disc with a circle, and A and B are deformation retracts of U and V. Also, $U \cap V$ is a "thickened" torus, i.e., homeomorphic to the product of $A \cap B$ and the open interval $(-\frac{1}{2}\varepsilon, \frac{1}{2}\varepsilon)$. We can now use the fact that

$$S^3 - N = (U - N) \cup (V - N)$$

and apply the Seifert-Van Kampen theorem to arrive at a presentation of $\pi(S^3 - N) \approx \pi(S^3 - K)$.

First, $U - N$ and $V - N$ both have the homotopy type of a circle; in fact, the center lines of A and B are deformation retracts of these two spaces. Therefore, their fundamental groups are infinite cyclic.

Secondly, the spaces $(U - N) \cap (V - N) = (U \cap V) - N$ and $(A - K) \cap (B - K) = (A \cap B) - K$ both have the same homotopy type. In fact, the set $(A - N) \cap (B - N) = (A \cap B) - N$ is a deformation retract of each of these spaces. We can readily see that $(A \cap B) - K$ is a subset of the torus $A \cap B$ homeomorphic to the product of a circle with an open interval. It is a strip wound spirally around the torus, like a bandage. Its fundamental group is infinite cyclic.

Finally, we must determine the homomorphisms

$$\varphi_1 : \pi(U \cap V - N) \to \pi(U - N),$$
$$\varphi_2 : \pi(U \cap V - N) \to \pi(V - N).$$

Here we leave the details to the reader. The result is that one of these homomorphisms is of degree m, and the other is of degree n. (We say a homomorphism of one infinite cyclic group into another is of degree m if the image of a generator of the first group is the mth power of a generator of the second group.) If we combine this result with Exercise 4.1(c) we obtain the following result:

Proposition 6.2 *The group G of a torus knot of type (m, n) has a presentation consisting of two generators, $\{\alpha, \beta\}$, and one relation, $\alpha^m\beta^n$.*

There remains the task of proving that these groups are nonisomorphic for different values of the pair (m, n). This we now do by a method due to O. Schreier. Consider the element $\alpha^m = \beta^{-n}$ in this group. This element commutes with α and β, and hence with every element; thus it belongs to the center. Let N denote the subgroup generated by this element; it is obviously a normal subgroup. Consider the quotient group

G/N. Let α' and β' denote the coset of α and β in G/N. Obviously, G/N is generated by the elements α' and β', and it has the following presentation:

$$\text{Generators: } \alpha', \beta'. \qquad \text{Relations: } \alpha'^m, \beta'^n.$$

From this presentation, it follows that G/N is the free product of a cyclic group of order m (generated by α') and a cyclic group of order n (generated by β'). The proof, which is not difficult, is left to the reader. We now apply Exercise III.4.1 to conclude that the center of G/N is $\{1\}$. Because the image of the center of G is contained in the center of G/N, it follows that N is the entire center of G. Thus, the quotient of G by its center is the free product of two cyclic groups (of order m and n). We can now apply the result of Exercise III.4.6 to conclude that the integers m, n are completely determined (up to their order) by G. Thus, we have proved the following:

Proposition 6.3 *If torus knots of types (m, n) and (m', n') are equivalent, then $m = m'$ and $n = n'$, or else $m = n'$ and $n = m'$. No torus knot is equivalent to an unknotted circle (assuming $m, n > 1$).*

Thus, by means of torus knots we have constructed an infinite family of nonequivalent knots.

Of course, most knots are not torus knots. The foregoing paragraphs should only be considered a brief introduction to the subject of knot theory. The reader who wishes to learn more about this subject can consult the books of Crowell and Fox [3], Neuwirth [5], or Rolfsen [9].

NOTES

Apparently a theorem along the lines of Theorem 2.1 was first proved by H. Seifert in 1931 in a paper entitled "Konstruktion dreidimensionaler geschlossener Räume" (*Ber. Sächs. Akad. Wiss.*, *83*, 1931, pp. 26–66). A little later a similar theorem was discovered and proved independently by E. R. Van Kampen ("On the connection between the fundamental groups of some related spaces," *Amer. J. Math.*, *55*, 1933, pp. 261–267). In spite of this, it is usually referred to as "Van Kampen's theorem" in American books and papers. Of course, the formulation of the theorem as the solution of a universal mapping problem came later. Our exposition is based on a paper by R. H. Crowell [2], which was apparently inspired by lectures of R. H. Fox at Princeton; see their joint textbook [3].

The reader who is familiar with the theory of simplicial complexes can easily derive Seifert's version of the Seifert-Van Kampen theorem (as stated in Section 52 of Seifert and Threlfall [6]) from Theorem 2.1 and Lemma 3.2 of this chapter. To do this, one makes use of the properties of a *regular neighborhood* of a subcomplex of a simplicial complex as outlined in Chapter II, Section 9 of the text by S. Eilenberg and N. Steenrod (*Foundations of Algebraic Topology*. Princeton, N.J.: Princeton University Press, 1951).

Free products with amalgamated subgroups

Let $\{W\} \cup \{V_i : i \in I\}$ be a covering of X by arcwise-connected open sets such that $V_i \cap V_j = W$ if $i \neq j$ and $x_0 \in W$ (see Exercise 3.1). Assume that, for each index i, the homomorphism $\pi(W, x_0) \to \pi(V_i, x_0)$ is a *monomorphism*. Then, the fundamental group $\pi(X, x_0)$, as specified by Theorem 2.2, has a structure that has been well studied by group theorists; it is called a "free product with amalgamated subgroup." It is a quotient group of the free product of the groups $\pi(V_i)$ obtained by "amalgamating" or identifying the various subgroups which correspond to $\pi(W, x_0)$ under the given monomorphisms. Every element of such a free product with amalgamated subgroups has a unique expression as a "word in canonical form." Such groups are important in certain aspects of group theory, and have also been used in topology. For further information on this subject, see the textbooks on group theory listed in the bibliography of Chapter III.

The Poincaré conjecture

It follows from the computations made in this chapter that any simply-connected, compact surface is homeomorphic to the 2-sphere S^2. H. Poincaré conjectured in the early 1900's that an analogous statement is true for 3-manifolds, namely, that a compact, simply-connected 3-manifold is homeomorphic to the 3-sphere S^3. In spite of the expenditure of much effort by many outstanding mathematicians over the years since Poincaré, it is still unknown whether or not this famous conjecture is true. It is easy to give examples of compact, simply-connected 4-manifolds which are not homeomorphic to S^4 (e.g., $S^2 \times S^2$). However, for all integers $n > 3$ there is an analog of the Poincaré conjecture, namely, that a compact n-manifold that has the homotopy type of an n-sphere is homeomorphic to S^n. This generalized Poincaré conjecture was proved for $n > 4$ by S. Smale in 1960 (see *Ann. Math.*, *74*, 1961, pp. 391–406). The case where $n = 4$ was proved by Michael Freedman in 1982.

Until the classical Poincaré conjecture (the case where $n = 3$) is settled, we cannot hope to have a classification theorem for compact 3-manifolds.

Homotopy type vs. topological type for compact manifolds

From the computations of the fundamental groups of compact surfaces in this chapter, the following fact emerges: If two compact surfaces are not homeomorphic, then they do not have the same homotopy type. The analogous statement for compact 3-manifolds is known to be false; there are fairly simple examples of compact 3-dimensional manifolds which are of the same homotopy type, but not homeomorphic (the so-called "lens spaces"). The proof of this fact is the culmination of the work of mathematicians in several countries over a period of years. The details are rather elaborate.

Higher dimensional examples of manifolds which are of the same homotopy type but not homeomorphic have been constructed by using a theorem of S. P. Novikov (topological invariance of rational Pontrjagin classes).

Fundamental group of a noncompact surface

The fundamental group of any noncompact surface (with a countable basis) is a free group on a countable or finite set of generators. Any simply-connected, noncompact surface is homeomorphic to the plane $\mathbf{R}^2$. For a proof of these facts, see Ahlfors and Sario [1], Chapter I, or the exercises of Section VI.5.

Sketch of the proof that any finitely presented group can be the fundamental group of a compact 4-manifold

First, note that the fundamental group of $S^1 \times S^3$ is infinite cyclic. Hence, by forming the connected sum of n copies of $S^1 \times S^3$, we obtain an orientable, compact 4-manifold whose fundamental group is a free group on n generators (see Exercise 3.7).

Next, suppose that M is a compact, orientable 4-manifold and C is a smooth, simple closed curve in M; it may be shown that any sufficiently small, closed tubular neighborhood N of C is homeomorphic to $S^1 \times E^3$ (this assertion would not be true if M were nonorientable). Also, the boundary of N is homeomorphic to $S^1 \times S^2$. Now $S^1 \times S^2$ is also the boundary of $E^2 \times S^2$, a 4-dimensional manifold with boundary. Let M' denote the complement of the interior of N. Form a quotient space of $M' \cup (E^2 \times S^2)$ by identifying corresponding points of the boundary of N and the boundary of $E^2 \times S^2$; denote the quotient space by M_1. Then, M_1 is readily seen to be a compact, orientable 4-manifold also; the process of obtaining M_1 from M is often called "surgery."

What is the fundamental group of M_1? We can answer this question by applying the Seifert-Van Kampen theorem twice. First, $M = M' \cup N$ and $M' \cap N$ is homeomorphic to $S^1 \times S^2$. It is readily seen that the homomorphism $\pi(M' \cap N) \to \pi(N)$ (induced by the inclusion) is an isomorphism; therefore by Exercise 4.1(a) the homomorphism $\pi(M') \to \pi(M)$ is also an isomorphism. Next, $M_1 = M' \cup (E^2 \times S^2)$ and $M' \cap (E^2 \times S^2) = M' \cap N$. Because $E^2 \times S^2$ is simply connected, Theorem 4.1 is applicable, and we can conclude that $\pi(M') \to \pi(M_1)$ is an epimorphism, and the kernel is the smallest normal subgroup containing the image of $\pi(M' \cap N) \to \pi(M')$; but it is readily seen that the images of $\pi(M' \cap N) \to \pi(M')$ and $\pi(C) \to \pi(M)$ are equivalent. (NOTE: Actually, each time we apply the Seifert-Van Kampen theorem, it is necessary to invoke Lemma 3.2, because M' and N are not open subsets of M.)

We can summarize the conclusion just obtained as follows: $\pi(M_1)$ is naturally isomorphic to the quotient of $\pi(M)$ by the smallest normal subgroup containing the image of $\pi(C) \to \pi(M)$. In other words, we have "killed off" the element α of $\pi(M)$ represented by the closed path C. If the group $\pi(M)$ is presented by means of generators and relations, then $\pi(M_1)$ has a presentation consisting of the same set of generators, and having one additional relation, namely, α.

It is not difficult to show that any element $\alpha \in \pi(M)$ can be represented by a smooth closed path C without any self-intersections, as required in the preceding argument. In fact, this is true for any orientable n-manifold M provided $n \geqq 3$. In a manifold of dimension $\geqq 3$ there is enough "room" to get rid of the self-intersections in any closed path by means of arbitrarily small deformations.

Now let G be a group which has a presentation consisting of n generators $x_1, \ldots, x_n$ and k relations, $r_1, r_2, \ldots, r_k$. Let M be the connected sum of n copies of $S^1 \times S^3$; then $\pi(M)$ is a free group on n generators, which we may denote by $x_1, \ldots, x_n$. We now perform surgery k times on M, killing off in succession the elements $r_1, \ldots, r_k$. The result will be a compact, orientable 4-manifold M_k such that $\pi(M_k) \approx G$, as required.[1]

This construction was utilized by A. A. Markov in his proof that there cannot exist any algorithm for deciding whether or not two given compact, orientable, triangulable 4-manifolds are homeomorphic. Markov's proof depends on the fact that there exists no general algorithm for deciding whether or not two given group presentations represent isomorphic groups (see *Proceedings of International Congress of Mathematicians*, 1958, pp. 300–306; also, W. Boone, W. Haken, and V. Poenaru, "On Recursively Unsolvable Problems in Topology and Their Classification" in *Contributions to Mathematical Logic*, edited by H. Schmidt, K. Schutte, and H.-J. Thiele, North-Holland Pub. Co., 1968, pp. 37–74).

Alternative proof of the Seifert-Van Kampen theorem

There is another method of proving the theorem of Seifert and Van Kampen, using the theory of covering spaces as described in the next chapter. Although this proof is not as long as that given in section 2, it uses more machinery and requires the assumption of additional hypotheses. An exposition of this proof is given in the French text Godbillon [7] and in the research paper Knill [8].

REFERENCES

1. Ahlfors, L. V., and L. Sario. *Riemann Surfaces*. Princeton, N.J.: Princeton University Press, 1960. Chapter I.
2. Crowell, R. H. "On the Van Kampen Theorem." *Pac. J. Math.*, *9*, 1959, pp. 43–50.
3. Crowell, R. H., and R. H. Fox. *Introduction to Knot Theory*. Boston: Ginn, 1963. (Reprinted by Springer-Verlag, New York, 1977.) Chapter V and Appendix III.
4. Hilton, P. J., and S. Wylie. *Homology Theory, an Introduction to Algebraic Topology*. Cambridge: The University Press, 1960. Chapter VI.
5. Neuwirth, L. P. *Knot Groups*. (Annals of Mathematics Studies No. 56) Princeton, N.J.: Princeton University Press, 1965.
6. Seifert, H., and W. Threlfall. *A Textbook of Topology*. New York: Academic Press, 1980. Chapter 7.
7. Godbillon, C. *Éléments de Topologie Algebrique*. Paris: Hermann, 1971. Chapters VI and X.
8. Knill, R. J. "The Seifert and Van Kampen Theorem via Regular Covering Spaces." *Pacific J. Math.*, *49*, 1973, pp. 149–160.
9. Rolfsen, D. *Knots and Links*. Berkeley: Publish or Perish, Inc., 1976.

[1] This result is due to Seifert and Threlfall [6], p. 180.

CHAPTER FIVE

Covering Spaces

1 Introduction

Let X be a topological space; *a covering space of* X consists of a space $\tilde{X}$ and a continuous map p of $\tilde{X}$ onto X which satisfies a certain very strong smoothness requirement. The precise definition is given below. The theory of covering spaces is important not only in topology, but also in related disciplines such as differential geometry, the theory of Lie Groups, and the theory of Riemann surfaces.

The theory of covering spaces is closely connected with the study of the fundamental group. Many basic topological questions about covering spaces can be reduced to purely algebraic questions about the fundamental groups of the various spaces involved. It would be practically impossible to give a complete exposition of either one of these two topics without also taking up the other.

2 Definition and some examples of covering spaces

In this chapter, we shall assume that *all spaces are arcwise connected and locally arcwise connected* (see Section II.2 for the definition) unless otherwise stated. To save words, we shall not keep repeating this assumption. On the other hand, it is not necessary to assume that the spaces we are dealing with satisfy any separation axioms.

Definition Let X be a topological space. A *covering space* of X is a pair consisting of a space $\tilde{X}$ and a continuous map $p : \tilde{X} \to X$ such that the following condition holds: Each point $x \in X$ has an arcwise-connected open neighborhood U such that each arc component of $p^{-1}(U)$ is mapped topologically onto U by p [in particular, it is assumed that $p^{-1}(U)$ is nonempty]. Any open neighborhood U that satisfies the condition just stated is called an *elementary neighborhood.* The map p is often called a *projection.*

To clarify this definition, we now give several examples. In some of the examples our discussion will be rather informal, which is often more helpful than a more rigorous and formal discussion in getting an intuitive feeling for the concept of covering space.

Examples

2.1 Let $p : \mathbf{R} \to S^1$ be defined by

$$p(t) = (\sin t, \cos t)$$

for any $t \in \mathbf{R}$. Then, the pair $(\mathbf{R}, p)$ is a covering space of the unit circle S^1. Any open subinterval of the circle S^1 can serve as an elementary neighborhood. This is one of the simplest and most important examples.

2.2 Let us use polar coordinates (r, θ) in the plane $\mathbf{R}^2$. Then, the unit circle S^1 is defined by the condition $r = 1$. For any integer n, positive or negative, define a map $p_n : S^1 \to S^1$ by the equation

$$p_n(1, \theta) = (1, n\theta).$$

The map p_n wraps the circle around itself n times. It is readily seen that, if $n \neq 0$, the pair (S^1, p_n) is a covering space of S^1. Once again, any proper open interval in S^1 is an elementary neighborhood.

2.3 If X is any space, and $i : X \to X$ denotes the identity map, then the pair (X, i) is a trivial example of a covering space of X. Similarly, if f is a homeomorphism of Y onto X, then (Y, f) is a covering space of X, which is also a rather trivial example. Later in this chapter, we shall prove that, if X is simply connected, then any covering space of X is one of these trivial covering spaces. Thus, we can only hope for nontrivial examples of covering spaces in the case of spaces that are not simply connected.

2.4 If $(\tilde{X}, p)$ is a covering space of X, and $(\tilde{Y}, q)$ is a covering space of Y, then $(\tilde{X} \times \tilde{Y}, p \times q)$ is a covering space of $X \times Y$ [the map $p \times q$ is defined by $(p \times q)(x, y) = (px, qy)$]. We leave the proof to the reader. It is clear that, if U is an elementary neighborhood of the point $x \in X$ and V is an elementary neighborhood of the point $y \in Y$, then $U \times V$ is an elementary neighborhood of $(x, y) \in X \times Y$.

Using this result and Examples 2.1 and 2.2, the reader can construct examples of covering spaces of the torus $T = S^1 \times S^1$. In particular, the plane $\mathbf{R}^2 = \mathbf{R} \times \mathbf{R}$, the cylinder $\mathbf{R} \times S^1$, or the torus itself can serve as a covering space of the torus. The reader should try to visualize the projection p involved in each of these cases.

2.5 In Section I.4, the projective plane P was defined as a quotient space of the 2-sphere S^2. Let $p : S^2 \to P$ denote the natural map. Then, it is readily seen that (S^2, p) is a covering space of P. We can take as an elementary neighborhood of any point $x \in P$ an open disc containing x.

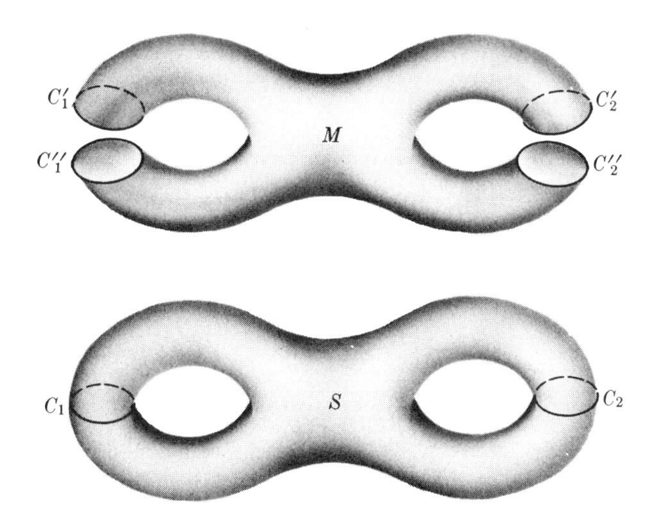

FIGURE 5.1 A surface of genus 2 as a quotient space of a bordered surface.

2.6 Let S be a compact, orientable surface of genus 2. We shall show how to construct a great variety of covering spaces of S. Note that we can regard S as a quotient space of a compact, bordered surface M, where M is orientable, of genus 0, and its boundary consists of four circles C_1', C_1'', C_2', and C_2''. The natural map $M \to S$ identifies the boundary circles in pairs (see Figure 5.1): C_i' and C_i'' are identified to a single circle C_i by means of a homeomorphism h_i of C_i' onto C_i'', $i = 1, 2$. We can also think of M as obtained from S by cutting along the circles C_1 and C_2.

Let D be the finite set $\{1, 2, 3, \ldots, n\}$ with the discrete topology and $q : M \times D \to M$, the projection of the product space onto the first factor. We can think of $M \times D$ as consisting of n disjoint copies of M, each of which is mapped homeomorphically onto M by q. We now describe how to form a quotient space of $M \times D$, which will be a connected 2-manifold $\tilde{S}$ and such that the map q will induce a map $p : \tilde{S} \to S$ of quotient spaces; i.e., so we will have a commutative diagram

$$\begin{array}{ccc} M \times D & \longrightarrow & \tilde{S} \\ {\scriptstyle q}\downarrow & & \downarrow{\scriptstyle p} \\ M & \longrightarrow & S \end{array}.$$

It will turn out that $(\tilde{S}, p)$ is a covering space of S. The identifications by which we form $\tilde{S}$ from $M \times D$ will all be of the following form: The circle $C_i' \times \{j\}$ is identified with the circle $C_i'' \times \{k\}$ by a homeomorphism which sends the point (x, j) onto the point $(h_i(x), k)$, where $i = 1$ or 2, and j and k are positive integers $\leq n$. We can carry out this identification of circles in pairs in many different ways, so long as we obtain a space $\tilde{S}$ which is connected. For example, in the case where $n = 3$, we could carry out the identifications according to the following scheme: Identify

$$\begin{aligned} C_1' \times \{1\} &\text{ with } C_1'' \times \{2\}, \\ C_1' \times \{2\} &\text{ with } C_1'' \times \{3\}, \\ C_1' \times \{3\} &\text{ with } C_1'' \times \{1\}, \\ C_2' \times \{1\} &\text{ with } C_2'' \times \{2\}, \\ C_2' \times \{2\} &\text{ with } C_2'' \times \{1\}, \\ C_2' \times \{3\} &\text{ with } C_2'' \times \{3\}. \end{aligned}$$

We leave it to the reader to concoct other examples, and to prove that in each case we actually obtain a covering space. Obviously, we could use a similar procedure to obtain examples of covering spaces of surfaces of higher genus.

2.7 Let X be a subset of the plane consisting of two circles tangent at a point:

$$\begin{aligned} C_1 &= \{(x, y) : (x - 1)^2 + y^2 = 1\}, \\ C_2 &= \{(x, y) : (x + 1)^2 + y^2 = 1\}, \\ X &= C_1 \cup C_2. \end{aligned}$$

We shall give two different examples of covering spaces of X. For the first example, let $\tilde{X}$ denote the set of all points $(x, y) \in \mathbf{R}^2$ such that x or y (or both) is an integer; $\tilde{X}$ is a union of horizontal and vertical straight lines. Define $p : \tilde{X} \to X$ by the formula

$$p(x, y) = \begin{cases} (1 + \cos(\pi - 2\pi x), \sin 2\pi x) & \text{if } y \text{ is an integer,} \\ (-1 + \cos 2\pi y, \sin 2\pi y) & \text{if } x \text{ is an integer.} \end{cases}$$

The map p wraps each horizontal line around the circle C_1 and each vertical line around the circle C_2.

For the second example, let D_n denote the circle $\{(x, y) \in \mathbf{R}^2 : (x - 1)^2 + (y - 3n)^2 = 1\}$ for any integer n, positive, negative, or zero, and let L denote the vertical line $\{(x, y) : x = 0\}$. The circles D_n are pairwise disjoint, and each is tangent to the line L. Define

$$\tilde{X}' = L \cup (\bigcup_{n \in \mathbf{Z}} D_n),$$

and $p' : \tilde{X}' \to X$ as follows: Let p' map each circle D_n homeomorphically onto C_1 by a vertical translation of the proper amount. Let p' wrap the line L around the circle C_2 in accordance with the formula

$$p'(0, y) = \left(-1 + \cos \frac{2\pi y}{3}, \sin \frac{2\pi y}{3}\right).$$

Then, $(\tilde{X}', p')$ is a covering space of X.

2.8 Here is an example for students who have at least a slight familiarity with the theory of functions of a complex variable. As usual, let

$$\exp(z) = \sum_{n=0}^{\infty} \frac{z^n}{n!}$$

denote the exponential function, where z is any complex number. The exponential function is a map, $\exp : \mathbf{C} \to \mathbf{C} - \{0\}$, where $\mathbf{C}$ denotes the complex plane. We assert that $(\mathbf{C}, \exp)$ is a covering space of $\mathbf{C} - \{0\}$, and that, for any $z \in \mathbf{C} - \{0\}$, the open disc

$$U_z = \{w \in \mathbf{C} : |w - z| < |z|\}$$

is an elementary neighborhood. To prove this, we would have to show that any component V of the inverse image of U_z is mapped homeomorphically onto U_z by exp; i.e., that there exists a continuous function $f : U_z \to V$ such that, for any $w \in U_z$,

$$\exp[f(w)] = w,$$

and, for any $v \in V$,

$$f(\exp v) = v.$$

Such a function f is called a "branch of the logarithm function in the disc U_z" in books on complex variables, and in the course of establishing the properties of the logarithm, the required facts are proved.

Recall that, if $z = x + iy$, then $\exp z = (\exp x) \cdot (\cos y + i \sin y)$, where $\exp x = e^x$ now refers to the more familiar real exponential function, $\exp : \mathbf{R} \to \{t \in \mathbf{R} : t > 0\}$. From this formula, the following fact emerges. We can regard $\mathbf{C} = \mathbf{R} \times \mathbf{R}$ and $\mathbf{C} - \{0\} = \{r \in \mathbf{R} : r > 0\} \times S^1$ (use polar coordinates). Then, we can consider the map $\exp : \mathbf{C} \to \mathbf{C} - \{0\}$ as a map $p \times q : \mathbf{R} \times \mathbf{R} \to \{r \in \mathbf{R} : r > 0\} \times S^1$, where $p(x) = e^x$ and $q(y) = (\cos y, \sin y)$. Compare Examples 2.1, 2.3, and 2.4.

2.9 We now give another example from the theory of functions of a complex variable. For any integer $n \neq 0$, let $p_n : \mathbf{C} \to \mathbf{C}$ be defined by $p_n(z) = z^n$. Then, $(\mathbf{C} - \{0\}, p_n)$ is a covering space of $\mathbf{C} - \{0\}$. The proof is given in books on complex variables when the existence and properties of the various "branches" of the function $\sqrt[n]{z}$ are discussed; the situation is analogous to that in Example 2.8. Note that it is necessary to omit 0 from the domain and range of the function p_n; otherwise we would not have a covering space. As in Example 2.8, we can consider $\mathbf{C} - \{0\} = \{r \in \mathbf{R} : r > 0\} \times S^1$ and decompose the covering space $(\mathbf{C} - \{0\}, p_n)$ into the Cartesian product of two covering spaces.

To clarify further the concept of covering space, we shall give some examples which are almost, but not quite, covering spaces.

Definition A continuous map $f : X \to Y$ is a *local homeomorphism* if each point $x \in X$ has an open neighborhood V such that $f(V)$ is open and f maps V topologically onto $f(V)$.

It is readily proved that, if $(\tilde{X}, p)$ is a covering space of X, then p is a local homeomorphism (the proof depends on the fact that in a locally arcwise connected space, the arc components of an open set are open). Also, the inclusion map of an open subset of a topological space into the whole space is a local homeomorphism. Finally, the composition of two local homeomorphisms is again a local homeomorphism. Thus, we can construct many examples of local homeomorphisms.

On the other hand, it is easy to construct examples of local homeomorphisms which are onto maps, but not covering spaces. For example, let p map the open interval $(0, 10)$ onto the circle S^1 as follows:

$$p(t) = (\cos t, \sin t).$$

Then, p is a local homeomorphism, but $((0, 10), p)$ is not a covering space of S^1. (Which points of S^1 fail to have an elementary neighborhood?) More generally, if $(\tilde{X}, p)$ is a covering space of X, and V is a connected, open, proper subset of $\tilde{X}$, then $p \mid V$ is a local homeomorphism, but $(V, p \mid V)$ is not a covering space of X. It is important to keep this distinction between covering spaces and local homeomorphisms in mind.

Note that a local homeomorphism is an open map (see Appendix A for the definition). In particular, if $(\tilde{X}, p)$ is a covering space of X, then p is an open map.

We next give a lemma which makes it possible to give many additional examples of covering spaces.

Lemma 2.1 *Let $(\tilde{X}, p)$ be a covering space of X, let A be a subspace of X which is arcwise connected and locally arcwise connected, and let $\tilde{A}$ be an arc component of $p^{-1}(A)$. Then, $(\tilde{A}, p \mid \tilde{A})$ is a covering space of A.*

The proof is immediate. The two covering spaces described in Example 2.7 can also be obtained by applying this lemma to the covering spaces $\mathbf{R}^2 = \mathbf{R} \times \mathbf{R}$ and $\mathbf{R} \times S^1$ of the torus $S^1 \times S^1$ described in Example 2.4 [choose A to be the following subset of $S^1 \times S^1 : A = (S^1 \times \{x_0\}) \cup (\{x_0\} \times S^1)$, where $x_0 \in S^1$].

We close this section by stating two of the principal problems in the theory of covering spaces:

(a) Give necessary and sufficient conditions for two covering spaces $(\tilde{X}_1, p_1)$ and $(\tilde{X}_2, p_2)$ of X to be isomorphic (by definition, they are isomorphic if and only if there exists a homeomorphism h of $\tilde{X}_1$ onto $\tilde{X}_2$ such that $p_2 h = p_1$).

(b) Given a space X, determine all possible covering spaces of X (up to isomorphism).

As we shall see, these problems have reasonable answers in terms of the fundamental groups of the spaces involved.

Exercises

2.1 Prove that the following four conditions on a topological space are equivalent:

(a) The arc components of any open subset are open.
(b) Every point has a basic family of arcwise-connected open neighborhoods.
(c) Every point has a basic family of arcwise-connected neighborhoods (they are not assumed to be open).
(d) For every point x and every neighborhood U of x, there exists a neighborhood V of x such that $V \subset U$ and any two points of V can be joined by an arc in U.

Thus, any one of these conditions could be taken as the definition of local arcwise connectivity.

2.2 Give an example of a local homeomorphism $f : X \to Y$ and a subset $A \subset X$ such that $f \mid A$ is not a local homeomorphism of A onto $f(A)$.

2.3 Prove that if X is compact and $f : X \to Y$ is a local homeomorphism, then, for any point $y \in Y$, $f^{-1}(y)$ is a finite set. If it is also assumed that Y is a connected Hausdorff space, then f maps X onto Y.

2.4 Assume X and Y are arcwise connected and locally arcwise connected, X is compact Hausdorff, and Y is Hausdorff. Let $f : X \to Y$ be a local homeomorphism; prove that (X, f) is a covering space of Y. (WARNING: This exercise is more subtle than it looks!)

3 Lifting of paths to a covering space

In this section, we prove some simple lemmas which provide the key to many of the results in this chapter. Let $(\tilde{X}, p)$ be a covering space of X, and let $g : I \to \tilde{X}$ be a path in $\tilde{X}$; then, pg is a path in X. Also, if g_0, $g_1 : I \to \tilde{X}$ and $g_0 \sim g_1$, then $pg_0 \sim pg_1$. We can now ask for a sort of converse result: If $f : I \to X$ is a path in X, does there exist a path $g : I \to \tilde{X}$ such that $pg = f$? If g_0, $g_1 : I \to \tilde{X}$ and $pg_0 \sim pg_1$, does it follow that $g_0 \sim g_1$? We shall see that the answer to both questions is *Yes*. This fact expresses one of the basic properties of covering spaces.

Lemma 3.1 *Let $(\tilde{X}, p)$ be a covering space of X, $\tilde{x}_0 \in \tilde{X}$, and $x_0 = p(\tilde{x}_0)$. Then, for any path $f : I \to X$ with initial point x_0, there exists a unique path $g : I \to \tilde{X}$ with initial point $\tilde{x}_0$ such that $pg = f$.*

PROOF: If the path f were contained in an elementary neighborhood U there would be no problem. For, if V denotes the arc component of $p^{-1}(U)$ which contains $\tilde{x}_0$, then, because p maps V topologically onto U, there would exist a unique g in V with the required properties.

Of course, f will not, in general, be contained in an elementary neighborhood U. However, we can always express f as the product of a finite number of "shorter" paths, each of which is contained in an elementary neighborhood, and then apply the argument in the preceding paragraph to each of these shorter paths in succession.

The details of this procedure may be described as follows. Let $\{U_i\}$ be a covering of X by elementary neighborhoods; then, $\{f^{-1}(U_i)\}$ is an open covering of the compact metric space I. Choose an integer n so large that $1/n$ is less than the Lebesgue number of this covering. Divide the interval I into the closed subintervals $[0, 1/n]$, $[1/n, 2/n]$, ..., $[(n-1)/n, 1]$. Note that f maps each subinterval into an elementary neighborhood in X. We now define g successively over these subintervals, starting with $[0, 1/n]$.

The uniqueness of the lifted path g is a consequence of the following more general lemma:

Lemma 3.2 *Let $(\tilde{X}, p)$ be a covering space of X and let Y be a space which is connected. Given any two continuous maps $f_0, f_1 : Y \to \tilde{X}$ such that $pf_0 = pf_1$,* the set *$\{y \in Y : f_0(y) = f_1(y)\}$ is either empty or all of Y.*

PROOF: Because Y is connected, it suffices to prove that the set in question is both open and closed. First we shall prove that it is closed. Let y be a point of the closure of this set, and let

$$x = pf_0(y) = pf_1(y).$$

Assume $f_0(y) \neq f_1(y)$; we will show that this assumption leads to a contradiction. Let U be an elementary neighborhood of x, and let V_0 and V_1 be the components of $p^{-1}(U)$ which contain $f_0(y)$ and $f_1(y)$ respectively. Since f_0 and f_1 are both continuous, we can find a neighborhood W of y such that $f_0(W) \subset V_0$ and $f_1(W) \subset V_1$. But it is readily seen that this contradicts the fact that any neighborhood W of y must meet the set in question.

An analogous argument enables us to show that every point of the set $\{y \in Y : f_0(y) = f_1(y)\}$ is an interior point. Q.E.D.

Lemma 3.3 *Let $(\tilde{X}, p)$ be a covering space of X and let $g_0, g_1 : I \to \tilde{X}$ be paths in $\tilde{X}$ which have the same initial point. If $pg_0 \sim pg_1$, then $g_0 \sim g_1$; in particular, g_0 and g_1 have the same terminal point.*

PROOF: The strategy of this proof is essentially the same as that of Lemma 3.1. Let $\tilde{x}_0$ be the initial point of g_0 and g_1. The hypothesis $pg_0 \sim pg_1$ implies the existence of a map $F : I \times I \to X$ such that

$$\begin{aligned} F(s, 0) &= pg_0(s), \\ F(s, 1) &= pg_1(s), \\ F(0, t) &= pg_0(0) = p(\tilde{x}_0), \\ F(1, t) &= pg_0(1). \end{aligned}$$

By an argument using the Lebesgue number, etc., we can find numbers $0 = s_0 < s_1 < \cdots < s_m = 1$ and $0 = t_0 < t_1 < \cdots < t_n = 1$ such that F maps each small rectangle $[s_{i-1}, s_i] \times [t_{j-1}, t_j]$ into some elementary neighborhood in X. We shall prove that there exists a unique map $G : I \times I \to \tilde{X}$ such that $pG = F$ and $G(0, 0) = \tilde{x}_0$. First, we define G over the small rectangle $[0, s_1] \times [0, t_1]$ so that the required properties hold; it is clear that this can be done, because F maps this small rectangle into an elementary neighborhood of the point $p(\tilde{x}_0)$. Then, we extend the definition of G successively over the rectangles $[s_{i-1}, s_i] \times [0, t_1]$ for $i = 2, 3, \ldots, m$, taking care that the definitions agree on the common edge of any two successive rectangles. Thus, G is defined over the strip $I \times [0, t_1]$. Next, G is defined over the rectangles in the strip $I \times [t_1, t_2]$, etc.

The uniqueness of G is assured by Lemma 3.2. Similarly, by the uniqueness assertion of Lemma 3.1, we see that $G(s, 0) = g_0(s)$, $G(0, t) = \tilde{x}_0$, $G(s, 1) = g_1(s)$, and that G maps $\{1\} \times I$ into a single point $\tilde{x}_1$ such that

$$p(\tilde{x}_1) = pg_0(1) = pg_1(1).$$

Thus, G defines an equivalence between the paths g_0 and g_1 as required.

Q.E.D.

As a corollary of these results on the lifting of paths, we shall prove the following lemma:

Lemma 3.4 *If $(\tilde{X}, p)$ is a covering space of X, then the sets $p^{-1}(x)$ for all $x \in X$ have the same cardinal number.*

PROOF: Let x_0 and x_1 be any two points of X. Choose a path f in X with initial point x_0 and terminal point x_1. Using the path f, we can define a mapping $p^{-1}(x_0) \to p^{-1}(x_1)$ by the following procedure. Given any point $y_0 \in p^{-1}(x_0)$, lift f to a path g in $\tilde{X}$ with initial point y_0 such that $pg = f$. Let y_1 denote the terminal point of g. Then, $y_0 \to y_1$ is the desired mapping. Using the inverse path $\bar{f}$ [defined by $\bar{f}(t) = f(1 - t)$], we can define in an analogous way a map $p^{-1}(x_1) \to p^{-1}(x_0)$. It is clear that these maps are the inverse of each other; hence, each is one-to-one and onto.

Q.E.D.

This common cardinal number of the sets $p^{-1}(x)$, $x \in X$, is called the *number of sheets* of the covering space $(\tilde{X}, p)$. For example, we speak of an n-sheeted covering, or an infinite-sheeted covering.

4 The fundamental group of a covering space

As a corollary of Lemma 3.3, we have the following fundamental result:

Theorem 4.1 *Let $(\tilde{X}, p)$ be a covering space of X, $\tilde{x}_0 \in \tilde{X}$, and $x_0 = p(\tilde{x}_0)$. Then, the induced homomorphism $p_* : \pi(\tilde{X}, \tilde{x}_0) \to \pi(X, x_0)$ is a monomorphism.*

This is a direct consequence of the special case of Lemma 3.3 in which g_0 and g_1 are assumed to be *closed* paths.

This theorem leads to the following question: Suppose $\tilde{x}_0$ and $\tilde{x}_1$ are points of X such that $p(\tilde{x}_0) = p(\tilde{x}_1) = x_0$. How do the images of the homomorphisms

$$p_* : \pi(\tilde{X}, \tilde{x}_0) \to \pi(X, x_0),$$

$$p_* : \pi(\tilde{X}, \tilde{x}_1) \to \pi(X, x_0),$$

compare? The answer is very simple. Choose a class γ of paths in $\tilde{X}$ from $\tilde{x}_0$ to $\tilde{x}_1$; this defines an isomorphism $u : \pi(\tilde{X}, \tilde{x}_0) \to \pi(\tilde{X}, \tilde{x}_1)$ by the formula $u(\alpha) = \gamma^{-1}\alpha\gamma$. Thus, we obtain the following commutative diagram (see the exercises in Section II.4):

$$\begin{array}{ccc} \pi(\tilde{X}, \tilde{x}_0) & \xrightarrow{p_*} & \pi(X, x_0) \\ \downarrow u & & \downarrow v \\ \pi(\tilde{X}, \tilde{x}_1) & \xrightarrow{p_*} & \pi(X, x_0) \end{array}$$

Here, $v(\beta) = (p_*\gamma)^{-1}\beta(p_*\gamma)$. But $p_*(\gamma)$ is a closed path, and, hence, an element of $\pi(X, x_0)$. Thus, we see that *the images of $\pi(\tilde{X}, \tilde{x}_0)$ and of $\pi(\tilde{X}, \tilde{x}_1)$ under p_* are conjugate subgroups of $\pi(X, x_0)$.*

Next, the question arises, can *every* subgroup in the conjugacy class of the subgroup $p_*\pi(\tilde{X}, \tilde{x}_0)$ be obtained as the image $p_*\pi(\tilde{X}, \tilde{x}_1)$ for some choice of the point $\tilde{x}_1 \in p^{-1}(x_0)$? Here the answer is *Yes*. To prove this, note that any subgroup in this conjugacy class is of the form $\alpha^{-1}[p_*\pi(\tilde{X}, \tilde{x}_0)]\alpha$ for some choice of the element $\alpha \in \pi(X, x_0)$. Choose a closed path $f : I \to X$ representing α. Apply Lemma 3.1 to obtain a path $g : I \to \tilde{X}$ covering α with initial point $\tilde{x}_0$. Let $\tilde{x}_1$ be the terminal point of this lifted path. Then, it is readily seen that

$$p_*\pi(\tilde{X}, \tilde{x}_1) = \alpha^{-1}[p_*\pi(\tilde{X}, \tilde{x}_0)]\alpha.$$

We can summarize what we have proved in the following theorem:

Theorem 4.2 *Let $(\tilde{X}, p)$ be a covering space of X and $x_0 \in X$. Then, the subgroups $p_*\pi(\tilde{X}, \tilde{x})$ for $\tilde{x} \in p^{-1}(x_0)$ are exactly a conjugacy class of subgroups of $\pi(X, x_0)$.*

The student who desires examples of this theorem can consider the various examples of covering spaces given in Section 2.

Exercise

4.1 Discuss the effect of changing the "base point" x_0 in the statement of Theorem 4.2 to a new base point $x_1 \in X$.

This conjugacy class of subgroups of $\pi(X, x_0)$ is an algebraic invariant of the covering space $(\tilde{X}, p)$. We shall later prove that it completely determines the covering space up to isomorphism!

5 Lifting of arbitrary maps to a covering space

In Section 3 we studied the "lifting" of paths in X to a covering space $\tilde{X}$. We now study the analogous problem for maps of any space Y into X. To discuss this question, we introduce the following notation: If X and Y are topological spaces, $x \in X$ and $y \in Y$, then the notation $f : (X, x) \to (Y, y)$ means f is a continuous map of X into Y and $f(x) = y$. With this notation, we can concisely state our main question as follows: Let $(\tilde{X}, p)$ be a covering space of X, $\tilde{x}_0 \in \tilde{X}$, $x_0 = p(\tilde{x}_0)$, $y_0 \in Y$, and $\varphi : (Y, y_0) \to (X, x_0)$. Under what conditions does there exist a map $\tilde{\varphi} : (Y, y_0) \to (\tilde{X}, \tilde{x}_0)$ such that the diagram

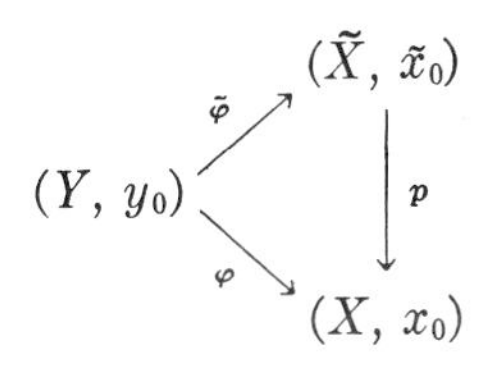

is commutative? If such a map $\tilde{\varphi}$ exists, we say that φ can be *lifted* to $\tilde{\varphi}$, or that $\tilde{\varphi}$ is a *lifting* of φ.

It is easy to obtain a necessary condition for the existence of such a lifting $\tilde{\varphi}$ by consideration of the fundamental groups of the spaces involved. For, if we assume such a map $\tilde{\varphi}$ exists, then we obtain the following commutative diagram of groups and homomorphisms:

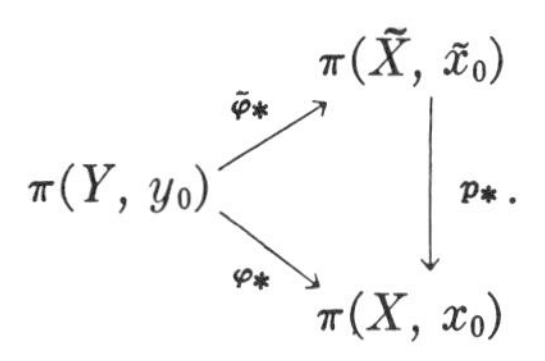

Because p_* is a monomorphism, the existence of a homomorphism $\tilde{\varphi}_* : \pi(Y, y_0) \to \pi(\tilde{X}, \tilde{x}_0)$, which makes this diagram commutative, is exactly equivalent to the condition that the image of φ_* be contained in the image of p_*. This is our desired necessary condition. The surprising thing is that this necessary condition is also sufficient.

Theorem 5.1 *Let $(\tilde{X}, p)$ be a covering space of X, Y a connected and locally arcwise-connected space, $y_0 \in Y$, $\tilde{x}_0 \in \tilde{X}$, and $x_0 = p(\tilde{x}_0)$. Given a map $\varphi : (Y, y_0) \to (X, x_0)$, there exists a lifting $\tilde{\varphi} : (Y, y_0) \to (\tilde{X}, \tilde{x}_0)$ if and only if $\varphi_*\pi(Y, y_0) \subset p_*\pi(\tilde{X}, \tilde{x}_0)$.*

PROOF: We have already proved the necessity of the given condition; it remains to prove it is sufficient. To do this, we must actually define the map $\tilde{\varphi}$. The following considerations show that there is an essentially unique way to define $\tilde{\varphi}$ if it exists at all. Assume that $\tilde{\varphi}$ exists; let y be any point of Y. Because Y is arcwise connected, we may choose a path $f : I \to Y$ with initial point y_0 and terminal point y. Consider the paths φf and $\tilde{\varphi} f$ in X and $\tilde{X}$, respectively. The path $\tilde{\varphi} f$ is a lifting of the path φf, and $\tilde{\varphi}(y)$ is the terminal point of the path $\tilde{\varphi} f$.

In view of these considerations, we *define* the map $\tilde{\varphi} : (Y, y_0) \to (\tilde{X}, \tilde{x}_0)$ as follows: Given any point $y \in Y$, choose a path $f : I \to Y$ with initial point y_0 and terminal point y. Then, φf is a path in X with initial point x_0. Apply Lemma 3.1 to obtain a path $g : I \to \tilde{X}$ such that the initial point of g is $\tilde{x}_0$ and $pg = \varphi f$. Define

$$\tilde{\varphi}(y) = \text{terminal point of } g.$$

To justify this definition, we must show that $\tilde{\varphi}(y)$ is independent of the choice of the path f. By using Lemma 3.3, we see that we can replace f by an equivalent path without altering the definition of $\tilde{\varphi}(y)$; i.e., $\tilde{\varphi}(y)$ only depends on the equivalence class α of the path f. Suppose that α and β are two different equivalence classes of paths in Y from y_0 to y.

Then, $\alpha\beta^{-1}$ is a closed path based at y_0; hence, $\alpha\beta^{-1} \in \pi(Y, y_0)$ and therefore by the hypothesis of the theorem, $\varphi_*(\alpha\beta^{-1}) \in p_*\pi(\tilde{X}, \tilde{x}_0)$. Thus, there is a class of loops based at $\tilde{x}_0$ in $\tilde{X}$ which projects onto $(\varphi_*\alpha)(\varphi_*\beta)^{-1}$, or, if $(\varphi_*\alpha)(\varphi_*\beta)^{-1}$ is "lifted" to a path in $\tilde{X}$ starting at $\tilde{x}_0$, the result is a closed path in $\tilde{X}$. Hence, if $\varphi_*\alpha$ and $\varphi_*\beta$ are each lifted to paths in $\tilde{X}$ starting at $\tilde{x}_0$, they have the same terminal point.

Next, we must prove that the function $\tilde{\varphi}$ thus defined is continuous. Let $y \in Y$ and let U be an arbitrary neighborhood of $\tilde{\varphi}(y)$. We must show that there exists a neighborhood V of y such that $\tilde{\varphi}(V) \subset U$. Choose an elementary neighborhood U' of $p\tilde{\varphi}(y) = \varphi(y)$ such that $U' \subset p(U)$. Let W be the arc component of $p^{-1}(U')$ which contains $\tilde{\varphi}(y)$, and let U'' be an elementary neighborhood of $\varphi(y)$ such that $U'' \subset p(U \cap W)$. Then it is easily shown that the arc component of $p^{-1}(U'')$, which contains $\tilde{\varphi}(y)$ is contained in U. Because φ is continuous, we can choose V such that $\varphi(V) \subset U''$. We can also choose V so that it is arcwise connected, because Y is locally arcwise connected. We leave it to the reader to verify that the neighborhood V thus chosen has the required properties.

It is obvious from our method of defining $\tilde{\varphi}$ that the required commutativity relation $p\tilde{\varphi} = \varphi$ holds. Q.E.D.

Remarks: 1. The map $\tilde{\varphi}$ is unique, in view of Lemma 3.2. The uniqueness of $\tilde{\varphi}$ is also clear from the proof of the theorem.

2. This theorem is a beautiful illustration of the general strategy of algebraic topology: A purely topological question (the existence of a continuous map satisfying certain conditions) is reduced to a purely algebraic question. In most cases in algebraic topology where such a reduction can be effected, the details are much more complicated than in Theorem 5.1.

Exercises

5.1 Let G be a topological space with a continuous multiplication $\mu : G \times G \to G$ with a unit e such that $\mu(e, x) = \mu(x, e) = x$ for any $x \in X$ (see Exercise II.7.5). Let $(\tilde{G}, p)$ be a covering space of G and $\tilde{e} \in \tilde{G}$ a point such that $p(\tilde{e}) = e$. Prove that there exists a unique continuous multiplication $\tilde{\mu} : \tilde{G} \times \tilde{G} \to \tilde{G}$ such that $\tilde{e}$ is a unit [i.e., $\tilde{\mu}(\tilde{e}, y) = \tilde{\mu}(y, \tilde{e}) = y$ for any $y \in \tilde{G}$ and p commutes with the multiplication in $\tilde{G}$ and G [i.e., $\mu(px, py) = p\tilde{\mu}(x, y)$]. (HINT: Use Theorem 5.1 together with the result of Example 2.4 and the exercise of Section II.7 referred to above.) Assume G is arcwise connected and locally arcwise connected as usual. Prove also that, if the multiplication μ is associative, then so is the multiplication $\tilde{\mu}$.

5.2 Let G be a connected, locally arcwise-connected topological group with unit e. Let $(\tilde{G}, p)$ be a covering space of G and $\tilde{e} \in \tilde{G}$ such that $p(\tilde{e}) = e$. Prove that there exists a unique continuous multiplication $\mu : \tilde{G} \times \tilde{G} \to \tilde{G}$ such that $\tilde{G}$ is a topological group with unit $\tilde{e}$ and p is a homomorphism. (HINT: Use the results of Exercises 5.1 and II.7.6 to show the existence of inverses in $\tilde{G}$.) Prove also that the kernel of p is a discrete normal subgroup of $\tilde{G}$ and hence is contained in the center of $\tilde{G}$.

5.3 Apply the considerations of Exercise 5.2 to the case in which $G = S^1$, the multiplicative group of all complex numbers of absolute value 1. Examples of covering spaces of S^1 were described in Section 2.

5.4 In Exercises 5.1 and 5.2, if the multiplication in G is commutative, prove that the multiplication in $\tilde{G}$ is also commutative.

6 Homomorphisms and automorphisms of covering spaces

We wish to obtain some information about the various possible covering spaces of a given space X. As we shall see, we can gain much insight into this problem by considering homomorphisms and automorphisms of covering spaces of X. This procedure is in accordance with the following semi-mystical principle which seems to help guide much present-day mathematical research: Whenever we wish to gain information about a certain class of mathematical objects, it is usually helpful to consider also the appropriate class of admissible maps and automorphisms of these objects.

Definition Let $(\tilde{X}_1, p_1)$ and $(\tilde{X}_2, p_2)$ be covering spaces of X. A *homomorphism* of $(\tilde{X}_1, p_1)$ into $(\tilde{X}_2, p_2)$ is a continuous map $\varphi : \tilde{X}_1 \to \tilde{X}_2$ such that the following diagram is commutative:

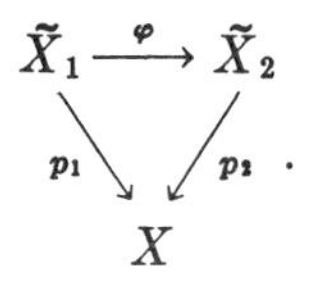

Note that the composition of two homomorphisms is again a homomorphism, and that, if $(\tilde{X}, p)$ is a covering space of X, then the identity map $\tilde{X} \to \tilde{X}$ is a homomorphism.

Definition A homomorphism φ of $(\tilde{X}_1, p_1)$ into $(\tilde{X}_2, p_2)$ is called an *isomorphism* if there exists a homomorphism ψ of $(\tilde{X}_2, p_2)$ into $(\tilde{X}_1, p_1)$ such that both compositions $\psi\varphi$ and $\varphi\psi$ are identity maps. Two covering spaces are said to be *isomorphic* if there exists an isomorphism of one onto the other. An *automorphism* is an isomorphism of a covering space onto itself; it may or may not be the identity map.

Automorphisms of covering spaces are usually called *covering transformations* in the literature (German: Deckbewegung). Note that a homomorphism of covering spaces is an isomorphism if and only if it is a homeomorphism in the usual sense. The set of all automorphisms of a covering space $(\tilde{X}, p)$ of X is obviously a group under the operation of composing maps. We shall use the notation $A(\tilde{X}, p)$ to denote this group.

We now derive some basic properties of homomorphisms and automorphisms of covering spaces.

Lemma 6.1 *Let φ_0 and φ_1 be homomorphisms of $(\tilde{X}_1, p_1)$ into $(\tilde{X}_2, p_2)$. If there exists any point $x \in \tilde{X}_1$ such that $\varphi_0(x) = \varphi_1(x)$, then $\varphi_0 = \varphi_1$.*

This is a special case of Lemma 3.2.

Corollary 6.2 *The group $A(\tilde{X}, p)$ operates without fixed points on the space $\tilde{X}$; i.e., if $\varphi \in A(\tilde{X}, p)$ and $\varphi \neq 1$, then φ has no fixed points.*

Lemma 6.3 *Let $(\tilde{X}_1, p_1)$ and $(\tilde{X}_2, p_2)$ be covering spaces of X and $\tilde{x}_i \in \tilde{X}_i$, $i = 1, 2$, points such that $p_1(\tilde{x}_1) = p_2(\tilde{x}_2)$. Then, there exists a homomorphism φ of $(\tilde{X}_1, p_1)$ into $(\tilde{X}_2, p_2)$ such that $\varphi(\tilde{x}_1) = \tilde{x}_2$ if and only if $p_{1*}\pi(\tilde{X}_1, \tilde{x}_1) \subset p_{2*}\pi(\tilde{X}_2, \tilde{x}_2)$.*

This is a special case of Theorem 5.1.

Corollary 6.4 *Under the hypotheses of Lemma 6.3, there exists an isomorphism φ of $(\tilde{X}_1, p_1)$ onto $(\tilde{X}_2, p_2)$ such that $\varphi(\tilde{x}_1) = \tilde{x}_2$ if and only if $p_{1*}\pi(\tilde{X}_1, \tilde{x}_1) = p_{2*}\pi(\tilde{X}_2, \tilde{x}_2)$.*

This is a direct consequence of Lemma 6.3, the definition of an isomorphism, and Corollary 6.2.

Corollary 6.5 *Let $(\tilde{X}, p)$ be a covering space of X and $\tilde{x}_1, \tilde{x}_2 \in p^{-1}(x_0)$, where $x_0 \in X$. There exists an automorphism $\varphi \in A(\tilde{X}, p)$ such that $\varphi(\tilde{x}_1) = \tilde{x}_2$ if and only if $p_*\pi(\tilde{X}, \tilde{x}_1) = p_*\pi(\tilde{X}, \tilde{x}_2)$.*

This is a special case of Corollary 6.4.

Theorem 6.6 *Two covering spaces $(\tilde{X}_1, p_1)$ and $(\tilde{X}_2, p_2)$ of X are isomorphic if and only if, for any two points $\tilde{x}_1 \in \tilde{X}_1$ and $\tilde{x}_2 \in \tilde{X}_2$ such that $p_1(\tilde{x}_1) = p_2(\tilde{x}_2) = x_0$, the subgroups $p_{1*}\pi(\tilde{X}_1, \tilde{x}_1)$ and $p_{2*}\pi(\tilde{X}_2, \tilde{x}_2)$ belong to the same conjugacy class in $\pi(X, x_0)$.*

PROOF: This follows directly from Corollary 6.4 and Theorem 4.2.

This theorem shows that the conjugacy class of subgroups mentioned in Theorem 4.2 completely determines a covering space up to isomorphism.

Lemma 6.7 *Let $(\tilde{X}_1, p_1)$ and $(\tilde{X}_2, p_2)$ be covering spaces of X, and let φ be a homomorphism of the first covering space into the second. Then, $(\tilde{X}_1, \varphi)$ is a covering space of $\tilde{X}_2$.*

PROOF: First, note that any point $x \in X$ has an open arcwise-connected neighborhood U which is an elementary neighborhood of x for both of the covering spaces simultaneously. We can obtain such a neighborhood by choosing open elementary neighborhoods U_1 and U_2 of x for the coverings $(\tilde{X}_1, p_1)$ and $(\tilde{X}_2, p_2)$, respectively, and then let U be the arc component of $U_1 \cap U_2$ which contains x.

Next, we prove that φ maps $\tilde{X}_1$ onto $\tilde{X}_2$. Let y be any point of $\tilde{X}_2$; we must show that there exists a point x of $\tilde{X}_1$ such that $\varphi(x) = y$. Choose a base point $x_1 \in \tilde{X}_1$, and let $x_2 = \varphi(x_1)$, $x_0 = p_1(x_1) = p_2(x_2)$. Choose a path f in $\tilde{X}_2$ with initial point x_2 and terminal point y, and let $g = p_2 f$ be the image path in X. By Lemma 3.1, there exists a unique path h in $\tilde{X}_1$ with initial point x_1 and such that $p_1 h = g$. Let x be the terminal point of h. Then the paths φh and f both have the same initial point and $p_2 \varphi h = g = p_2 f$, hence $\varphi h = f$ by the uniqueness assertion of of Lemma 3.1. Therefore $\varphi(x) = y$, as required.

It should now be clear how to choose an elementary neighborhood of any point $z \in \tilde{X}_2$. Choose a neighborhood U of $x = p_2(z)$ which is elementary for both coverings, and let W be the component of $p_2^{-1}(U)$ which contains z. The proof that W has the required properties is easy. Q.E.D.

Let $(\tilde{X}, p)$ be a covering space of X such that $\tilde{X}$ is simply connected. If $(\tilde{X}', p')$ is any other covering space of X, then, by Lemma 6.3, there exists a homomorphism φ of $(\tilde{X}, p)$ onto $(\tilde{X}', p')$, and, by the lemma just proved, $(\tilde{X}, \varphi)$ is a covering space of $\tilde{X}'$; i.e., $\tilde{X}$ can serve as a covering space of any covering space of X. For this reason a simply connected covering space, such as $(\tilde{X}, p)$, is called a *universal* covering space. By Theorem 6.6, any two universal covering spaces of X are isomorphic.

Exercises

6.1 Prove that, if X is a simply connected space and $(\tilde{X}, p)$ is a covering space of X, then p is a homeomorphism of $\tilde{X}$ onto X.

6.2 Determine all covering spaces (up to isomorphism) of each of the following spaces: S^1, the circle; P, the projective plane; the subset $\{(x, y) \in \mathbf{R}^2 : 1 \leqq x^2 + y^2 \leqq 4\}$ of the plane. Exhibit an explicit covering space $(\tilde{X}, p)$ from each isomorphism class. (SUGGESTION: Consider the examples in Section 2.)

6.3 Let X be a topological space whose fundamental group is abelian and which has a universal covering space. If $(\tilde{X}_1, p_1)$ and $(\tilde{X}_2, p_2)$ are covering spaces of X, define $(\tilde{X}_1, p_1) \geqq (\tilde{X}_2, p_2)$ if and only if there exists a homomorphism of $(\tilde{X}_1, p_1)$ onto $(\tilde{X}_2, p_2)$. Prove that this relation is transitive, reflexive, and, if $(\tilde{X}_1, p_1) \leqq (\tilde{X}_2, p_2)$ and $(\tilde{X}_2, p_2) \leqq (\tilde{X}_1, p_1)$, then $(\tilde{X}_1, p_1)$ is isomorphic to $(\tilde{X}_2, p_2)$. Finally, prove that any two covering spaces of X have a least upper bound and a greatest lower bound with respect to this partial ordering relation. [NOTE: This result is definitely not true if the hypothesis that $\pi(X)$ is abelian is omitted.] (SUGGESTION: Use Lemma 10.1.)

6.4 Let

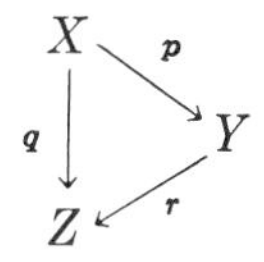

be a commutative diagram of spaces and continuous maps. Assume that (X, p) is a covering space of Y and (X, q) is a covering space of Z. Prove that (Y, r) is a covering space of Z. [HINT: Let $U \subset Z$ be an elementary neighborhood for the covering space (X, q), and let V be an arc component of $r^{-1}(U)$. Apply Lemma 2.1 to V considered as a subspace of Y.]

6.5 Let X be a space which has a universal covering space. If (X_1, p_1) is a covering space of X and (X_2, p_2) is a covering space of X_1, then (X_2, p_1p_2) is a covering space of X.

7 The action of the group $\pi(X, x)$ on the set $p^{-1}(x)$

To study further the group of automorphisms of a covering space $(\tilde{X}, p)$ of X, we define an action of the group $\pi(X, x)$ on the set $p^{-1}(x)$ for any $x \in X$; i.e., we make $\pi(X, x)$ operate on the right on the set $p^{-1}(X)$. The definition is very natural and simple; it depends on Lemmas 3.1 and 3.3 on the lifting of paths.

Definition Let $(\tilde{X}, p)$ be a covering space of X and $x \in X$. For any point $\tilde{x} \in p^{-1}(x)$ and any $\alpha \in \pi(X, x)$, define $\tilde{x} \cdot \alpha \in p^{-1}(x)$ as follows. By Lemmas 3.1 and 3.3, there exists a unique path class $\tilde{\alpha}$ in $\tilde{X}$ such that $p_*(\tilde{\alpha}) = \alpha$ and the initial point of $\tilde{\alpha}$ is the point $\tilde{x}$. Define $\tilde{x} \cdot \alpha$ to be the terminal point of the path class $\tilde{\alpha}$.

We leave it to the reader to verify the formulas:

$$(\tilde{x} \cdot \alpha) \cdot \beta = \tilde{x} \cdot (\alpha \cdot \beta), \tag{5.7-1}$$

$$\tilde{x} \cdot 1 = \tilde{x}. \tag{5.7-2}$$

These are exactly the conditions needed for $\pi(X, x)$ to be a group of right operators on the set $p^{-1}(x)$ (see Appendix B). We assert that the group $\pi(X, x)$ operates *transitively* on the set $p^{-1}(x)$. To prove this, let $\tilde{x}_0$ and $\tilde{x}_1 \in p^{-1}(x)$; because $\tilde{X}$ is assumed to be arcwise connected, there exists a path class $\tilde{\alpha}$ in $\tilde{X}$ with initial point $\tilde{x}_0$ and terminal point $\tilde{x}_1$. Let $\alpha = p_*(\tilde{\alpha})$. Then, α is an equivalence class of closed paths, and obviously $\tilde{x}_0 \cdot \alpha = \tilde{x}_1$ as was to be proved.

Thus, the set $p^{-1}(x)$ is a homogeneous right $\pi(X, x)$-space (as defined in Appendix B). From the definition, we see immediately *that, for any point $\tilde{x} \in p^{-1}(x)$, the isotropy subgroup corresponding to this point is precisely the subgroup $p_*\pi(\tilde{X}, \tilde{x})$ of $\pi(X, x)$*. Hence, as a right $\pi(X, x)$-space, $p^{-1}(x)$ is isomorphic to the space of cosets, $\pi(X, x)/p_*\pi(\tilde{X}, \tilde{x})$, and the *number of sheets of the covering is equal to the index of the subgroup $p_*\pi(\tilde{X}, \tilde{x})$.*

We now have the following important result, which establishes a connection between the group of automorphisms of a covering space and the action of $\pi(X, x)$ on $p^{-1}(x)$.

Proposition 7.1 *For any automorphism $\varphi \in A(\tilde{X}, p)$, any point $\tilde{x} \in p^{-1}(x)$, and any $\alpha \in \pi(X, x)$,*

$$\varphi(\tilde{x} \cdot \alpha) = (\varphi\tilde{x}) \cdot \alpha;$$

i.e., each element $\varphi \in A(\tilde{X}, p)$ induces an automorphism of the set $p^{-1}(x)$ considered as a right $\pi(X, x)$-space.

PROOF: The proof is simple. Lift α to a path $\tilde{\alpha}$ in $\tilde{X}$ with initial point $\tilde{x}$ and such that $p_*(\tilde{\alpha}) = \alpha$; then $\tilde{x} \cdot \alpha$ is the terminal point of $\tilde{\alpha}$. Now consider the path $\varphi_*(\tilde{\alpha})$ in $\tilde{X}$. Its initial point is $\varphi(\tilde{x})$, and its terminal point is $\varphi(\tilde{x} \cdot \alpha)$. Next, observe that

$$p_*[\varphi_*(\tilde{\alpha})] = (p\varphi)_*(\tilde{\alpha}) = p_*(\tilde{\alpha}) = \alpha;$$

i.e., $\varphi_*(\tilde{\alpha})$ is a lifting of the path α also. Hence, by definition, $(\varphi\tilde{x}) \cdot \alpha$ is the terminal point of the path $\varphi_*(\tilde{\alpha})$; i.e., $(\varphi\tilde{x}) \cdot \alpha = \varphi(\tilde{x} \cdot \alpha)$, as required.

We can now completely determine the structure of the automorphism group $A(\tilde{X}, p)$.

Theorem 7.2 *Let $(\tilde{X}, p)$ be a covering space of X. Then, the group of automorphisms, $A(\tilde{X}, p)$, is naturally isomorphic to the group of automorphisms of the set $p^{-1}(x)$, $x \in X$, considered as a right $\pi(X, x)$-space.*

PROOF: If φ is any automorphism of $(\tilde{X}, p)$, then the restriction $\varphi \mid p^{-1}(x)$ is an automorphism of $p^{-1}(x)$ as a right $\pi(X, x)$-space, in view of Proposition 7.1. Moreover, it follows from Corollary 6.2 that each automorphism φ is completely determined by its restriction, $\varphi \mid p^{-1}(x)$. In other words, the mapping $\varphi \to \varphi \mid p^{-1}(x)$ is a monomorphism of $A(\tilde{X}, p)$ into the group of automorphisms of the right $\pi(X, x)$-space $p^{-1}(x)$. Next, it follows from Lemma 2.1 of Appendix B and Corollary 6.5 that the mapping $\varphi \to \varphi \mid p^{-1}(x)$ is an epimorphism of $A(\tilde{X}, p)$ onto the group of automorphisms of $p^{-1}(x)$. Hence, we have the theorem. Q.E.D.

Corollary 7.3 *For any point $x \in X$ and any $\tilde{x} \in p^{-1}(x)$, the automorphism group $A(\tilde{X}, p)$ is isomorphic to the quotient group $N[p_*\pi(\tilde{X}, \tilde{x})]/p_*\pi(\tilde{X}, \tilde{x})$, where $N[p_*\pi(\tilde{X}, \tilde{x})]$ denotes the normalizer of the subgroup $p_*\pi(\tilde{X}, \tilde{x})$ in $\pi(X, x)$.*

This corollary is obtained by applying Theorem 2.2 of Appendix B to Theorem 7.2.

An especially important class of covering spaces consists of those for which $p_*\pi(\tilde{X}, \tilde{x})$ is a *normal* subgroup of $\pi(X, x)$. [Note that this condition is independent of the choice of the point $\tilde{x} \in p^{-1}(x)$.] Such a covering space is called *regular.*

Corollary 7.4 *If $(\tilde{X}, p)$ is a regular covering space of X, then $A(\tilde{X}, p)$ is isomorphic to the quotient group $\pi(X, x)/p_*\pi(\tilde{X}, \tilde{x})$ for any $x \in X$ and any $\tilde{x} \in p^{-1}(x)$.*

This follows from Corollary 7.3, because $N[p_*\pi(\tilde{X}, \tilde{x})] = \pi(X, x)$ in this case.

This corollary applies in particular to the universal covering space:

Corollary 7.5 *Let $(\tilde{X}, p)$ be a universal covering space of X. Then, $A(\tilde{X}, p)$ is isomorphic to $\pi(X)$, and the order of the group $\pi(X)$ is equal to the number of sheets of the covering space $(\tilde{X}, p)$.*

Examples

7.1 Consider the covering space $(\mathbf{R}, p)$ of the circle S^1 defined by $p(t) = (\sin t, \cos t)$ for any $t \in \mathbf{R}$ (see Example 2.1). Because the real line $\mathbf{R}$ is contractible, it is simply connected. Therefore, $(\mathbf{R}, p)$ is a universal covering space of S^1, and Corollary 7.5 is applicable. Let us determine the group of automorphisms of this covering space. From the known periodicity of the functions $\sin t$ and $\cos t$, it is clear that the "translation" $T_n : \mathbf{R} \to \mathbf{R}$ defined by $T_n(t) = t + 2n\pi$ is an automorphism for any integer n. Moreover, it is clear that, if x is any point of S^1 and t_1 and t_2 are any two points of $p^{-1}(x)$, then there exists an integer n such that $T_n(t_1) = t_2$. It follows that every automorphism of the covering space $(\mathbf{R}, p)$ is such a translation (see Lemma 6.1 and Corollary 6.2).

Because the group of all such translations $\{T_n : n \in \mathbf{Z}\}$ is obviously infinite cyclic, we have once again proved that $\pi(S^1)$ is infinite cyclic. The second half of the proof of this fact in Section II.5 actually used some of the ideas of the theory of covering spaces in disguise. Indeed, the notion of the angle $a(z)$ for any $z \in S^1$ implicitly involves the covering space $(\mathbf{R}, p)$ of S^1; and the argument given to define the degree of a closed path in S^1 amounts to the lifting of such a path in S^1 to the covering space $(\mathbf{R}, p)$.

7.2 Let $p : S^2 \to P$ denote the natural map of the 2-sphere onto its quotient space, the projective plane; then, (S^2, p) is a covering space of P (see Example 2.5), and, because S^2 is simply connected, it is a universal covering space. Because it is a 2-sheeted covering space, the fundamental group $\pi(P)$ and the automorphism group must both be of order 2. It is clear that the automorphism group is generated by the antipodal map $T : S^2 \to S^2$, $T(x, y, z) = (-x, -y, -z)$.

Exercises

7.1 Let $p : \tilde{G} \to G$ be a continuous homomorphism of topological groups such that $(\tilde{G}, p)$ is a covering space of G. (It is assumed, of course, that both G and $\tilde{G}$ are connected and locally arcwise connected.) Let K denote the kernel of p; then K is a discrete subgroup of $\tilde{G}$ which is contained in the center (see the exercises in Section 5). For each element $k \in K$, define a map $\varphi_k : \tilde{G} \to \tilde{G}$ by $\varphi_k(x) = x \cdot k = k \cdot x$. Prove that the mapping $k \to \varphi_k$ is an isomorphism of K onto $A(\tilde{G}, p)$.

7.2 Determine the group of automorphisms of the covering spaces described in Examples 2.2, 2.4, 2.7, 2.8, and 2.9.

8 Regular covering spaces and quotient spaces

Let $(\tilde{X}, p)$ be a covering space of X; because p is an open map, X has the quotient topology induced by p (see Appendix A). Thus, we can regard X as being obtained from $\tilde{X}$ by a process of identifying certain points: For any point $x \in X$, all the points of the set $p^{-1}(x)$ are to be identified to a single point. Recall that the automorphism group $A(\tilde{X}, p)$ permutes the points of the set $p^{-1}(x)$ among themselves. However, it is *not* true, in general, that the quotient space $\tilde{X}/A(\tilde{X}, p)$ is naturally homeomorphic to X, because there may exist distinct points $\tilde{x}_1, \tilde{x}_2 \in p^{-1}(x)$ such that there is no automorphism $\varphi \in A(\tilde{X}, p)$ for which $\varphi(\tilde{x}_1) = \tilde{x}_2$; in other words, the automorphism group $A(\tilde{X}, p)$ need not operate transitively on $p^{-1}(x)$. Indeed, we have the following lemma:

Lemma 8.1 *Let $(\tilde{X}, p)$ be a covering space of X. The automorphism group $A(\tilde{X}, p)$ operates transitively on $p^{-1}(x)$, $x \in X$, if and only if $(\tilde{X}, p)$ is a regular covering space of X.*

This is an immediate consequence of Theorem 4.2 and Corollary 6.5.

As a result, we see that if $(\tilde{X}, p)$ is a regular covering space of X, then X is naturally homeomorphic to the quotient space $\tilde{X}/A(\tilde{X}, p)$. This leads to the following rather natural question: Let Y be a topological space, and let G be a group of homeomorphisms of Y. Let $p : Y \to Y/G$ denote the natural map of Y onto its quotient space. Under what conditions is (Y, p) a regular covering space of Y/G with $A(Y, p) = G$? First, it is clear that there are some necessary conditions which must be satisfied. For example, if $(\tilde{X}, p)$ is a regular covering space of X, then $A(\tilde{X}, p)$ acts on $\tilde{X}$ without fixed points (this is the content of Corollary 6.2). Also, the orbit of any point $\tilde{x} \in \tilde{X}$ under the action of the group $A(\tilde{X}, p)$ [i.e., the set of points $\{\varphi(\tilde{x}) : \varphi \in A(\tilde{X}, p)\}$] is a discrete, closed subset of $\tilde{X}$. In fact, the following even stronger condition holds: Every point $\tilde{x} \in \tilde{X}$ has a neighborhood U such that the sets $\varphi(U)$, $\varphi \in A(\tilde{X}, p)$, are pairwise disjoint (we can choose U to be a component of the inverse image of an appropriate elementary neighborhood in X). A group of homeomorphisms satisfying this condition is said to be *properly discontinuous*. Note that a properly discontinuous group of homeomorphisms is fixed point free. It turns out that these necessary conditions are also sufficient.

Proposition 8.2 *Let Y be a connected, locally arcwise-connected topological space and let G be a properly discontinuous group of homeomorphisms of Y. Let $p : Y \to Y/G$ denote the natural projection of Y onto its quotient space. Then, (Y, p) is a regular covering space of Y/G, and $G = A(Y, p)$.*

PROOF: Let $x \in Y/G$; we must show that x has an elementary neighborhood. Choose a point $y \in Y$ such that $p(y) = x$. By hypothesis, there exists a neighborhood N of y such that the sets $\varphi(N)$, $\varphi \in G$, are pairwise disjoint. Because Y is locally arcwise connected, there exists an open, arcwise-connected neighborhood V of y such that $V \subset N$. Let $U = p(V)$. We assert that U is an elementary neighborhood of x. Because p is an open map (see Section 1 of Appendix A), U is an open set, and it is clearly arcwise connected. It is also clear that p maps V in a one-to-one, continuous fashion onto U; and because p is an open map, it is a homeomorphism of V onto U. If W is any component of $p^{-1}(U)$ different from V, then there exists a $\varphi \in G$ such that $W = \varphi(V)$. Because φ is a homeomorphism of V onto W, and $p = p\varphi$, it follows that p also maps W homeomorphically onto U. Thus, U is an elementary neighborhood of x, and (Y, p) is a covering space of Y/G. It is obvious that every $\varphi \in G$ is an automorphism of (Y, p); thus, $G \subset A(Y, p)$. The assumption that G is a *proper* subgroup of $A(Y, p)$ is readily seen to imply that $A(Y, p)$ has elements with fixed points. Hence, $G = A(Y, p)$. Finally, it follows from Lemma 8.1 that (Y, p) is a regular covering space of Y/G. Q.E.D.

We shall now give some simple examples of this theorem.

Examples

8.1 Let $Y = \mathbf{R}$, the real line, and, for each integer n, define $\varphi_n : \mathbf{R} \to \mathbf{R}$ by $\varphi_n(x) = x + n$. Let $G = \{\varphi_n : n \in \mathbf{Z}\}$. Then, G is a properly discontinuous group of homeomorphisms of $\mathbf{R}$; indeed, for any $x \in \mathbf{R}$, if we let U be the open interval $(x - \frac{1}{3}, x + \frac{1}{3})$, then the neighborhoods $\varphi_n(U)$ are pairwise disjoint. Hence, by the proposition just proved, $\mathbf{R}$ is a regular covering space of the quotient space $\mathbf{R}/G$. By the results of Section 4 of Appendix A, it follows that $\mathbf{R}/G$ is homeomorphic to the quotient space of the closed unit interval $[0, 1]$ obtained by identifying the two end points of the interval. Thus, $\mathbf{R}/G$ is a circle. Once again we have proved that the universal covering space of a circle is the real line, and that the group of automorphisms is infinite cyclic (see Example 7.1).

8.2 Let $Y = S^n$, the unit n-sphere in Euclidean $(n + 1)$-space, and let $T : S^n \to S^n$ be the antipodal map defined by $T(x) = -x$ for any $x \in S^n$. Clearly, T^2 is the identity transformation; hence, T generates a group G of homeomorphisms of S^n, which is cyclic of order 2. It is obvious that G is a properly discontinuous group of homeomorphisms; therefore, S^n is a covering space of S^n/G, which is a real projective n-space. Because S^n is simply connected, it is a universal covering space, and the fundamental group of a real projective n-space is cyclic of order 2 (see Example 7.2 for the case where $n = 2$).

Exercises

8.1 Let Y be a Hausdorff space and let G be a *finite* group of homeomorphisms of Y such that each element $\varphi \neq 1$ of G has no fixed points. Prove that G is a properly discontinuous group of homeomorphisms.

8.2 Let Y be a topological group and let G be a discrete subgroup of Y. Prove that there exists a neighborhood U of the identity such that the sets $g \cdot U$ for $g \in G$ are pairwise disjoint (NOTE: $g \cdot U = \{g \cdot x : x \in U\}$). HINT: Choose a neighborhood V of the identity such that $V \cap G = \{1\}$. Then prove that there exists a neighborhood U of the identity such that $\{x \cdot y^{-1} : x, y \in U\} \subset V$.

8.3 Let Y be a topological group and let G be a discrete subgroup. Let Y/G denote the space of cosets $\{G \cdot y : y \in Y\}$ with the quotient space topology, and $p : Y \to Y/G$ the natural projection. Prove that (Y, p) is a regular covering space of Y/G with $A(Y, p) = G$, where G operates on Y by multiplication on the left. (HINT: Use the result of Exercise 8.2 and Proposition 8.2.) Note that Example 8.1 is a special case of this exercise.

8.4 Let X be a regular topological space, and $(\tilde{X}, p)$ a covering space of X. Prove that for any compact set $C \subset \tilde{X}$, the set $\{\varphi \in A(\tilde{X}, p) : \varphi(C) \cap C \neq \varnothing\}$ is finite.

8.5 Let Y be a locally compact Hausdorff space, and let G be a group of homeomorphisms of Y such that each element $\varphi \neq 1$ of G has no fixed points, and for any compact set $C \subset Y$, the set $\{\varphi \in G : \varphi(C) \cap C \neq \varnothing\}$ is finite. Prove that G is a properly discontinuous group of homeomorphisms, and that the quotient space Y/G is locally compact and Hausdorff.

We shall conclude this section with two examples to illustrate some of the possibilities inherent in Proposition 8.2. In the first example, it is shown that the quotient space Y/G need not be Hausdorff, even though the space Y satisfies all the separation axioms. In fact, we give such an example where Y is the Euclidean plane $\mathbf{R}^2$, and G is an infinite cyclic, properly discontinuous group of homeomorphisms of $\mathbf{R}^2$.

Examples

8.3 We start by considering the following simple system of ordinary differential equations in the (x, y) plane:

$$\begin{cases} \dfrac{dx}{dt} = \cos^2 x, \\ \dfrac{dy}{dt} = \sin x. \end{cases}$$

It is easily seen that the integral curves are the curves

$$y = \sec x + C$$

for various values of the constant of integration C, and the vertical lines

$$x = (n + \tfrac{1}{2})\pi$$

for all integers n. We can consider this system of differential equations the equations of motion of a particle in the plane; t represents the time, and (x, y) are the coordinates of the particle at time t. The particle must move along one of the integral curves. Which curve it will move along depends on its initial position.

Using this differential equation, we shall define an operation of the additive group of real numbers, $\mathbf{R}$, on the Euclidean plane. For any real number t and any point (x, y) of the plane, we define $t \cdot (x, y)$ to be the position at time t of a particle which was at the point (x, y) at time 0. It is clear that

$$s \cdot [t \cdot (x, y)] = (s + t) \cdot (x, y),$$
$$0 \cdot (x, y) = (x, y).$$

Also, the map $\mathbf{R} \times \mathbf{R}^2 \to \mathbf{R}^2$ defined by $(t, (x, y)) \to t \cdot (x, y)$ is continuous (it is even differentiable). This is a consequence of standard theorems on differential equations. It is also clear that this action of $\mathbf{R}$ on the plane is fixed point free.

We now consider the action of the subgroup $\mathbf{Z}$ of $\mathbf{R}$ on the plane; this will give our desired example.

We shall first prove that the action of $\mathbf{Z}$ on $\mathbf{R}^2$ is properly discontinuous. Given any point $P = (x, y)$, let C be the unique integral curve passing through P. Let C_1 and C_2 be two integral curves near by, one on each side of C. Let T_0 be a smooth curve through P which is orthogonal to all the integral curves between C_1 and C_2. For any real number t, let $T_t = t \cdot T_0$. Let U be the neighborhood of P bounded by $T_{-1/3}$, $T_{+1/3}$, and the curves C_1 and C_2. Then, it is readily seen that the successive "translates" of U,

$$\{n \cdot U : n \in \mathbf{Z}\}$$

are pairwise disjoint.

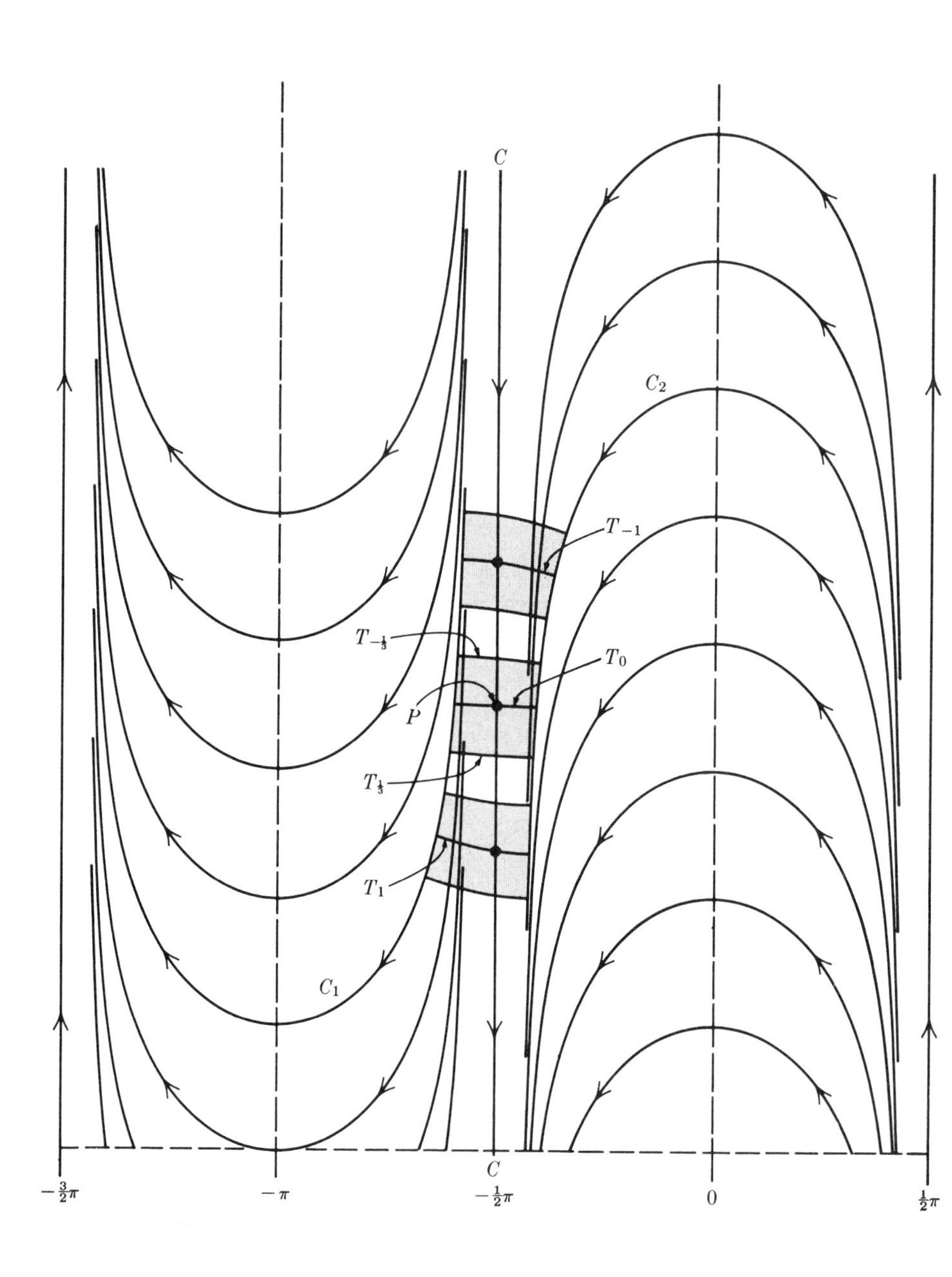

FIGURE 5.2 Diagram for Examples 8.3 and 8.4.

Next, we shall prove that the quotient space is non-Hausdorff. Consider the points

$$P_1 = \left(\frac{\pi}{2}, 0\right), \qquad P_2 = \left(-\frac{\pi}{2}, 0\right)$$

in the plane. We shall prove that their images in the quotient space $\mathbf{R}^2/\mathbf{Z}$ do not have disjoint neighborhoods. For this purpose, it suffices to prove that given *any* neighborhoods N_1 of P_1 and N_2 of P_2, there exists a point in N_1 equivalent to a point of N_2 under the action of the group $\mathbf{Z}$. To do this, consider for any small number $a > 0$ the two points $((\pi/2) - a, 0)$ and $(-(\pi/2) + a, 0)$. These two points are obviously on the same integral curve. How long would it take a particle located at the point $(-(\pi/2) + a, 0)$ to move along its integral curve to the point $((\pi/2) - a, 0)$? To compute this, it obviously suffices to compute how long it would take its projection on the x axis to move from the first position to the second. Because $dx/dt = \cos^2 x$, the time in question is given by the integral

$$I_a = \int_{-(\pi/2)+a}^{(\pi/2)-a} \frac{dx}{\cos^2 x} = \Big[\tan x\Big]_{-(\pi/2)+a}^{(\pi/2)-a}$$

$$= 2 \tan\left(\frac{\pi}{2} - a\right).$$

From this formula for the elapsed time, we can draw several conclusions:

(1) The elapsed time is a continuous function of a.
(2) As $a \to 0$, the elapsed time I_a approaches $+\infty$.
(3) For any number $\varepsilon > 0$, there are infinitely many values of a such that $0 < a < \varepsilon$ and I_a is an integer.

Recall that the points $((\pi/2) - a, 0)$ and $(-(\pi/2) + a, 0)$ are equivalent if and only if the elapsed time I_a is an integer.

From this, the desired conclusion readily follows.

8.4 We next give an example[1] of an infinite cyclic group of homeomorphisms acting without fixed points on a nice space in such a way that the "orbit" of each point is a closed, discrete subspace, but the action is not properly discontinuous! This example shows the strength of the requirement that G be properly discontinuous in Proposition 8.2.

Consider the action of the group of integers $\mathbf{Z}$ on the Euclidean plane $\mathbf{R}^2$ just described. The infinite strip

$$S = \left\{(x, y) : -\frac{\pi}{2} \leqq x \leqq +\frac{\pi}{2}\right\}$$

is mapped into itself by every element of the group $\mathbf{Z}$. We now form a quotient space of S by identifying the points $(\pi/2, y)$ and $(-\pi/2, -y)$ for any real number y. The quotient space is a Möbius strip without boundary (a noncompact sur-

---

[1] This example was suggested to the author by Joseph Auslander.

face). Moreover, the action of the group $\mathbf{Z}$ on S is readily seen to be compatible with the identifications, so it also acts on the quotient space. It is clear that the action of $\mathbf{Z}$ on this open Möbius strip is without fixed points, and that the orbit of any point x (i.e., the set of all points $n \cdot x$ for $n \in \mathbf{Z}$) is a discrete, closed subset. The argument given in the last example to show that the quotient space is non-Hausdorff may be applied to this example to show that the point $(\pi/2, 0)$ [which is identified with $(-\pi/2, 0)$] does not have any neighborhood U such that the sets $n \cdot U$ for $n \in \mathbf{Z}$ are pairwise disjoint. Hence, the action of the group on the Möbius strip is not properly discontinuous.

9 Application: The Borsuk-Ulam theorem for the 2-sphere

As usual, let S^n denote the unit n-sphere in $\mathbf{R}^{n+1}$:

$$S^n = \{x \in \mathbf{R}^{n+1} : |x| = 1\}.$$

For any positive integers m and n, let us agrée to call a map $f : S^m \to S^n$ *antipode preserving* in case $f(-x) = -f(x)$ for any $x \in S^m$. The following well-known theorem, due to the Polish mathematicians K. Borsuk and S. Ulam, has many interesting consequences.

Theorem 9.1 *There does not exist any continuous, antipode-preserving map* $f : S^n \to S^{n-1}$ $(n > 0)$.

We will prove this theorem only for $n \leqq 2$. Before giving the proof, we indicate and prove some interesting corollaries.

Corollary 9.2 *Assume that* $f : S^n \to \mathbf{R}^n$ *is a continuous map such that* $f(-x) = -f(x)$ *for any* $x \in S^n$. *Then, there exists a point* $x \in S^n$ *such that* $f(x) = 0$.

PROOF: Assume to the contrary that $f(x) \neq 0$ for all $x \in S^n$. For any $x \in S^n$, define

$$g(x) = \frac{f(x)}{|f(x)|}.$$

Then, g is a continuous map $S^n \to S^{n-1}$, which is antipode preserving, contrary to Theorem 9.1.

Corollary 9.3 *Assume* $f : S^n \to \mathbf{R}^n$ *is a continuous map. Then, there exists a point* $x \in S^n$ *such that* $f(x) = f(-x)$. *In particular,* f *is not one-to-one.*

PROOF: Assume to the contrary that, for every point $x \in S^n$, $f(x) \neq f(-x)$. Define $g(x) = f(x) - f(-x)$. Then, $g(-x) = -g(x)$, and $g(x) \neq 0$ for all x, which contradicts Corollary 9.2.

Corollary 9.4 *No subset of* $\mathbf{R}^n$ *is homeomorphic to* S^n.

This is an obvious consequence of Corollary 9.3.

There is another interpretation of Corollary 9.3 which is interesting. If $f : S^n \to \mathbf{R}^n$ is a continuous map, we can write

$$f(x) = (f_1(x), \ldots, f_n(x))$$

where $f_1(x), \ldots, f_n(x)$ are continuous real-valued functions on S^n. Thus, we may reword the corollary as follows: *Let* $f_1, f_2, \ldots, f_n$ *be continuous real-valued functions on* S^n. *Then, there exists a point* $x \in S^n$ *such that* $f_i(x) = f_i(-x)$ *for* $i = 1, \ldots, n$. For example, if $f_1(x)$ and $f_2(x)$ denote the temperature and barometric pressure at a certain instant at any point x on the earth's surface, and we assume that the temperature and barometric pressure both vary continuously over the earth's surface, then we conclude that there exists a pair of antipodal points on the surface of the earth which simultaneously have the same temperature and pressure! This is a topological theorem par excellence; only topological hypotheses are involved in the statement and proof.

PROOF OF THEOREM 9.1: For $n \leqq 2$: The case where $n = 1$ is trivial, because S^1 is connected, but S^0 is not connected. Therefore, we concentrate on the case where $n = 2$. The proof is by contradiction; assume that there exists a continuous antipode-preserving map $f : S^2 \to S^1$. Consider the quotient spaces of S^2 and S^1 obtained by identifying diametrically opposite points. These spaces are the real projective plane P_2, and a space which is again homeomorphic to S^1, respectively. We denote by $p_2 : S^2 \to P_2$ and $p_1 : S^1 \to S^1$ the natural maps of each space onto its quotient space. Because f is antipode preserving, it induces a continuous map $g : P_2 \to S^1$ such that the following diagram is commutative:

$$\begin{array}{ccc} S^2 & \xrightarrow{f} & S^1 \\ \downarrow p_2 & & \downarrow p_1 \\ P_2 & \xrightarrow[g]{} & S^1 \end{array}$$

Note that (S^2, p_2) and (S^1, p_1) are 2-sheeted covering spaces of P_2 and S^1, respectively; this is a consequence of Proposition 8.2 (with G a cyclic group of order 2). We shall now reach a contradiction by an argument involving the induced homomorphism

$$g_* : \pi(P_2) \to \pi(S^1)$$

of the fundamental groups.

On the one hand, we know that $\pi(P_2)$ is cyclic of order 2, and $\pi(S^1)$ is infinite cyclic. Therefore, the homomorphism g_* must be trivial for purely algebraic reasons.

On the other hand, let α denote an equivalence class of paths on S^2 such that the end points of α are antipodal points of S^2. Because f is antipode preserving, the end points of $f_*(\alpha)$ are antipodal points of S^1. Now, $p_{2*}(\alpha)$ and $p_{1*}f_*(\alpha)$ are closed paths on P_2 and S^1, and hence represent elements of the fundamental groups, $\pi(P_2)$ and $\pi(S^1)$. We assert that $p_{2*}(\alpha) \neq 1$ and $p_{1*}f_*(\alpha) \neq 1$; this follows by considering the action of the fundamental groups $\pi(P_2, x_0)$ and $\pi(S^1, y_0)$ on the sets $p_2^{-1}(x_0)$ and $p_1^{-1}(y_0)$, respectively (see Section 7). It follows from the definitions that $p_{2*}(\alpha)$ and $p_{1*}f_*(\alpha)$ operate nontrivially on these sets. Next, by commutativity of the diagram above,

$$g_*p_{2*}(\alpha) = p_{1*}f_*(\alpha).$$

Therefore, g_* sends $p_{2*}(\alpha)$ onto $p_{1*}f_*(\alpha)$, contradicting the fact that g_* is trivial.

Q.E.D.

It is clear that to prove the Brouwer fixed point theorem (see Chapter II) and the Borsuk-Ulam theorem in the cases where $n > 2$, we need higher dimensional analogs of the fundamental group. The fundamental group is essentially a 1-dimensional invariant of a space and will not suffice for this purpose. This is one of the main purposes of the subject of algebraic topology: to develop a complete theory of such higher dimensional analogs of the fundamental group, and to use them to prove theorems like those of Brouwer and Borsuk-Ulam.

Exercise

9.1 Generalize the argument used for proving the Borsuk-Ulam theorem as follows: Let X and Y be spaces which are connected and locally arcwise connected, let G be a group which operates on the left on X and Y such that it is a properly discontinuous group of homeomorphisms of each, and let $f : X \to Y$ be a continuous G-equivariant map (see Appendix B for the definition). Let $p : X \to X/G$ and $q : Y \to Y/G$ denote the natural maps, and let $g : X/G \to Y/G$ denote the map induced by f. Prove that the homomorphism $g_* : \pi(X/G) \to \pi(Y/G)$ induces an *isomorphism* of quotient groups: $\pi(X/G)/p_*\pi(X) \approx \pi(Y/G)/q_*\pi(Y)$.

9.2 Prove the following corollary of the general Borsuk-Ulam theorem: There does not exist a continuous 1–1 map $f : \mathbf{R}^{n+1} \to \mathbf{R}^n$ for any $n \geqq 1$.

9.3 Does there exist a continuous antipode preserving map $f : S^1 \to S^1$ of *even* degree? Prove your answer.

10 The existence theorem for covering spaces

We have proved that a covering space $(\tilde{X}, p)$ of X is determined up to isomorphism by the conjugacy class of the subgroup $p_*\pi(\tilde{X}, \tilde{x})$ of $\pi(X, x)$. This fact gives rise to the following question: Suppose X is a topological space and we are given a conjugacy class of subgroups of $\pi(X, x)$. Does there exist a covering space $(\tilde{X}, p)$ of X such that $p_*\pi(\tilde{X}, \tilde{x})$ belongs to the given conjugacy class? We shall show that this question can be answered affirmatively, provided X satisfies a slight additional hypothesis.

First, we prove that it suffices to consider this problem for the special case where the given conjugacy class of subgroups consists of the trivial subgroup $\{1\}$.

Lemma 10.1 *Let X be a topological space which has a universal covering space. Then, for any conjugacy class of subgroups of $\pi(X, x)$, there exists a covering space $(\tilde{X}, p)$ of X such that $p_*\pi(\tilde{X}, \tilde{x})$ belongs to the given conjugacy class.*

PROOF: Let (Y, q) be a universal covering space of X; i.e., Y is simply connected. According to Section 7, $\pi(X, x)$ operates transitively on the right on the set $q^{-1}(x)$, and, since Y is simply connected, it operates without any fixed points. Also, the group of automorphisms $A(Y, q)$ is isomorphic to $\pi(X)$, and it operates transitively without fixed points on the left on the set $q^{-1}(x)$. Choose a point $y \in q^{-1}(x)$ and a subgroup G of $\pi(X, x)$, which belongs to the given conjugacy class. Let H be the subgroup of $A(Y, q)$ defined as follows: $\varphi \in H$ if and only if there exists an element $\alpha \in G$ such that $\varphi(y) = y \cdot \alpha$. It is readily seen that G and H are isomorphic under the following correspondence: $\varphi \leftrightarrow \alpha$ if and only if $\varphi(y) = y \cdot \alpha$.

Because H is a subgroup of $A(Y, q)$, it is a properly discontinuous group of homeomorphisms of Y. Let $\tilde{X}$ denote the quotient space Y/H, $r : Y \to \tilde{X}$ the natural projection, and $p : \tilde{X} \to X$ the map induced by $q : Y \to X$. Then, we have the following commutative diagram:

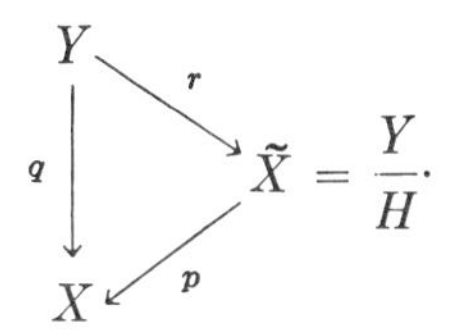

By assumption, (Y, q) is a covering space of X, and (Y, r) is a covering space of $\tilde{X}$ by Proposition 8.2. It follows by an easy argument that $(\tilde{X}, p)$ is a covering space of X (see Exercise 6.4). Since $(\tilde{X}, p)$ is a covering space of X, the group $\pi(X, x)$ operates on the right on the set $p^{-1}(x)$. Let $\tilde{x} = r(y) \in p^{-1}(x)$. By our construction of $\tilde{X}$, it is clear that the isotropy subgroup of $\pi(X, x)$ corresponding to the point $\tilde{x}$ is precisely the subgroup G. But this is exactly equivalent to the assertion that $p_*\pi(\tilde{X}, \tilde{x}) = G$ (see Section 7). Q.E.D.

We now consider the following problem: Given a topological space X, does X have a universal covering space? First, we derive a rather simple necessary condition. Let $(\tilde{X}, p)$ be a universal covering space of X, let x be an arbitrary point of X, let $\tilde{x}$ be a point of $p^{-1}(x)$, let U be an elementary neighborhood of x, and let V be the component of $p^{-1}(U)$ which contains the point $\tilde{x}$. We then have the following commutative diagram involving fundamental groups:

$$\begin{array}{ccc} \pi(V, \tilde{x}) & \longrightarrow & \pi(\tilde{X}, \tilde{x}) \\ {\scriptstyle (p|V)_*}\downarrow & & \downarrow{\scriptstyle p_*} \\ \pi(U, x) & \underset{i_*}{\longrightarrow} & \pi(X, x) \end{array}.$$

Because $p \mid V$ is a homeomorphism of V onto U, $(p \mid V)_*$ is an isomorphism. Note also that, by hypothesis, $\pi(\tilde{X}, \tilde{x}) = \{1\}$. From these two facts and the commutativity of this diagram, it follows that i_* is a trivial homomorphism; i.e., image $i_* = \{1\}$. Thus, we conclude that the space X has the following property: *Every point $x \in X$ has a neighborhood U such that the homomorphism $\pi(U, x) \to \pi(X, x)$ is trivial.* A space which has this property is called *semilocally simply connected.*[2] This definition can also be phrased as follows: A space X is semilocally simply connected if and only if every point $x \in X$ has a neighborhood U such that any loop in U can be shrunk to a point in X.

The following is a simple example of a space which is connected and locally arcwise connected, but not semilocally simply connected. For any positive integer n, let

$$C_n = \left\{(x, y) \in \mathbf{R}^2 : \left(x - \frac{1}{n}\right)^2 + y^2 = \frac{1}{n^2}\right\};$$

i.e., C_n is a circle of radius $1/n$ with center at the point $(1/n, 0)$. Let X denote the union of the circles C_n for all positive integers n. Then, X is not semilocally simply connected; the point $(0, 0)$ does not have the required kind of neighborhood.

[2] This name is rather long and awkward, but it is an accurate description of the property in question. It lies between ordinary simple connectivity and true local simple connectivity (which we do not consider in this book). Moreover, this name is sanctioned by several years of common acceptance.

Fortunately, most of the topological spaces that arise in problems from other branches of mathematics where covering spaces are involved are semilocally simply connected. For example, all manifolds and manifolds with boundary have this property.

We shall now prove that this necessary condition for the existence of a universal covering space is also sufficient.

Theorem 10.2 *Let X be a topological space which is connected, locally arcwise connected, and semilocally simply connected. Then, given any conjugacy class of subgroups of $\pi(X, x)$, there exists a covering space $(\tilde{X}, p)$ of X corresponding to the given conjugacy class* [i.e., *such that $p_*\pi(\tilde{X}, \tilde{x})$ belongs to the given conjugacy class*].

PROOF: In view of Lemma 10.1, it suffices to prove that X has a universal covering space. This we will do by a direct construction. To motivate this construction, we shall try to describe how an early topologist might have discovered it.

Let us assume for the moment that X has a universal covering space $(\tilde{X}, p)$. Choose a base point $\tilde{x}_0 \in \tilde{X}$, and let $x_0 = p(\tilde{x}_0)$. Given any point $y \in \tilde{X}$, there exists a path class α with initial point $\tilde{x}_0$ and terminal point y, because $\tilde{X}$ is arcwise connected. Because $\tilde{X}$ is simply connected, this path class is unique. Now consider the function which assigns to the point y the path class $p_*(\alpha)$ in X. It follows from Lemmas 3.1 and 3.3 that this is a one-to-one map of Y onto the set of path classes in X which have x_0 as initial point. Thus, we can identify the points of $\tilde{X}$ with the path classes in X which start at the point x_0. This simple observation is the basis of the following construction.

Choose a base point $x_0 \in X$ and *define* $\tilde{X}$ to be the set of all equivalence classes of paths α in X which have x_0 as their initial point. Define a function $p : \tilde{X} \to X$ by setting $p(\alpha)$ equal to the terminal point of the path class α. We shall now show how to topologize $\tilde{X}$ so that it is a simply connected space and $(\tilde{X}, p)$ is a covering space of X.

Observe that our hypotheses imply that the topology on X has a basis consisting of open sets U with the following properties: U is arcwise connected and the homomorphism $\pi(U) \to \pi(X)$ (induced by the inclusion map) is trivial. Equivalently, every closed path in U is equivalent (in X) to a constant path. For brevity let us agree to call such an open set U *basic*. Note that, if x and y are any two points in a basic open set U, then any two paths f and g in U with initial point x and terminal point y are equivalent (in X).

Given any path $\alpha \in \tilde{X}$ and any basic open set U which contains the end point $p(\alpha)$, denote by (α, U) the set of all paths $\beta \in \tilde{X}$ such that, for some path class α' in U, $\beta = \alpha \cdot \alpha'$. Then (α, U) is a subset of $\tilde{X}$. We topologize $\tilde{X}$ by choosing as a basic family of open sets the family of all such sets (α, U). In order that the family of all sets of the form (α, U) can be a basis for some topology on $\tilde{X}$, it is necessary to prove the following statement: If $\gamma \in (\alpha, U) \cap (\beta, V)$, then there exists a basic open set W such that $(\gamma, W) \subset (\alpha, U) \cap (\beta, V)$. However, the proof of this statement is easy: We choose W to be any basic open set such that $p(\gamma) \in W \subset U \cap V$.

Before proceeding with the proof that $(\tilde{X}, p)$ is a universal covering space of X, it is convenient to make the following two simple observations:

(a) Let $\alpha \in \tilde{X}$, and let U be a basic open neighborhood of $p(\alpha)$. Then, $p \mid (\alpha, U)$ is a one-to-one map of (α, U) onto U.

(b) Let U be any basic open set, and let x be any point of U. Then,

$$p^{-1}(U) = \bigcup_{\lambda} (\alpha_\lambda, U),$$

where $\{\alpha_\lambda\}$ denotes the totality of all path classes in X with initial point x_0 and terminal point x. Moreover, the sets (α_λ, U) are pairwise disjoint.

The proof of these two observations is easy, and can be left to the reader.

Note that it follows from (b) that p is continuous. Hence, $p \mid (\alpha, U)$ is a one-to-one continuous map of (α, U) onto U, by (a). We assert that $p \mid (\alpha, U)$ is an open map of (α, U) onto U. For, any open subset of (α, U) is a union of sets of the form (β, V), where $V \subset U$, and hence the fact that $p \mid (\alpha, U)$ is open also follows from (a). Thus, p maps (α, U) homeomorphically onto U. Since U is arcwise connected, so is (α, U). Because the sets (α_λ, U) occurring in statement (b) are pairwise disjoint, it follows that any basic open set $U \subset X$ has all the properties required of an elementary neighborhood.

Next, we shall prove that the space $\tilde{X}$ is arcwise connected. Let $\tilde{x}_0 \in \tilde{X}$ denote the equivalence class of the constant path at x_0. Given any point $\alpha \in \tilde{X}$, it suffices to exhibit an arc joining the points $\tilde{x}_0$ and α. For this purpose, choose a path $f : I \to X$ belonging to the equivalence class α. For any real number $s \in I$, define $f_s : I \to X$ by $f_s(t) = f(st)$, $t \in I$. Then, $f_1 = f$ and $f_0 =$ constant path at x_0. Let α_s denote the equivalence class of the path f_s. We assert that the map $s \to \alpha_s$ is a continuous map $I \to \tilde{X}$, i.e., a path in $\tilde{X}$. To prove this assertion, we must check that, for any $s_0 \in I$ and any basic neighborhood U of $f(s_0)$, there exists a real number $\delta > 0$ such that if $|s - s_0| < \delta$, then $\alpha_s \in (\alpha_{s_0}, U)$. For this purpose, we choose δ so that, if $|s - s_0| < \delta$, then $f(s) \in U$; such a number δ exists because f is continuous. Thus, $s \to \alpha_s$ is a path in $\tilde{X}$ with initial point $\tilde{x}_0$ and terminal point α, as required.

Finally, we must prove that $\tilde{X}$ is simply connected. Now $p_*\pi(\tilde{X}, \tilde{x}_0)$

is the isotropy subgroup corresponding to the point $\tilde{x}_0$ for the action of $\pi(X, x_0)$ on $p^{-1}(x_0)$ (see Section 7). Thus, we must determine $\tilde{x}_0 \cdot \alpha$ for any $\alpha \in \pi(X, x_0)$. Choose a closed path $f : I \to X$ belonging to the equivalence class α, and, by the method of the preceding paragraph, define the path $s \to \alpha_s$ in $\tilde{X}$. This path in $\tilde{X}$ has $\tilde{x}_0$ as initial point, $\alpha \in \tilde{X}$ as terminal point, and is obviously a lifting of the path f. Hence, $\tilde{x}_0 \cdot \alpha = \alpha$, by the definition of the action of $\pi(X, x_0)$ on $p^{-1}(x_0)$. Therefore, $\tilde{x}_0 \cdot \alpha = \tilde{x}_0$ if and only if $\alpha = 1$; hence, the isotropy subgroup consists of the element 1 alone, as required. Q.E.D.

Exercise

10.1 Prove that for any positive integer n there exists a noncompact surface S and a properly discontinuous group G of homeomorphisms of S such that G is a free abelian group of rank $2n$ and S/G is a compact, orientable surface of genus n.

11 The induced covering space over a subspace

Let $(\tilde{X}, p)$ be a covering space of X, let A be a subspace of X which is connected and locally arcwise connected, and let $\tilde{A}$ be an arc component of $p^{-1}(A)$. Then, according to Lemma 2.1, $(\tilde{A}, p \mid \tilde{A})$ is a covering space of A. It is natural to ask, to which conjugacy class of subgroups of $\pi(A)$ does this covering space correspond? Under what conditions is $p^{-1}(A)$ connected, i.e., $\tilde{A} = p^{-1}(A)$? As we shall see, these questions have relatively simple answers. To fix the notation, let $\tilde{a} \in \tilde{A}$, $a = p(\tilde{a})$, $p' = p \mid \tilde{A} : \tilde{A} \to A$, and let $i : A \to X$ denote the inclusion map.

Proposition 11.1 *Under the above hypotheses,*

$$p'_*\pi(\tilde{A}, \tilde{a}) = i_*^{-1}[p_*\pi(\tilde{X}, \tilde{a})].$$

PROOF: First we prove that $p'_*\pi(\tilde{A}, \tilde{a}) \subset i_*^{-1}[p_*\pi(\tilde{X}, \tilde{a})]$. This is a direct consequence of the commutativity of the following diagram:

$$\begin{array}{ccc} \pi(\tilde{A}, \tilde{a}) & \longrightarrow & \pi(\tilde{X}, \tilde{a}) \\ {\scriptstyle p_*'}\downarrow & & \downarrow{\scriptstyle p_*} \\ \pi(A, a) & \xrightarrow{i_*} & \pi(X, a) \end{array}$$

Next, we prove the opposite inclusion, $p'_*\pi(\tilde{A}, \tilde{a}) \supset i_*^{-1}[p_*\pi(\tilde{X}, \tilde{a})]$. Let $\alpha \in i_*^{-1}[p_*\pi(\tilde{X}, \tilde{a})]$; i.e., there exists an element $\beta \in \pi(\tilde{X}, \tilde{a})$ such that $i_*(\alpha) = p_*(\beta)$. Choose a closed path $f : I \to A$ representing α. By

Lemma 3.1, there exists a unique path $g : I \to \tilde{A}$ with initial point $\tilde{a}$ such that $pg = f$. On account of uniqueness, g must belong to the equivalence class $\beta \in \pi(\tilde{X}, \tilde{a})$; i.e., g is a closed path. Let $\gamma \in \pi(\tilde{A}, \tilde{a})$ denote the equivalence class of g. Then, $p'_*(\gamma) = \alpha$ as required. Q.E.D.

Proposition 11.2 *Under the above hypotheses,* $p^{-1}(A)$ *is connected* [i.e., $\tilde{A} = p^{-1}(A)$] *if and only if the subgroup* $i_*\pi(A, a)$ *meets every coset of the subgroup* $p_*\pi(\tilde{X}, \tilde{a})$.

PROOF: We make use of the considerations of Section 7. The set $p^{-1}(a)$ is a homogeneous right $\pi(X, a)$-space, and $p_*\pi(\tilde{X}, \tilde{a})$ is the isotropy subgroup corresponding to the point $\tilde{a}$. Similarly, the set $p'^{-1}(a) = \tilde{A} \cap p^{-1}(a)$ is a homogeneous right $\pi(A, a)$-space with $p_*\pi(\tilde{A}, \tilde{a})$ as the isotropy subgroup corresponding to the point $\tilde{a}$. From the definition of the action of the groups $\pi(X, a)$ and $\pi(A, a)$ on these two sets, it is readily seen that, for any $x \in p'^{-1}(a)$ and any $\alpha \in \pi(A, a)$,

$$x \cdot \alpha = x \cdot (i_*\alpha).$$

Now observe that, if $p^{-1}(A)$ is connected, then $p^{-1}(a) = p'^{-1}(a)$. Conversely, if $p^{-1}(a) = p'^{-1}(a)$, then we assert that $p^{-1}(A)$ is connected; for, if x is any point of $p^{-1}(A)$, let $f : I \to A$ be a path with initial point $p(x)$ and terminal point a. By Lemma 3.1, there exists a path $g : I \to \tilde{X}$ such that $pg = f$ and the initial point of g is x. Then, the terminal point of g is a point of $p^{-1}(a) = p'^{-1}(a)$, hence, a point of $\tilde{A}$. Because pg is a path in A, g is a path in $p^{-1}(A)$. Thus, we have proved that every point of $p^{-1}(A)$ can be joined to some point of $\tilde{A}$ by a path in $p^{-1}(A)$. Thus, $p^{-1}(A)$ is connected.

In view of this, we must investigate under what condition is $p^{-1}(a) = p'^{-1}(a)$. As a homogeneous right $\pi(X, a)$-space, $p^{-1}(a)$ is isomorphic to the coset space $\pi(X, a)/p_*\pi(\tilde{X}, \tilde{a})$; similarly, $p'^{-1}(a)$ is isomorphic to $\pi(A, a)/p'_*\pi(\tilde{A}, \tilde{a})$ as a right $\pi(A, a)$-space (see Section 2 of Appendix B). The inclusion map $p'^{-1}(a) \to p^{-1}(a)$ is equivalent under these isomorphisms to the map of coset spaces

$$\frac{\pi(A, a)}{p'_*\pi(\tilde{A}, \tilde{a})} \to \frac{\pi(X, a)}{p_*\pi(\tilde{X}, \tilde{a})},$$

induced by $i_* : \pi(A, a) \to \pi(X, a)$, in view of the considerations of the first paragraph of this proof. Thus, $p'^{-1}(a) = p^{-1}(a)$ if and only if this map of coset spaces is onto. From this we immediately derive the condition in the statement of the proposition. Q.E.D.

We will now consider some special cases and examples of this theorem. We keep the same notation.

Examples

11.1 Assume that $(\tilde{X}, p)$ is a regular covering space of X; then, $(\tilde{A}, p')$ is a regular covering space of A. For, if $p_*\pi(\tilde{X}, \tilde{a})$ is a normal subgroup of $\pi(X, a)$, then $i_*^{-1}[p_*\pi(\tilde{X}, \tilde{a})]$ is a normal subgroup of $\pi(A, a)$. Note that the group of automorphisms of $(\tilde{A}, p')$ may be considered a subgroup of the group of automorphisms of $(\tilde{X}, p)$. In this case, $p^{-1}(A)$ is connected if and only if the homomorphism of quotient groups

$$\frac{\pi(A, a)}{p'_*\pi(\tilde{A}, \tilde{a})} \longrightarrow \frac{\pi(X, a)}{p_*\pi(\tilde{X}, \tilde{a})}$$

induced by i_* is an epimorphism (it is always a monomorphism). Note that it may happen that $(\tilde{A}, p')$ is a regular covering space of A even though $(\tilde{X}, p)$ is *not* a regular covering space of X.

11.2 Assume that $(\tilde{X}, p)$ is a universal covering space of X. Then, $p'_*\pi(\tilde{A}, \tilde{a})$ is the kernel of i_*. Hence, $\tilde{A}$ is simply connected if and only if i_* is a monomorphism. By Proposition 11.2, $\tilde{A} = p^{-1}(A)$ if and only if i_* is an epimorphism. Thus, $p^{-1}(A)$ is a simply connected, covering space of A if and only if i_* is an isomorphism of $\pi(A)$ onto $\pi(X)$.

11.3 Assume that $i_* : \pi(A) \longrightarrow \pi(X)$ is the trivial homomorphism; i.e., image $i_* = \{1\}$. Then, $p' : \tilde{A} \rightarrow A$ is a trivial covering space; i.e., p' is a homeomorphism of $\tilde{A}$ onto A. Thus, in this case, $p^{-1}(A)$ splits up into arc components, each of which is mapped homeomorphically onto A by p. We can apply this in particular to the case where A is an open, arcwise-connected set: We conclude that, if the homomorphism $i_* : \pi(A) \rightarrow \pi(X)$ is trivial, then A is an elementary neighborhood in X for *any* covering space $(\tilde{X}, p)$ of X. This fact motivates the definition of the topology on $\tilde{X}$ in the construction used to prove Theorem 10.2.

11.4 If i_* is an epimorphism of $\pi(A)$ onto $\pi(X)$, then $p^{-1}(A)$ is connected for any covering space $(\tilde{X}, p)$ of X.

11.5 If i_* is an isomorphism of $\pi(A)$ onto $\pi(X)$, then given *any* covering space $(\tilde{A}, p')$ of A, there exists a covering space $(\tilde{X}, p)$ of X such that $(\tilde{A}, p')$ is isomorphic to the covering space $(p^{-1}(A), p \mid (p^{-1}A))$ (provided X is semilocally simply connected). In other words, under these conditions every covering space of A can be considered the part over A of some covering space of X. Thus, there is a natural one-to-one correspondence between covering spaces of X and covering spaces of A.

11.6 Let $X = S^1 \times S^1$ be a torus and let $(\mathbf{R}^2, p)$ be the universal covering space of X defined by

$$p(x, y) = (\cos 2\pi x, \sin 2\pi x, \cos 2\pi y, \sin 2\pi y)$$

for any point $(x, y) \in \mathbf{R}^2$. Let A be the subset of $S^1 \times S^1$ consisting of all points (u, v) such that $u = (1, 0)$ or $v = (1, 0)$. Then, A is the union of two circles with a single point in common, and $\pi(A)$ is a free group on two generators. In Chapter IV we proved that $i_* : \pi(A) \rightarrow \pi(X)$ is an epimorphism, and its kernel is the commutator subgroup of $\pi(A)$. In this case $\tilde{A} = p^{-1}(A)$ is the union of the horizontal lines $y =$ an integer and the vertical lines $x =$ an integer. Thus, $(\tilde{A}, p)$ is essentially the covering space corresponding to the first part of Example

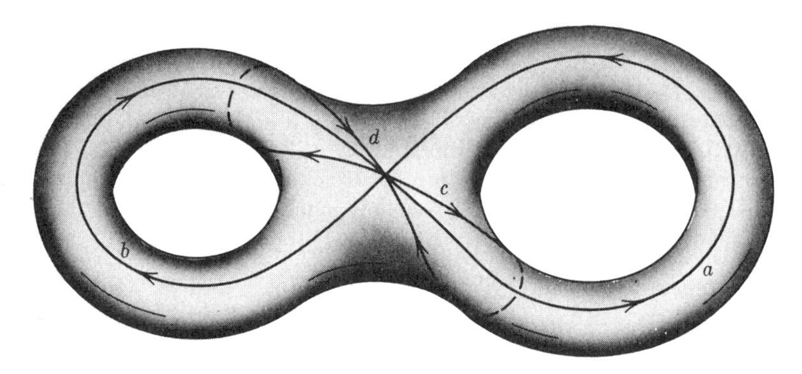

FIGURE 5.3 Diagram for Example 11.7.

2.7. We see by Proposition 11.1 that $(\tilde{A}, p)$ is the regular covering space of A corresponding to the commutator subgroup of $\pi(A)$.

This example could be modified in several ways: (a) Instead of taking the universal covering space of the torus, we could take some other covering space. (b) Instead of taking $X = S^1 \times S^1$, we could take X to be the product of n copies of S^1, and A to be the subset of X consisting of all points having $(n - 1)$ coordinates equal to $(1, 0)$. Then, A is the union of n circles having a single point in common. We leave the details of these modified examples to the reader.

11.7 Let X be a compact set which is the closure of the region in $\mathbf{R}^3$ bounded by an orientable surface of genus 2 (imbedded in the standard way) and let A be the bounding surface (see Figure 5.3). It is geometrically evident that the union of the two circles a and b in Figure 5.3 is a deformation retract of X. Hence, $\pi(X)$ is a free group generated by the equivalence classes of the closed paths a and b. In Chapter IV we showed that $\pi(A)$ is a group which is generated by the equivalence classes of the four loops a, b, c, and d subject to one relation. Thus, $i_* : \pi(A) \to \pi(X)$ is an epimorphism, and the kernel is the least normal subgroup containing the equivalence classes of the paths c and d. If $(\tilde{X}, p)$ is any covering space of X, we can now apply Propositions 11.1 and 11.2 to the induced covering space $(\tilde{A}, p')$ of A. Because the union of the two circles a and b is a deformation retract of X, the problem of constructing a covering space of X is equivalent to the problem of constructing a covering space of the union of two circles with a single point in common, by Example 11.5. It is comparatively easy to visualize such covering spaces; we leave it to the student to construct examples.

12 Point set topology of covering spaces

In this section we consider some results on the point set topology of covering spaces, which are useful in some contexts and are not exactly obvious. We assume, as in the rest of this chapter, that all spaces involved are connected and locally arcwise connected.

Proposition 12.1 *Let X be a space which is semilocally simply connected and has a countable basis of open sets. Then, any covering space of X also has a countable basis of open sets.*

PROOF: First, we make the following two simple observations that will be useful in the course of the proof:

(a) Any open subset of X has at most a countable number of arc components.
(b) Let U be any arcwise-connected, open subset of X such that the natural homomorphism $\pi(U) \to \pi(X)$ is trivial; if x_0, $x_1 \in U$, then any two paths in U joining x_0 and x_1 are equivalent (in X).

The first step in the proof is to prove that the fundamental group of X is countable. For this purpose, note that our hypotheses imply that X is a Lindelöf space; i.e, any open covering of X has a countable subcovering. Hence, we may choose a countable open covering $\{U_1, U_2, U_3, \ldots\}$ of X such that each U_i is arcwise connected and the natural homomorphism $\pi(U_i) \to \pi(X)$ is trivial.

If $U_i \cap U_j \neq \phi$, then $U_i \cap U_j$ has at most a countable number of components. [See remark (a).] In each component of the intersection $U_i \cap U_j$ choose a point. Do this for all pairs (i, j) such that $U_i \cap U_j \neq \phi$. Such points will be called *preferred points.* There are at most a countable number of preferred points.

For each pair of distinct preferred points which lie in the same set U_i, choose a path in U_i connecting these two points. Such a path will be called a *preferred path.* Note that there are at most a countable number of preferred paths. Also note that it follows from remark (b) that the equivalence class of any particular preferred path is unique.

Choose a preferred point x_0 as base point. We shall complete the proof by showing that any element of $\pi(X, x_0)$ is equivalent to a (finite) product of preferred paths. Let

$$f : I \to X$$

be a closed path representing an element $\alpha \in \pi(X, x_0)$. We readily prove (by use of the Lebesgue number) that there exists a partition of the unit interval $I = [0, 1]$,

$$t_0 = 0 < t_1 < t_2 < \cdots < t_n = 1$$

such that, for each subinterval $[t_i, t_{i+1}]$, $f([t_i, t_{i+1}])$ is contained in one of the open sets U_j; to be precise, for $i = 1, 2, \ldots, n$, we choose an open set

$U_{a(i)}$ such that

$$f([t_{i-1}, t_i]) \subset U_{a(i)}, \qquad i = 1, 2, \ldots, n.$$

We may assume that $U_{a(i)}$ and $U_{a(i+1)}$ are different; if they were the same, we could consolidate the intervals $[t_{i-1}, t_i]$ and $[t_i, t_{i+1}]$ into the single interval $[t_{i-1}, t_{i+1}]$ by omitting the point t_i from the subdivision.

Let α_i denote the equivalence class of the path $f \mid [t_{i-1}, t_i]$ for $i = 1, 2, \ldots, n$. Then, α_i is a path class in the set $U_{a(i)}$, and

$$\alpha = \alpha_1\alpha_2 \ldots \alpha_n.$$

Each point $f(t_i)$, $i = 1, \ldots, n - 1$, lies in a certain component of the intersection $U_{a(i)} \cap U_{a(i+1)}$; call this component C_i. Choose a path class β_i in C_i from the point $f(t_i)$ to a preferred point in C_i. Then,

$$\alpha = \alpha_1\beta_1\beta_1^{-1}\alpha_2\beta_2\beta_2^{-1}\alpha_3 \cdots \alpha_{n-1}\beta_{n-1}\beta_{n-1}^{-1}\alpha_n$$

and the successive paths

$$\alpha_1\beta_1, \ \beta_1^{-1}\alpha_2\beta_2, \ \ldots, \ \beta_{n-2}^{-1}\alpha_{n-1}\beta_{n-1}, \ \beta_{n-1}^{-1}\alpha_n$$

join preferred points in the sets $U_{a(1)}, U_{a(2)}, \ldots, U_{a(n)}$; therefore, each is equivalent to a preferred path [by remark (b)], and thus we have shown that α can be expressed as a product of preferred paths.

Exercise

12.1 Prove that, if X is semilocally simply connected and compact metric, then the fundamental group of X is finitely generated.

Because $\pi(X)$ is countable, any subgroup of $\pi(X)$ has countable index, and hence any covering space of X has countably many sheets. Therefore, to complete the proof, it suffices to prove the following lemma.

Lemma 12.2 *Let X be a space which is semilocally simply connected, and has a countable basis of open sets. If $(\tilde{X}, p)$ is a covering space of X having a countable number of sheets, then $\tilde{X}$ has a countable basis of open sets also.*

PROOF: Choose a countable basis of open sets $U_1, U_2, U_3, \ldots$ for X such that each U_i is arcwise connected and $\pi(U_i) \to \pi(X)$ is a trivial homomorphism. It is readily shown that such a basis exists. By Example 11.3, for each integer i, $p^{-1}(U_i)$ has a countable number of components, each of which is mapped homeomorphically onto U_i by p; denote these components by $U_{i1}, U_{i2}, \ldots$. Then, $\{U_{ij}\}$ is a countable family of open sets, and may be readily proved to be a basis for the topology of $\tilde{X}$.

Q.E.D.

Next, we prove a proposition regarding the topology of covering spaces which will be used in the following chapters. The reader can omit this proposition until it is needed.

We consider the following problem. Let $(\tilde{Y}, p)$ be a covering space of Y. As usual, we assume that Y and $\tilde{Y}$ are connected and locally arcwise connected. Moreover, we assume that Y is a regular Hausdorff space (i.e., satisfies the T_3 separation axiom of Alexandroff and Hopf). This implies that $\tilde{Y}$ satisfies the same separation axiom.

Let $\{X_\lambda : \lambda \in \Lambda\}$ be a family of compact Hausdorff spaces which are locally arcwise connected, connected, and simply connected, and let $\{f_\lambda : X_\lambda \to Y : \lambda \in \Lambda\}$ be a family of continuous maps. Our hypotheses imply that each of the maps f_λ can be lifted (in various ways) to maps $f_{\lambda i} : X_\lambda \to \tilde{Y}$ such that $f_\lambda = pf_{\lambda i}$. Let $\{f_{\lambda i} : i \in M_\lambda\}$ denote the set of *all* such liftings of f_λ.

With these preliminaries out of the way, we can state our problem as follows: Suppose that Y has the largest topology which makes all the maps f_λ continuous [i.e., a set $U \subset Y$ is open if and only if $f_\lambda^{-1}(U)$ is open in X_λ for all λ]. Does $\tilde{Y}$ have the largest topology making all the maps $f_{\lambda i}$ continuous? The following lemma, due essentially to J. H. C. Whitehead, shows that under slight additional hypotheses, the answer is affirmative.

Lemma 12.3 *Using the above hypotheses and notation, assume that* $(\tilde{Y}, p)$ *is a regular covering space of* Y, *or, alternatively, that* Y *is semilocally simply connected. If* Y *has the largest topology which makes all the maps* f_λ *continuous, then* $\tilde{Y}$ *has the largest topology which makes all the maps* $f_{\lambda i}$ *continuous.*

PROOF: First, we define some terminology. As usual, by an *elementary* neighborhood of a point in Y, we mean an arcwise-connected open neighborhood U such that each component of $p^{-1}(U)$ is mapped topologically onto U by p. By a *basic* open neighborhood V of a point in Y, we mean an arcwise-connected open neighborhood such that $\bar{V}$ is contained in some elementary neighborhood U. The hypothesis that Y is regular assures us that each point of Y contains arbitrarily small basic neighborhoods. If V is any basic neighborhood in Y, we shall call each component W of $p^{-1}(V)$ a *basic* neighborhood in $\tilde{Y}$.

The idea is to prove this lemma first for the case where $(\tilde{Y}, p)$ is a regular covering, and then to prove it for the remaining case. Therefore, we assume that $(\tilde{Y}, p)$ is a regular covering space of Y, and we denote the group of automorphisms of $(\tilde{Y}, p)$ by G.

First, we make two assertions which will be needed in the proof.

Assertion 1 A subset $A \subset \tilde{Y}$ is closed if and only if $A \cap \bar{W}$ is closed for every basic neighborhood W in $\tilde{Y}$ (where basic neighborhood is defined as above).

Assertion 2 If X is a compact space, $f : X \to \tilde{Y}$ is a continuous map, and W is a basic neighborhood in $\tilde{Y}$, then only a finite number of the sets

$$\{f^{-1}\varphi^{-1}(\bar{W}) : \varphi \in G\}$$

are nonempty.

PROOF OF ASSERTION 1 Because the basic neighborhoods cover $\tilde{Y}$, it follows that $\tilde{Y} - A$ is the union of the sets $W - A$ for all basic neighborhoods W. But $W - A = W - (A \cap \bar{W})$ is obviously open. Therefore $\tilde{Y} - A$ is open, and A is closed.

PROOF OF ASSERTION 2 Let $V = p(W)$; then, V is a basic neighborhood in Y. Hence, there exists an elementary neighborhood U in Y such that $\bar{V} \subset U$. Let $\tilde{U}$ denote the component of $p^{-1}(U)$ which contains $\bar{W}$. For any $\varphi \in G$, let

$$U_\varphi = f^{-1}\varphi^{-1}(\tilde{U}).$$

Note that the sets U_φ are pairwise disjoint. Next, note that

$$C = \bigcup_{\varphi \in G} f^{-1}\varphi^{-1}(\bar{W}) = f^{-1}p^{-1}(\bar{V})$$

is a closed subset of X, which must therefore be compact. Moreover, $\{U_\varphi : \varphi \in G\}$ is an open covering of C, which must therefore have a finite refinement. But the only way this could be true is for all but a finite number of the sets $f^{-1}\varphi^{-1}(\bar{W})$ to be empty.

We can now proceed to the proof of the lemma. Let A be a subset of $\tilde{Y}$ such that $f_{\lambda i}^{-1}(A)$ is a closed subset of X_λ for every map $f_{\lambda i}$; we must prove that A is closed. From Assertion 1, we see that it suffices to prove the following implication: For any subset $A \subset \tilde{Y}$ and any basic neighborhood W in $\tilde{Y}$, if $f_{\lambda i}^{-1}(A)$ is closed for any map $f_{\lambda i}$, then $\bar{W} \cap A$ is closed. Replacing A by $A \cap \bar{W}$ if necessary, it is clear that it suffices to prove this implication in the special case where $A \subset \bar{W}$; this we shall do. Let $p(A) = B \subset Y$. Because p maps $\bar{W}$ homeomorphically onto its image, it suffices to prove that B is closed. To prove that B is closed, we must show that $f_\lambda^{-1}(B)$ is closed for any $\lambda \in \Lambda$. It is readily seen that

$$f_\lambda^{-1}(B) = \bigcup_i f_{\lambda i}^{-1}(A). \tag{5.12-1}$$

Choose a particular index j; then,

$$\bigcup_i f_{\lambda_i}^{-1}(A) = \bigcup_{\varphi \in G} f_{\lambda_j}^{-1}\varphi^{-1}(A).$$

By Assertion 2, $f_{\lambda_j}^{-1}\varphi^{-1}(\bar{W})$ is nonempty for only finitely many $\varphi \in G$; hence, $f_{\lambda_j}^{-1}\varphi^{-1}(A)$ is nonempty for only finitely many $\varphi \in G$. Thus, in Equation (5.12-1), the right-hand side is a *finite* union of closed sets. Hence, the left-hand side is closed, as desired.

Now we shall prove the lemma in the case where $(\tilde{Y}, p)$ is *not* a regular covering; in this case, we assume Y is semilocally simply connected. Hence, there exists a universal covering space $(\hat{Y}, q)$ of Y, and a map $r : \hat{Y} \to \tilde{Y}$ such that $(\hat{Y}, r)$ is a covering space of $\tilde{Y}$; hence, $\tilde{Y}$ has the quotient topology determined by r. We can now apply what we have just proved to conclude that the covering space $\hat{Y}$ has the largest topology which makes certain lifted maps $X_\lambda \to \hat{Y}$ continuous. We then apply Lemma 2.4 of Appendix A to the map $r : \hat{Y} \to \tilde{Y}$ to complete the proof.

Q.E.D.

Example

12.1 Any compact surface M is a quotient space of a polygonal disc D under a map

$$f : D \to M$$

which identifies certain edges of D in pairs. If $(\tilde{M}, p)$ is any covering space of M, then there exist liftings

$$f_i : D \to \tilde{M}, \qquad pf_i = f,$$

of f. Then, $\tilde{M}$ has the largest topology which makes all the f_i's continuous by Lemma 12.3. Note that the images $f_i(D)$ cover $\tilde{M}$. Each such image is called a *fundamental domain* in $\tilde{M}$. (See also Example 2.6.)

Exercises

12.2 Let X be a space which is connected and locally arcwise connected, and let $(\tilde{X}, p)$ be a covering space of X. Prove that, if X has any of the following properties, then so does $\tilde{X}$:

(a) Hausdorff.
(b) Regular.
(c) Completely regular.
(d) Locally compact.

12.3 Let X be a space which is connected and locally arcwise connected, and let $(\tilde{X}, p)$ be a covering space of X. Prove that $\tilde{X}$ is compact if and only if X is compact and the covering has only a finite number of sheets.

12.4 Let X be a space which is connected, locally arcwise connected, semi-locally simply connected, and separable metric. Prove that any covering space of X is also separable metric. (HINT: Use the metrization theorem of Urysohn and Tychonoff; i.e., a space is separable metric if and only if it is regular and has a countable basis of open sets.)

NOTES

Branched covering spaces

The Riemann surface of a so-called "multiple-valued" analytic function is usually not a covering space of the domain of definition of the function because of the existence of the "branch points." It is an example of a "branched covering space." For the general theory of branched covering spaces, see the article by R. H. Fox entitled "Covering Spaces with Singularities" in the book *Algebraic Geometry and Topology: A Symposium in Honor of S. Lefschetz* (Princeton, N.J.: Princeton University Press, 1957), pp. 243–257; also, E. Michael, *Proc. Kon. Ned. Akad. Weten. Amsterdam* (ser. *A*) *66*, 1963, pp. 629–633. There is no extensive general theory of branched covering spaces, although a great deal is known in various special cases, e.g., for 2-dimensional manifolds.

Covering spaces without any assumption of local connectedness

It is natural to ask whether the assumption of local arcwise connectedness which was assumed throughout this chapter can be weakened or omitted. This problem has been considered by several authors. For example, in the book by Chevalley [1], the assumption of local arcwise connectedness is reduced to ordinary local connectedness. Other papers considering this subject are the following:

B. Banaschewski. *Math. Nachr.*, *15*, 1956, pp. 175–180.
J. Dugundji. *Proc. Nat. Acad. Sci. U.S.A.*, *36*, 1950, pp. 141–143.
J. Gil de Lamadrid and J. P. Jans. *Proc. Amer. Math. Soc.*, *10*, 1959, pp. 710–715.
C. N. Lee. *Duke Math. J.*, *24*, 1957, pp. 547–554.
R. S. Novosad. *Trans. Amer. Math. Soc.*, *79*, 1955, pp. 216–228.
S. Lubkin. *Trans. Amer. Math. Soc.*, *104*, 1962, pp. 205–238.

These authors, in turn, refer to earlier papers on the subject. Most of these papers are brief, and there is no attempt to reconstruct the entire theory under weakened hypotheses.

So far the ideas developed in these papers have not proved to be important, because in the situations where the theory of covering spaces finds natural application, the spaces involved satisfy all the local conditions one could desire. Or course, it may happen that sometime in the future somebody will need to apply the theory of covering spaces to spaces which are bad locally.

The reader's attention should be called to the example due to E. C. Zeeman on p. 258 of Hilton and Wylie [2]. This example shows the necessity of the assumption that Y be locally connected in Theorem 5.1.

Partial ordering of isomorphism classes of covering spaces of a given space

In Exercise 6.3 it was shown that, if X is a space with abelian fundamental group $\pi(X)$, then there is a natural partial ordering of the isomorphism classes of covering spaces of X: We define $(\tilde{X}_1, p_1) \geqq (\tilde{X}_2, p_2)$ if and only if there exists a homomorphism of $(\tilde{X}_1, p_1)$ onto $(\tilde{X}_2, p_2)$. Under this partial ordering, the isomorphism classes of coverings constitute a lattice isomorphic to the lattice of all subgroups of $\pi(X)$.

This result remains true if we make the slightly weaker assumption that the fundamental group of X is Hamiltonian; i.e., every subgroup of $\pi(X)$ is normal. For the structure of Hamiltonian groups, see M. Hall, *The Theory of Groups* (New York: Macmillan, 1959), p. 190.

If we try to extend this result to more general groups, we encounter serious troubles. For example, it can happen that there exist covering spaces $(\tilde{X}_1, p_1)$ and $(\tilde{X}_2, p_2)$ of X such that $(\tilde{X}_1, p_1) \geqq (\tilde{X}_2, p_2)$ and $(\tilde{X}_2, p_2) \geqq (\tilde{X}_1, p_1)$, but $(\tilde{X}_1, p_1)$ and $(\tilde{X}_2, p_2)$ are nonisomorphic. This could occur as follows. Choose a base point $x \in X$, and assume there exist points $x_1, x_1' \in p_1^{-1}(x)$ and $x_2, x_2' \in p_2^{-1}(x)$ with the following properties:

(a) The subgroups $p_{1*}\pi(\tilde{X}_1, x_1)$ and $p_{2*}\pi(\tilde{X}_2, x_2)$ are *not* conjugate subgroups of $\pi(X, x)$.

(b) The following inclusion relations hold between subgroups:

$$p_{1*}\pi(\tilde{X}_1, x_1) \supset p_{2*}\pi(\tilde{X}_2, x_2'),$$

$$p_{2*}\pi(\tilde{X}_2, x_2) \supset p_{1*}\pi(\tilde{X}_1, x_1').$$

It then follows by the results of Section 6 that the stated relations between $(\tilde{X}_1, p_1)$ and $(\tilde{X}_2, p_2)$ hold.

It is not a difficult algebraic problem to exhibit a group having four subgroups with the desired properties; and in Chapter VII we shall show that any group can be "realized" as the fundamental group of a space X which is connected, locally arcwise connected, and semilocally simply connected.

Covering spaces as fibre spaces or fibre bundles

The reader who is familiar with the theory of fibre spaces and fibre bundles will recognize that a covering space as we have defined the term is a locally trivial fibre space with a discrete fibre. Thus, the theory of covering spaces may be considered as a chapter in the general theory of fibre spaces. We may also consider that a covering space $(\tilde{X}, p)$ of X is a fibre bundle with $\pi(X)$ as structural group and the discrete homogeneous space $\pi(X)/p_*\pi(\tilde{X})$ as the fibre. The regular covering spaces then correspond to the principal fibre bundles. This topic is discussed in the book *The Topology of Fibre Bundles* by N. E. Steenrod (Princeton, N.J.: Princeton University Press, 1951); especially in Sections 13 and 14.

Higher homotopy groups of a covering space

For any space X and point $x_0 \in X$, the notation $\pi_n(X, x_0)$ denotes the set of all homptopy classes of maps $(S^n, y_0) \to (X, x_0)$; here $y_0 \in S^n$ and it is understood that all homotopies are relative to the chosen base point y_0 (see Section II.4 for the definition of relative homotopy). Note that, for $n = 1$, $\pi_1(X, x_0) = \pi(X, x_0)$ is just the fundamental group. It is also possible to define in a natural way an addition in $\pi_n(X, x_0)$ for $n > 1$, so that it becomes an abelian group, called *the* nth *homotopy group of* X. We assert that, *if* $(\tilde{X}, p)$ *is a covering space of* X, *the projection* p *induces an isomorphism of* $\pi_n(\tilde{X}, x)$ *onto* $\pi_n(X, p(x))$ *for any point* $x \in \tilde{X}$ *and any integer* $n > 1$. The proof is a simple application of Theorem 5.1. If $\varphi : (S^n, y_0) \to (X, p(x))$ is any continuous map, then there exists a unique map $\tilde{\varphi} : (S^n, y_0) \to (\tilde{X}, x)$ such that $p\tilde{\varphi} = \varphi$. Moreover, two such maps, $\varphi_0, \varphi_1 : (S^n, y_0) \to (X, p(x))$ are homotopic (relative to y_0) if and only if the corresponding lifted maps $\tilde{\varphi}_0, \tilde{\varphi}_1 : (S^n, y_0) \to (\tilde{X}, x)$ are homotopic.

This result is often useful in the study of higher homotopy groups.

Determination of all covering spaces with a finite number of sheets

In general, we probably cannot hope for an *effective* procedure for determining all covering spaces of a given space X [or equivalently, of determining all conjugacy classes of subgroups of $\pi(X)$]. However, if the fundamental group $\pi(X)$ is finitely presented, then for any given integer n there is an effective procedure for finding all n-sheeted covering spaces of X. This procedure is illustrated on pp. 201–203 of Seifert and Threlfall [3].

The Jordan curve theorem

A proof of the Jordan curve theorem using the theory of fundamental groups and covering spaces may be found in Munkres, [6], pp. 374–386. This book also has exercises which contain an outline of a proof of the Brouwer theorem on "Invariance of Domain" in the plane.

REFERENCES

1. Chevalley, C. *The Theory of Lie Groups I*. Princeton, N.J.: Princeton University Press, 1946. Chapter II, Sections VI–X.
2. Hilton, P. J., and S. Wylie. *Homology Theory, An Introduction to Algebraic Topology*. Cambridge: The University Press, 1960. Chapter 6.
3. Seifert, H., and W. Threlfall. *A Textbook of Topology*. New York: Academic Press, 1980. Chapter 8.
4. Kinoshita, S. "Notes on Covering Transformation Groups." *Proc. Amer. Math. Soc.*, *19*, 1968, pp. 421–424.
5. Gray, W. J. "A Note on Covering Transformations." *Indag. Math.*, *31*, 1969, pp. 283–284.
6. Munkres, J. R. *Topology: A First Course*. Englewood Cliffs, N.J.: Prentice Hall, Inc., 1975. Chapter 8.

CHAPTER SIX

The Fundamental Group and Covering Spaces of a Graph. Applications to Group Theory

1 Introduction

A *graph* is a topological space which consists of a collection of points, called *vertices*, and a collection of *edges*. Each edge is either homeomorphic to an interval of the real line and joins two distinct vertices, or it is homeomorphoric to a circle and joins a given vertex to itself. It is assumed that any two distinct edges are either disjoint, or else intersect in a common end point.

Graphs are objects of fairly common occurrence; e.g., wiring diagrams as used by electrical engineers can be considered to be graphs. Many problems from various branches of mathematics and related disciplines, especially problems of a combinatorial nature, can be translated into problems involving graphs. As a result, within the past century, an extensive theory has grown up. It should be emphasized, however, that this theory is mostly concerned with the combinatorial properties of a graph, i.e., properties having to do with the relations between the edges and vertices. The topological properties are completely neglected more often than not. There are several texts on graph theory; see [1] or [2].

Our object in this chapter will be to study the fundamental group and covering spaces of a graph; thus we shall emphasize the topological properties of graphs. We shall then apply the results obtained to group theory to prove certain classical theorems. Although these group theoretic results can be given purely algebraic proofs, the topological proofs that we give are more transparent and better motivated. In fact, many of these theorems were probably originally discovered by the consideration of the fundamental group and covering spaces of a graph.

If a graph has only a finite number of vertices and edges, there is no problem in the definition of its topology: There is one rather obvious way to topologize it. For an infinite graph, the problem is more delicate. It is necessary for us to consider this problem in detail, however, because a covering space of a finite graph may very well be an infinite graph.

A graph, as we define it, is a 1-dimensional CW-complex. Thus, the material of this chapter may be considered an introduction to the theory of the CW-complexes of J. H. C. Whitehead.

2 Definition and examples

A *graph* is a pair consisting of a Hausdorff space X and a subspace X^0 (called "the set of vertices of X") such that the following conditions hold:

(a) X^0 is a discrete, closed subspace of X. Points of X^0 are called "vertices."
(b) $X - X^0$ is the disjoint union of open subsets e_i, where each e_i is homeomorphic to an open interval of the real line. The sets e_i are called "edges."
(c) For each edge e_i, its boundary $\bar{e}_i - e_i$ is a subset of X^0 consisting of one or two points. If $\bar{e}_i - e_i$ consists of two points, then the pair $(\bar{e}_i, e_i)$ is homeomorphic to the pair $([0, 1], (0, 1))$; if $\bar{e}_i - e_i$ consists of one point, then the pair $(\bar{e}_i, e_i)$ is homeomorphic to the pair $(S^1, S^1 - \{1\})$, where S^1 is the unit circle in the plane.
(d) X has the so-called weak topology: A subset $A \subset X$ is closed (open) if and only if $A \cap \bar{e}_i$ is closed (open) for all edges e_i.

Of these conditions, (d) is undoubtedly the hardest to comprehend. It is automatically satisfied in the case where X has only a finite number of edges. Thus, it is only of interest when X has an infinite number of edges. We shall clarify the condition by giving some examples.

Examples

2.1 For any integer $n > 0$, let C_n denote the circle in the xy plane with center at the point $(1/n, 0)$ and radius $1/n$. It is tangent to the y axis at the origin. Let X denote the union of the circles C_n for all $n > 0$. Let $X^0 = \{(0, 0)\}$; i.e., there is only one vertex. Then, conditions (a), (b), and (c) hold, but condition (d) does not hold. The student should find a set A which is not closed in X, but such that $A \cap \bar{e}_i$ is closed for all edges e_i.

2.2 Using polar coordinates (r, θ) in the plane, let

$$e_n = \left\{(r, \theta) : \theta = \frac{1}{n} \quad \text{and} \quad 0 < r < 1\right\}$$

for any integer $n > 0$. Let

$$X = \bigcup_{n=1}^{\infty} \bar{e}_n,$$

where $\bar{e}_n$ denotes the closure of e_n in the topology of the plane. Define X^0 to consist of the origin and the points $(r, \theta) = (1, 1/n)$ for $n > 0$. Once again, conditions (a), (b), and (c) hold, but condition (d) fails.

2.3 Let $X = \mathbf{R}$, the real line, and let X^0 be the subset of integral points. Then, conditions (a)–(d) all hold. The verification of (d) is left as an exercise for the reader.

2.4 For each integer n, positive or negative, let

$$A_n = \{(x, y) \in \mathbf{R}^2 : x = n\},$$

$$B_n = \{(x, y) \in \mathbf{R}^2 : y = n\}.$$

Then, each A_n is a vertical line in the plane, and each B_n is a horizontal line. Let X be the union of all these lines, and let X^0 be the set of all points of intersection of horizontal and vertical lines, i.e., the set of points in the plane having integral coordinates. Then, conditions (a)–(d) hold.

In Examples 2.1 and 2.2, we can define a new topology on the space X in each case so that condition (d) will hold. In Example 2.1, the topology of each circle C_n is left unchanged; then, a set $A \subset X$ is defined to be closed if and only if $A \cap C_n$ is closed in C_n for all n. We can follow similar procedure in Example 2.2, using the edges $\bar{e}_n$ in place of the circles C_n. The reader should verify that X is a Hausdorff space in each case.

We leave it to the reader to construct examples of graphs with a finite number of vertices and edges. It can be proved that every such finite graph is homeomorphic to a subset of Euclidean 3-space. Because we shall not make any use of this fact, we shall not prove it.

Note that, if X is the space of a graph, it is possible to give X the structure of a graph in several different ways. For example, if $X = [0, 1]$, we may take $X^0 = \{0, \frac{1}{2}, 1\}$, or we may choose $X^0 = \{0, \frac{1}{3}, \frac{2}{3}, 1\}$; i.e., we may subdivide the unit interval into two or three subintervals. In general, if X^0 and X'^0 are two different choices of sets of vertices on the space X, we say that (X, X'^0) is a *subdivision* of (X, X^0) if $X^0 \subset X'^0$. It is clear that any graph can be subdivided by inserting new vertices in some or all of the edges.

Exercises

2.1 Let X be a graph and let Y be a topological space. Prove that a function $f : X \to Y$ is continuous if and only if $f \mid \bar{e}_i$ is continuous for each edge e_i of X.

2.2 Let X be a graph, and let I be the unit interval, $[0, 1]$. Prove that a subset $A \subset X \times I$ is closed (open) if and only if $A \cap (\bar{e} \times I)$ and $A \cap (v \times I)$ are closed (open) for every edge e and every vertex v of X.

2.3 Let X be a graph and let Y be a topological space. Prove that a function $f : X \times I \to Y$ is continuous if and only if $f \mid \bar{e} \times I$ and $f \mid v \times I$ are continuous for all edges e and vertices v of X.

3 Basic properties of graphs

A graph may be either *connected* or *disconnected*, as simple examples show. In fact, there is nothing in our definition to preclude the existence of vertices that are isolated points, i.e., points that do not belong to the closure of any edge.

We shall say that a graph is *finite* if it has only a finite number of vertices and edges. A finite graph is compact, because it is the union of a finite number of compact subsets. Conversely, if a graph is compact, then it is finite; the proof is easy. A graph is *locally finite* if each vertex is incident with only a finite number of edges (a vertex v and an edge e are *incident* if and only if $v \in \bar{e}$). A graph is locally compact if and only if it is locally finite. Once again, the proof is easy. Note that, if we did not impose condition (d) in the definition, these propositions would not be true, as Example 2.1 illustrates.

A subset A of a graph (X, X^0) is called a *subgraph* if the pair (A, A^0) is a graph, where $A^0 = A \cap X^0$. This will be true if and only if A is the exact union of a collection of vertices and closed edges of X. It follows from this that A is a closed subset of X.

The next lemma shows that locally a graph is as "nice" as could be desired.

Lemma 3.1 *Every point of a graph has a basic family of contractible neighborhoods.*

PROOF: This is clear for points that are interior points of an edge, or for isolated vertices. Let v be a nonisolated vertex, and let U be an open set containing v. We must exhibit a neighborhood V of v which is contractible and such that $V \subset U$. For each edge e incident with v, $U \cap \bar{e}$ is an open neighborhood of v in $\bar{e}$. Choose V so that $V \cap \bar{e}$ is an open, contractible neighborhood of v in $\bar{e}$, $V \cap \bar{e} \subset U \cap \bar{e}$, and for any edge e' not incident with v, $V \cap \bar{e}' = \phi$. Clearly, such a choice is possible by condition (c). By condition (d), V is an open set. It remains to prove that V is contractible. For each edge e incident with v, choose a contracting homotopy

$$\varphi_e : (V \cap \bar{e}) \times I \to V \cap \bar{e}$$

such that, for any $x \in V \cap \bar{e}$ and $t \in I$,

$$\varphi_e(x, 0) = x,$$

$$\varphi_e(x, 1) = v,$$

$$\varphi_e(v, t) = v.$$

Then, define $f : V \times I \to V$ by

$$f \mid (V \cap \bar{e}) \times I = \varphi_e.$$

The proof that f is continuous depends on the fact that each of the maps φ_e is continuous, and that a graph has the weak topology [i.e., condition (d) holds]; see Exercise 2.3. Q.E.D.

It follows that a graph is locally arcwise connected, and semilocally simply connected. Thus, the entire theory of covering spaces is applicable. Also, if a graph is connected, it is arcwise connected. In particular, any two vertices in a connected graph can be joined by a path.

An edge e of a graph is *oriented* by placing an arrow on it to indicate a choice of positive direction along the edge. Every edge has two possible orientations. This intuitive idea can be made precise as follows. By definition, an edge e is homeomorphic to the open interval $(0, 1)$. Define two homeomorphisms

$$h_0, h_1 : e \to (0, 1)$$

to be *equivalent* if the composite homeomorphism $h_0 h_1^{-1} : (0, 1) \to (0, 1)$ is a monotone increasing map. It is readily seen that there are two equivalence classes of homeomorphisms $e \to (0, 1)$. To *orient* the edge e is to choose one of these equivalence classes as the preferred class of homeomorphisms.

Let e be an edge of a graph incident with two vertices. Assume that e is oriented; then, it should be clear what we mean by the *initial* vertex and *terminal* vertex of the oriented edge e. For completeness, we now give precise mathematical definitions of these concepts. By assumption, the pair $(\bar{e}, e)$ is homeomorphic to the pair $([0, 1], (0, 1))$. We may choose a definite homeomorphism $h : \bar{e} \to [0, 1]$ such that $h \mid e$ belongs to the preferred equivalence class determined by the choice of orientation of e. Then, the initial vertex of e is $h^{-1}(0)$ and the terminal vertex is $h^{-1}(1)$.

In the case of an edge e incident with only one vertex [i.e., the pair $(\bar{e}, e)$ is homeomorphic to $(S^1, S^1 - \{1\})$], we make the convention that the one vertex is both the initial vertex and the terminal vertex.

An *edge path* in a graph is a finite sequence of *oriented* edges $(e_1, e_2, \ldots, e_n)$ $(n \geqq 1)$ such that the terminal vertex of e_{i-1} is the initial vertex of e_i for $1 < i \leqq n$. The edge path $(e_1, \ldots, e_n)$ is called *reduced* if it is *not* the case that e_{i-1} and e_i are the same edge with opposite orientations for any $i = 2, 3, \ldots, n$. Usually we shall only be interested in reduced edge paths, and unless the contrary is explicitly stated, we shall assume that all edge paths under discussion are reduced. If $(e_1, e_2, \ldots, e_n)$ is an edge path, then the initial vertex of e_1 is called the *initial vertex of the edge path*, and the terminal vertex of e_n is called the *terminal vertex of the edge path*. The edge path is said to *join* its initial and terminal vertices; it is called a *closed* edge path if its initial and terminal vertices are the same. A closed edge path is also called a *circuit* or *closed circuit*.

Exercises

3.1 Prove that any compact subset of a graph is contained in a finite subgraph.

3.2 Prove that a graph is connected if and only if every pair of vertices can be joined by an edge path.

3.3 Prove that the union or intersection of any collection of subgraphs of a graph X is also a subgraph.

3.4 Let X be a graph which is *not* locally finite and v a vertex of X incident with an infinite number of edges. Prove that v does not have a countable basis of neighborhoods. Deduce from this that, if a graph is metrizable, it is locally finite.

4 Trees

A *tree* is a connected graph that contains no closed (reduced) edge paths. For example, a graph consisting of a single edge e and two vertices incident with it, or a graph consisting of a single vertex and no edges, is a tree. Example 2.3 is a tree, as is Example 2.2 if it is given the weak topology as explained in Section 2. On the other hand, any graph which contains an edge e incident with only one vertex (i.e., $\bar{e}$ is homeomorphic with S^1) is not a tree.

We leave to the reader the proof of the following two properties of trees:

(a) Any connected subgraph of a tree is also a tree.
(b) In a tree, any two distinct vertices can be joined by a unique reduced edge path.

The main theorem about the topology of a tree is the following.

Theorem 4.1 *Any tree is contractible.*

PROOF: First, we prove this for finite trees by induction on the number of edges. It is obvious for the cases in which a tree has zero edges (i.e., a graph consisting of a single vertex) or exactly one edge. Assume we have proved the theorem for all trees having less than n edges, and let T be a tree having n edges, $n > 1$. We assert that there exists a vertex v of T such that v is incident with only one edge of T. This assertion is proved by contradiction; if every vertex in a finite, connected graph is incident with two or more edges, it is easy to construct a closed edge path. Let e be the unique edge with which v is incident, and let $T' = T - (e \cup \{v\})$. Then, T' is a connected subgraph of T, as is easily proved. Hence, T' is also a tree. By the inductive hypothesis, T' is contractible. To complete the proof we show that T' is a deformation retract of T. We leave it to the reader to describe the obvious deformation retraction.

Now let T be an arbitrary tree. Choose a vertex $v_0 \in T$; we shall define a contracting homotopy

$$f : T \times I \to T$$

such that $f(x, 0) = x$, $f(x, 1) = v_0$, and $f(v_0, t) = v_0$ for any $x \in T$ and $t \in I$. We construct the map f in successive steps as follows. For each vertex $v \in T$, choose a finite connected subgraph $T(v)$ of T which contains the vertices v_0 and v. Such a choice is possible because T is connected. For each such v, choose a path $t \to f(v, t)$, $0 \leqq t \leqq 1$, in $T(v)$ with initial point v and terminal point v_0. We also define $f(v_0, t) = v_0$ for all t. We have now defined the function f over the set $T^0 \times I$, where T^0 denotes the set of vertices of T.

Next let e be any edge of T; we shall show how to extend the map f over $\bar{e} \times I$. Let v_1 and v_2 be the vertices incident with e. Then, $T(v_1) \cup T(v_2) \cup e$ is a finite, connected subgraph of T, and, hence, a tree. By the first part of the proof, this finite tree is contractible, hence, simply connected. The set $\bar{e} \times I$ is homeomorphic to a square. The map f is already defined on the two sides $\{v_1\} \times I$ and $\{v_2\} \times I$ of this square. We have no choice as to the definition of f on the other two sides of the square. On the side $\bar{e} \times \{0\}$, we must have

$$f(x, 0) = x,$$

whereas on the side $\bar{e} \times \{1\}$ we must have

$$f(x, 1) = v_0.$$

Thus, we have the four edges of the square $\bar{e} \times I$ mapped into the simply connected space $T(v_1) \cup T(v_2) \cup e$; we now invoke Lemma II.8.1 to conclude that the map f can be extended over the interior of the square.

By this process, we can extend the map $f : T^0 \times I \to T$ to a map $f : T \times I \to I$. It remains to prove that the map thus defined is continuous. This follows from the fact that the map is continuous over $\bar{e} \times I$ for every edge e, and the fact that T has the weak topology (compare Exercise 2.3). Q.E.D.

Remark: In the second part of the proof, the vertex v_0 was chosen arbitrarily; thus, we actually proved a slightly stronger result, namely, *a tree T may be contracted homotopically to any preassigned vertex v_0 in such a way that the image of v_0 remains fixed during the contracting homotopy.* This remark will be used in the proof of Theorem 5.2.

Any graph contains subgraphs which are trees, e.g., a subgraph consisting of a single vertex. The set of all such trees contained in a given graph is partially ordered by inclusion.

Theorem 4.2 *Let X be a graph; then any tree*[1] *contained in X is contained in a maximal tree*[1] *in X.*

PROOF: If X is a finite graph, then there are only a finite number of subgraphs of X and the theorem is obvious. To handle the case where X is an infinite graph, we invoke Zorn's lemma. To do this, we must prove the following assertion: Let $\{T_\lambda : \lambda \in \Lambda\}$ be a family of trees contained in X which is linearly ordered by inclusion. Then, the union

$$\bigcup_{\lambda \in \Lambda} T_\lambda$$

is a subgraph of X which is also a tree. The proof of this assertion is left to the reader.

The truth of the theorem now follows directly from Zorn's lemma.

Q.E.D.

The following theorem gives further insight into the nature of the maximal trees in a graph.

Theorem 4.3 *Let X be a connected graph and let T be a subgraph of X which is a tree. Then, T is a maximal tree if and only if T contains all the vertices of X.*

PROOF: Assume that T is a maximal tree which does not contain all the vertices of X. Because X is connected, we can find an edge path $(e_1, \ldots, e_n)$ in X whose initial vertex is contained in T and whose terminal vertex is not in T. An easy argument shows that one of edges occurring in this edge path, say e_i, must have its initial vertex in T and its final vertex not in T. Hence, $\bar{e}_i$ is not contained in T. However, $T \cup \bar{e}_i$ is a connected subgraph of X which is easily proven to be a tree. This contradicts the maximality of T. Thus, half the theorem is proved.

To prove the other half, assume that T contains all the vertices of X. If e is any edge of X not contained in T, then the vertices of e are con-

[1] In this theorem it is tacitly assumed that we are only considering trees contained in X which are subgraphs of X.

tained in T; hence, $T \cup e$ is a connected subgraph of X. It is readily seen that $T \cup e$ must contain a closed edge path; hence, it is not a tree. Because this argument applies for any edge e not in T, we conclude that T is a maximal tree. Q.E.D.

5 The fundamental group of a graph

We are now ready to prove one of the main theorems of this chapter.

Theorem 5.1 *The fundamental group of any connected graph X is a free group.*

The theorem is obvious if X is a tree, because then the fundamental group is trivial (to avoid special cases, we may as well regard the trivial group as a free group on an empty set of generators). For the case where X is not a tree, we shall prove the following more explicit theorem.

First, note that, although an edge path in a graph is a different kind of object from a path as defined in Chapter II, there is a rather obvious relation between the two concepts: An edge path $(e_1, \ldots, e_n)$ in the graph X joining the vertices v_0 and v_1 determines a unique equivalence class of paths in the topological space X joining the points v_0 and v_1, as follows. For each oriented edge e_i, choose a map $f_i : I \to \bar{e}_i$ such that $f_i \mid (0, 1)$ is a homeomorphism of $(0, 1)$ onto e_i whose inverse belongs to the preferred equivalence class determined by the orientation of e_i. Let α_i denote the equivalence class of the path f_i. Then, the product $\alpha_1\alpha_2 \ldots \alpha_n$ is uniquely determined by the edge path $(e_1, \ldots, e_n)$.

In the remainder of this chapter, it will be convenient to permit ourselves a certain sloppiness of terminology regarding edge paths and their associated equivalence classes of paths. It will usually be clear from the context which is meant, and no confusion will result.

Let X be a connected graph, let v_0 be a vertex of X, and let T be a maximal tree in X containing v_0. Let $\{e_\lambda : \lambda \in \Lambda\}$ denote the set of edges of X not contained in T. Choose a definite orientation for each of the edges e_λ; let a_λ and b_λ denote the initial and terminal vertices of e_λ (of course, it may happen that $a_\lambda = b_\lambda$). To each edge e_λ we associate an element $\alpha_\lambda \in \pi(X, v_0)$ as follows. There is a unique reduced edge path A_λ in T from v_0 to a_λ, and a unique edge path B_λ in T from b_λ to v_0. Then, α_λ is the path class associated with the edge path $(A_\lambda, e_\lambda, B_\lambda)$. If $a_\lambda = v_0$, we omit A_λ; similarly, if $b_\lambda = v_0$, we omit B_λ.

Theorem 5.2 *The fundamental group $\pi(X, v_0)$ is a free group on the set of generators $\{\alpha_\lambda \mid \lambda \in \Lambda\}$.*

PROOF: First, we prove this theorem for the case where the index set Λ contains only one element; i.e., there is only one edge of X not contained in T. We denote this edge by e_1. Because X is not a tree, there exist closed edge paths in X, and it is clear that any such closed edge path must involve the edge e_1. Give the edge e_1 a definite orientation. Then, there must exist reduced closed edge paths in X starting with e_1, i.e., edge paths of the form $(e_1, \ldots, e_n)$. By choosing the "shortest" among all such closed edge paths, we can obtain a *simple* closed edge path; i.e., one in which no vertex or edge occurs twice. Denote this simple closed edge path by $(e_1, \ldots, e_m)$. Let

$$C = \bigcup_{i=1}^{m} \bar{e}_i.$$

Then, C is a subgraph of X homeomorphic to a circle.

Consider the complementary set $X - C$; let $\{Y_i\}$ denote the set of components of $X - C$. Each $\bar{Y}_i$ is a subgraph of T; hence, it is a tree. An easy argument shows that $\bar{Y}_i$ has exactly one vertex in common with C. By the remark following the proof of Theorem 4.3, each of the trees $\bar{Y}_i$ can be contracted to this vertex. This shows C is a deformation retract of X; hence, the inclusion map $C \to X$ induces an isomorphism of fundamental groups. This shows that $\pi(X)$ is an infinite cyclic group. We leave it to the reader to verify that it is generated as stated in Theorem 5.2.

We now use the general form of the Seifert-Van Kampen theorem to prove the general case of Theorem 5.2.

Choose a point $x_\lambda \in e_\lambda$ for each $\lambda \in \Lambda$. The set $\{x_\lambda : \lambda \in \Lambda\}$ is closed and discrete because X has the weak topology. Let U denote the complement of $\{x_\lambda : \lambda \in \Lambda\}$ in X. Then, T is a deformation retract of U; hence, U is contractible. For any index λ, let

$$V_\lambda = U \cup \{x_\lambda\}.$$

Then, $V_\lambda \supset U$ for all λ, and, if $\lambda \neq \mu$,

$$V_\lambda \cap V_\mu = U.$$

Clearly, V_λ has $T \cup e_\lambda$ as a deformation retract; thus, the fundamental group $\pi(V_\lambda, v_0)$ is infinite cyclic and is generated by α_λ.

We can now apply Theorem IV.2.2 to the open covering of X by the sets V_λ and U. We conclude that $\pi(X, v_0)$ is the free product of the groups $\pi(V_\lambda, v_0)$ (see Exercise IV.3.1). Q.E.D.

Exercises

5.1 Let T be a triangle in the plane $\mathbf{R}^2$ with vertices a, b, and c. (a) Prove that the closed segment $\overline{ab}$ is a deformation retract of T. (b) Prove that the union of the two closed segments $\overline{ab}$ and $\overline{bc}$ is a deformation retract of T.

5.2 Let S be a compact, bordered surface with a given triangulation. Prove that there is a finite, connected graph $X \subset S$ such that X is a union of edges and vertices of the triangulation, X is a deformation retract of S, and X has the same Euler characteristic as S. Deduce from this that $\pi(S)$ is a free group on $1 - \chi(S)$ generators. [HINT: Let T be a triangle of the given triangulation such that T has at least one of its edges contained in the boundary. Show that $\overline{S - T}$ is a deformation retract of T. (There are two cases, depending on whether one or two edges of T are contained in the boundary.) Use the result of the previous exercise. By successive repetitions of this process, we can get rid of all the triangles. Let X denote what is left at the end.]

5.3 Let X be a connected graph, let Y be a connected subgraph, and let v be a vertex of Y. Prove that it is possible to choose generators $\{a_i\}$ for $\pi(Y, v)$ and $\{b_j\}$ for $\pi(X, v)$ such that $\pi(X, v)$ and $\pi(Y, v)$ are free groups on the sets $\{b_j\}$ and $\{a_i\}$, respectively, and the homomorphism $i_* : \pi(Y, v) \rightarrow \pi(X, v)$ maps the set of generators $\{a_i\}$ into the set of generators $\{b_j\}$ in a one-to-one fashion. In particular, i_* is a monomorphism. (HINT: Choose a maximal tree in Y and enlarge it to a maximal tree in X.)

5.4 Let S and S' be compact, bordered surfaces such that S' is contained in the interior of S, each connected component of $S - S'$ meets the boundary of S, and there exists a triangulation of S such that S' is a union of certain triangles of this triangulation. Prove that it is possible to choose generators $\{a_i\}$ for $\pi(S', x)$ and $\{b_j\}$ for $\pi(S, x)$ such that $\pi(S', x)$ and $\pi(S, x)$ are freely generated by the sets $\{a_i\}$ and $\{b_j\}$, respectively, and the homomorphism $i_* : \pi(S', x) \rightarrow \pi(S, x)$ maps the set of generators $\{a_i\}$ into the set of generators $\{b_j\}$ in a one-to-one fashion. (HINT: Carry out the process of collapsing triangles described in Exercise 5.2 in such a way that all the triangles of $S - S'$ are collapsed first before those of S'. By this method, we reduce the problem to that of Exercise 5.3.)

Remark: It can be shown that the hypothesis assumed about the existence of a certain triangulation of S is always satisfied automatically.

5.5 Let S be a noncompact surface with a given triangulation. Prove that there exists a sequence $S_1, S_2, S_3, \ldots$ of compact, bordered surfaces contained in S having the following properties: (a) S_n is contained in the interior of S_{n+1} for all n. (b)

$$S = \bigcup_{n=1}^{\infty} S_n.$$

(c) Each component of $S_{n+1} - S_n$ meets the boundary of S_{n+1}. (d) Each surface S_n can be triangulated so that S_{n-1} is a union of triangles of the given triangulation. [HINT: Construct S_n by induction on n. Take S_1 to be a single triangle of the triangulation. Assume inductively that $S_1, \ldots, S_n$ have been constructed satisfying the given conditions. Construct S_{n+1} by adjoining triangles and pieces of triangles to S_n. To satisfy condition (c) at the next stage of the construction, choose S_{n+1} so it satisfies the following condition: The closure of each component of $S - S_{n+1}$ is noncompact. Note that, because there are only a countable number of triangles in any triangulation of S, it is not difficult to satisfy condition (b).]

5.6 Use Exercises 5.4 and 5.5 to prove that the fundamental group of any noncompact surface (which can be triangulated) is a free group on a finite or countable set of generators. (HINT: Observe that in the construction of Exercise 5.5, if C is a compact subset of S, then there exists an integer n such that $C \subset S_n$. Hence, the result of Exercise II.4.11 is applicable.)

5.7 Assume that S is a noncompact, triangulable surface which is simply connected. Prove that, if the construction of Exercise 5.5 is applied to S, then each S_n is homeomorphic to the unit disc E^2. (HINT: Use Exercise 5.6 and the classification theorem for compact, bordered surfaces in Section I.10.)

Remark: It follows from this that S is homeomorphic to the plane $\mathbf{R}^2$. The proof depends on arguments involving the classical Schönfliess theorem.

6 The Euler characteristic of a finite graph

Exactly as in the case of a compact 2-manifold, we define the *Euler characteristic* of a finite graph X [denoted by $\chi(X)$] to be the number of vertices of X minus the number of edges of X. We leave it to the reader to verify that the Euler characteristic is invariant under subdivision.

Proposition 6.1 *If T is a finite tree, then $\chi(T) = +1$.*

PROOF: The proof is by induction on the number of edges in the tree. If there are 0 edges and 1 vertex, or 1 edge and 2 vertices, the result is obvious. To make the inductive step, we use the argument in the first part of the proof of Theorem 4.1. The details may be left to the reader.

Theorem 6.2 *Let X be a finite, connected graph. Then, the fundamental group $\pi(X)$ is a free group on $1 - \chi(X)$ generators.*

PROOF: Let T be a maximal tree in X, and let $e_1, \ldots, e_k$ be the edges not contained in T. Using Proposition 6.1, we see that

$$\chi(X) = 1 - k.$$

On the other hand, by Theorem 5.2, $\pi(X)$ is a free group on k generators. From this the theorem follows.

Corollary 6.3 *If two finite, connected graphs X and Y have the same homotopy type, then $\chi(Y) = \chi(X)$; i.e., the Euler characteristic is a homotopy type invariant.*

The converse of this corollary is also true, although we will not take up the matter.

Corollary 6.4 *If X is a finite, connected graph, and $\chi(X) = 1$, then X is a tree.*

Corollary 6.5 *If X is a finite, connected graph, then $\chi(X) \leqq 1$.*

Note, however, that, if the hypothesis that X be connected is dropped, then $\chi(X)$ may be arbitrarily large.

7 Covering spaces of a graph

As we remarked earlier in this chapter, a connected graph has sufficiently nice local properties so that the entire theory of covering spaces is applicable. We now show that any covering space of a graph is also a graph in a natural way.

Theorem 7.1 *Let X be a connected graph with vertex set X^0, let (Y, p) be a covering space of X, and let $Y^0 = p^{-1}(X^0)$. Then, Y is a graph with vertex set Y^0.*

PROOF: It is clear that Y^0 is a closed, discrete subset of Y. Let e be an edge of X. Then, by Lemma V.2.1, each component of $p^{-1}(e)$ is a covering space of e. But, because e is simply connected, this means each component of $p^{-1}(e)$ is mapped homeomorphically onto e. Also, each component of $p^{-1}(e)$ is open in $p^{-1}(e)$, by local connectivity. Thus, condition (b) of Section 2 holds. Similarly, it is easy to verify condition (c); if $\bar{e}$ is homeomorphic to $[0, 1]$, then each component of $p^{-1}(\bar{e})$ is mapped homeomorphically. If $\bar{e}$ is homeomorphic to S^1, then we apply what we know about the possible covering spaces of a circle. Finally, condition (d) is a direct consequence of Lemma V.12.3, because there is a map $f_i : I \to X$, corresponding to any edge e_i, such that $f_i(I) = \bar{e}_i$ and f_i maps the open interval $(0, 1)$ homeomorphically onto e_i. Q.E.D.

There is a simple way to indicate by a drawing all the relevant information regarding a covering space of a graph. Let X be a connected graph, preferably a finite one. Label the edges $a, b, c, \ldots$, give each edge an orientation, and indicate the chosen orientations by placing a small arrow on each edge. Label the vertices $U, V, W, \ldots$, etc. Let $(\tilde{X}, p)$ be a covering space of X. The mapping p can be indicated in a diagram of $\tilde{X}$ as follows. Label the edges of $p^{-1}(a)$ with the symbols $a_1, a_2, a_3, \ldots,$

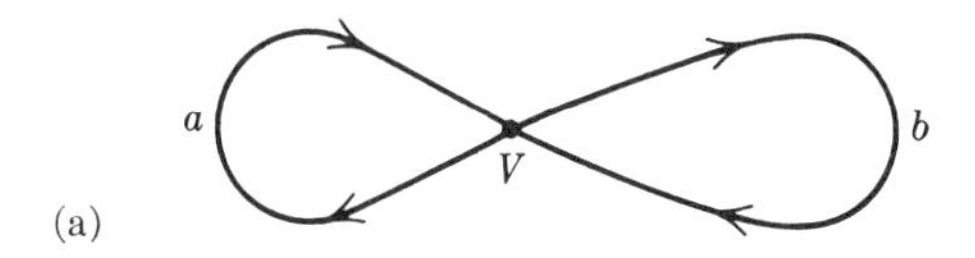

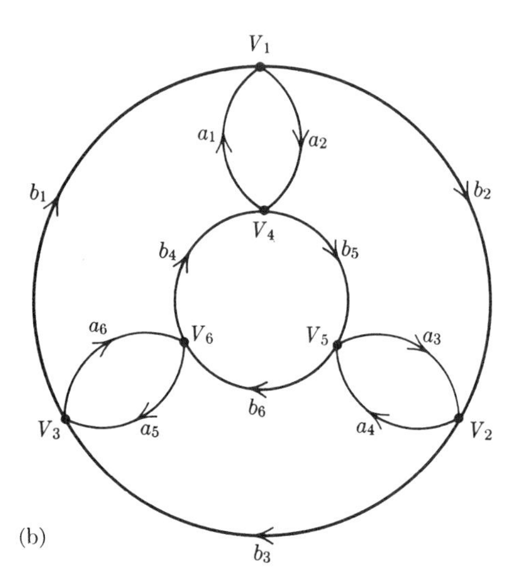

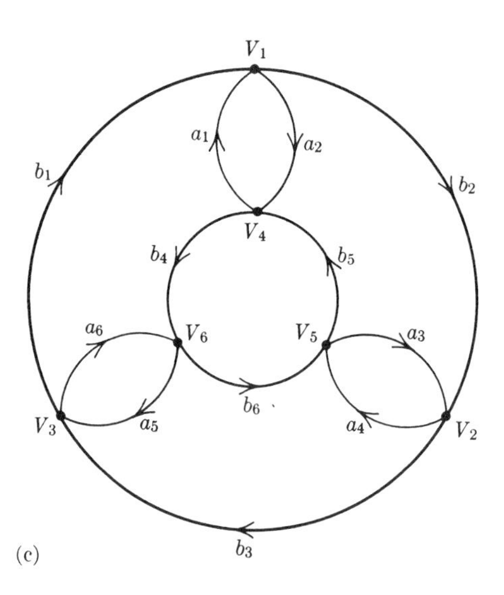

FIGURE 6.1 Example 7.1, two different 6-sheeted regular covering spaces of a graph. (a) The base space. (b) First covering space. (c) Second covering space.

and place a small arrow on each edge a_i indicating an orientation which agrees with that of a under the map p. Do the same thing with the edges of $p^{-1}(b)$, $p^{-1}(c)$, etc. Label the vertices of $p^{-1}(U)$ with the symbols $U_1, U_2, U_3, \ldots$, etc.

Examples

7.1 In Figure 6.1(a) is shown a simple graph labeled as described above, and in Figures 6.1(b) and 6.1(c) are shown two different covering spaces labeled according to the method described. Although both are 6-sheeted regular covering spaces, they are not isomorphic. It can be proved that the automorphism group of the covering space in Figure 6.1(b) is cyclic of order 6, whereas that in Figure 6.1(c) is a non-abelian group of order 6. We suggest that the reader try to discover the various automorphisms in each case. This example shows the importance of the arrows to indicate orientations of the edges. Note also that it is easy to read off from these diagrams the operations of the group $\pi(X, V)$ on the set of vertices $p^{-1}(V)$.

7.2 In Figure 6.2 is shown another covering space of the graph of Figure 6.1(a). The reader should prove that the group of automorphisms of this covering space consists of the identity element alone.

A direct consequence of Theorem 7.1 is the following very important group theoretic result:

Theorem 7.2 *Any subgroup of a free group is a free group.*

PROOF: Let F be a free group on the set S and let F' be a subgroup of F. Construct a connected graph X such that $\pi(X) \approx F$. This may be

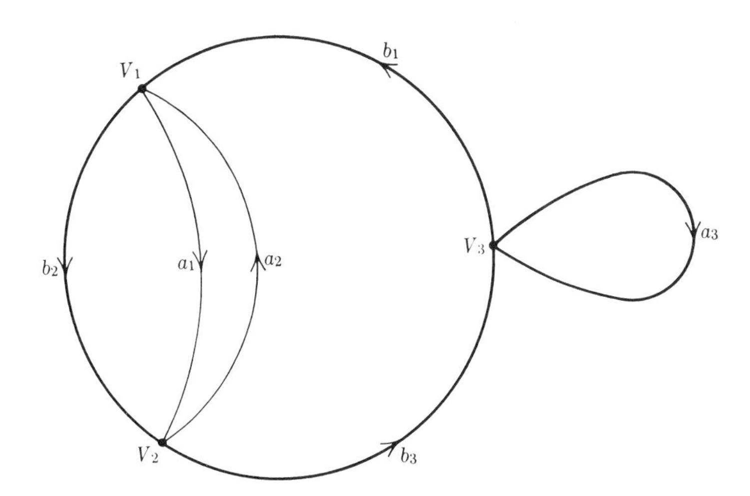

FIGURE 6.2 Example 7.2, a 3-sheeted nonregular covering space of a graph.

done as follows: Let X have one vertex and a set of edges in one-to-one correspondence with the elements of S; i.e., X is the union of a collection of circles, any two of which have the unique vertex of X in common. Topologize X by giving it the weak topology. By the theory of covering spaces, there exists a covering space $(\tilde{X}, p)$ of X such that $\pi(\tilde{X}) \approx F'$. By Theorem 7.1, $\tilde{X}$ is a graph. Hence, F' is a free group. Q.E.D.

Later we shall consider the problem of describing a set of free generators for the subgroup F'. For the time being, we shall concentrate on finitely generated free groups.

Theorem 7.3 *Let F be a free group on k generators, $k \geqq 1$, and let F' be a subgroup of index n. Then, F' is a free group on $n(k - 1) + 1$ generators.*

PROOF: Let X be a finite graph such that $\pi(X) \approx F$. Then,

$$\chi(X) = 1 - k$$

by Theorem 6.2. Let $\tilde{X}$ be a covering space of X corresponding to the subgroup F'. Then, $\tilde{X}$ is an n-sheeted covering of X. Hence,

$$\chi(\tilde{X}) = n \cdot \chi(X) = n - nk.$$

Therefore, F' is a free group on

$$1 - \chi(\tilde{X}) = 1 - n + nk$$

generators. Q.E.D.

By comparison with the theory of free abelian groups, or of finite groups, this result seems paradoxical. If $k > 1$, then as the index increases, the number of generators also increases. On the other hand, for finite groups, the larger the index of a subgroup, the smaller the order of the subgroup. If A is a free abelian group on k generators, then any subgroup of finite index is also a free abelian group of rank k.

8 Generators for a subgroup of free group

In the preceding section we proved that every subgroup of a free group was also a free group by using the fact that any covering space of a graph is also a graph, and that the fundamental group of a graph is a free group. However, our theorem on the fundamental group of a graph actually proved even more: It gives an explicit description of a set of generators for the fundamental group. It seems reasonable to expect that we should be able to use this more explicit result to actually specify generators for a subgroup of a free group. We now show how this can be done.

Because the choice of generators for the fundamental group of a graph involved the choice of a maximal tree, our first task is to show how a maximal tree in the covering space $\tilde{X}$ of a graph X is determined by the choice of certain elements in the fundamental group of X.

Let F be a free group on the set S, and let F' be a subgroup of F. We wish to determine a set S' of generators for F' such that F' is a free group on the set S'. As in the proof of Theorem 7.2, let X be a connected graph having only one vertex v_0 such that $\pi(X, v_0) \approx F$; the edges of X are in one-to-one correspondence with the elements of the set S. Each edge is assumed to have a definitely chosen orientation. Let $(\tilde{X}, p)$ be a covering space of X such that $p_*\pi(\tilde{X}, v_1) = F'$ for some vertex v_1 of X. We now intend to apply Theorem 5.2 to determine generators for the fundamental group $\pi(\tilde{X}, v_1)$. Let T be a maximal tree in $\tilde{X}$. Corresponding to each vertex v in $\tilde{X}$, there is a unique reduced edge path γ_v in T with initial vertex v_1 and terminal vertex v. Let $p_*(\gamma_v)$ denote the edge path in X which is the image of γ_v under the projection p; it is a closed edge path in X, and its associated equivalence class of paths is an element of $\pi(X, v_0)$, which we denote by the same notation, $p_*(\gamma_v)$. We now make the following assertions about this set of closed paths:

(1) The collection of closed edge paths $\{p_*(\gamma_v) : v \text{ a vertex of } \tilde{X}\}$ completely determines the tree T (obvious).

(2) Each coset $F' \cdot \alpha$ of the subgroup F' contains exactly one of the path classes $p_*(\gamma_v)$. (This follows easily from the considerations of Chapter V.)

(3) The set $\{p_*(\gamma_v) : v \text{ a vertex of } \tilde{X}\}$ is a *Schreier system* in the free group $F = \pi(X, v_0)$. By a *Schreier system* in the free group F on a set of generators S is meant a nonempty subset of F which satisfies the following condition. Assume $g \neq 1$ belongs to the Schreier system. Express g as a reduced word in the generators:

$$g = s_1^{n_1} s_2^{n_2} \ldots s_k^{n_k}.$$

Let

$$g' = \begin{cases} g s_k & \text{if } n_k < 0, \\ g s_k^{-1} & \text{if } n_k > 0. \end{cases}$$

Then, we demand that g' also belong to the system (g' may be thought of as the reduced word obtained from g by dropping off the last letter). Note that 1 belongs to every Schreier system.

We leave it to the reader to prove that $\{p_*(\gamma_v) : v \text{ a vertex of } \tilde{X}\}$ is actually a Schreier system in F; the proof is not difficult.

Conversely, if G is any Schreier system in the free group F such that each coset $F' \cdot \alpha$ of the subgroup F' contains at most one element of G, then there exists a unique tree T in $\tilde{X}$ which contains v_1 and has the property that

$$\{p_*(\gamma_v) : v \text{ a vertex of } T\} = G,$$

where γ_v denotes the unique edge path in T from v_1 to v. If every coset $F' \cdot \alpha$ of F' contains an element of G, then this unique tree T is a maximal tree of $\tilde{X}$.

These facts, all of which are easily verified, show that there is a natural one-to-one correspondence between maximal trees in $\tilde{X}$ and Schreier systems G in F such that each coset of F' contains exactly one element of G. We now exploit this one-to-one correspondence to determine a set of generators for F'. In our notation, a set of generators for $\pi(\tilde{X}, v_1)$ is obtained as follows. Let e be an edge of $\tilde{X}$ not contained in T; orient e so that its orientation agrees with that of $p(e)$ under the projection p, and let v and v' denote its initial and terminal vertices. Then, $\gamma_v e(\gamma_{v'})^{-1}$ is a typical generator of $\pi(\tilde{X}, v_1)$, and we obtain a complete set of generators by taking the typical generator corresponding to each edge of $\tilde{X}$ not in T. If we apply the isomorphism p_*, we obtain a set of free generators for the subgroup F'. Now

$$p_*[\gamma_v e(\gamma_{v'})^{-1}] = (p_*\gamma_v)(p_*e)(p_*\gamma_{v'})^{-1}. \qquad (6.8\text{-}1)$$

In this expression, $p_*(\gamma_v)$ and $p_*(\gamma_{v'})$ are both elements of G, and $p_*(e)$ is a generator of F (i.e., an element of the set S). Note that, because the entire expression represents an element of the subgroup F', it follows that $(p_*\gamma_v)(p_*e)$ and $(p_*\gamma_{v'})$ belong to the same coset of F'.

We can put this expression in an algebraically more convenient form, as follows. For any element $a \in F$, let $\Phi(a)$ denote the unique element of G which belongs to the same coset of F'. Then, the right-hand side of (6.8-1) is an expression of the following form:

$$gs[\Phi(gs)]^{-1}, \qquad (6.8\text{-}2)$$

where $g \in G$, and $s \in S$. Thus, we have shown that the subgroup F' of F is a free group generated by certain elements of the form (6.8-2). In general, not every element of the form (6.8-2) is one of the free generators of F' given by formula (6.8-1). We leave it to the reader to verify that there are only two alternatives: For any $g \in G$ and any $s \in S$, either $gs[\Phi(gs)]^{-1} = 1$ or $gs[\Phi(gs)]^{-1}$ is one of the generators given by formula (6.8-1). Thus, we have proved the following theorem.

Theorem 8.1 *Let F be a free group on the set S, let F' be a subgroup of F, and let G be a Schreier system which contains exactly one element from* *each coset of F'. Then, F' is a free group on the following set of generators:*

$$\{gs[\Phi(gs)]^{-1} : g \in G,\ s \in S,\ gs[\Phi(gs)]^{-1} \neq 1\}.$$

(Here the function Φ assigns to each element of F the unique element of G in the same coset of F'.)

Example

8.1 Let F be a free group on a set S consisting of two elements: $S = \{s_1, s_2\}$; let F' be the commutator subgroup of F. In this case the graph X is a figure "8" curve, consisting of two circles with a point in common, and $\tilde{X}$ is a graph which consists of the following two families of parallel lines in the plane: the vertical lines $x = m$ and the horizontal line $y = n$ where m and n range over all integers (see Example 2.4). The projection $p : \tilde{X} \to X$ wraps each horizontal line around the circle s_1 and each vertical line around the circle s_2. For the proof that $(\tilde{X}, p)$ is the covering space of X corresponding to the commutator subgroup of $\pi(X)$, see Example V.11.6.

The quotient group F/F' is a free abelian group on two generators s_1 and s_2. Hence, we can take as coset representatives the elements $s_1^m s_2^n$, where m and n are integers. This set is clearly a Schreier system. The corresponding maximal tree in the graph $\tilde{X}$ consists of the union of the x axis and all the vertical lines $x = m$. The elements of the form $gs[\Phi(gs)]^{-1}$ are the following:

$$(s_1^m s_2^n) s_1 (s_1^{m+1} s_2^n)^{-1}, \tag{6.8-3}$$

$$(s_1^m s_2^n) s_2 (s_1^m s_2^{n+1})^{-1}. \tag{6.8-4}$$

These elements are obtained by taking $g = s_1^m s_2^n$ and $s = s_1$ or s_2. It is readily seen that the element given by formula (6.8-4) is the unit, whereas the element given by formula (6.8-3) is equal to $s_1^m s_2^n s_1 s_2^{-n} s_1^{-m-1}$. Thus, we have proved that *the commutator subgroup of a free group on two generators $\{s_1, s_2\}$ is a free group on the generators $s_1^m s_2^n s_1 s_2^{-n} s_1^{-m-1}$, where (m, n) ranges over all pairs of integers $\neq (0, 0)$.*

We recommend that the reader study this example, because it is readily visualized, and illustrates the general case quite well. In particular, the reader should satisfy himself that the generators we have obtained for the commutator subgroup correspond exactly to the generators of the fundamental group of the graph $\tilde{X}$ obtained by using Theorem 5.2. The reader should also consider the possibility of using other maximal trees in the graph $\tilde{X}$, and determine the corresponding Schreier systems in F.

Exercises

8.1 Determine a set of free generators for the commutator subgroup of a free group on n generators $\{s_1, s_2, \ldots, s_n\}$.

8.2 Let F be a free group on two generators s_1 and s_2. How many subgroups of index 2 has F? (HINT: Every subgroup of index 2 is normal, and the quotient group is cyclic of order 2.) For each subgroup of index 2, determine a set of free generators. Interpret your results by means of graphs.

8.3 Let F be a free group and let N be a nontrivial normal subgroup of infinite index. Prove that N is not finitely generated.

8.4 Let F be a free group on two generators a and b, and let S_3 denote the symmetric group of degree 3 considered as the set of all permutations of the set $\{1, 2, 3\}$. Define an epimorphism $f : F \to S_3$ by the rule $f(a) = (1, 2)$ and $f(b) = (1, 2, 3)$, where $(1, 2)$ and $(1, 2, 3)$ are the usual notations for cyclic permutations. Use Theorem 8.1 to find a set of free generators of the kernel of f. Interpret your results in terms of a covering space of a graph.

8.5 Let F be a free group on two generators a and b, and let H denote a cyclic group of order 6 with generator x. Define an epimorphism $f : F \to H$ by the rules $f(a) = x^3$ and $f(b) = x^2$. Use Theorem 8.1 to find a set of free generators of the kernel of f, and interpret your results in terms of a covering space of a graph. Compare the covering spaces which occur in Exercises 8.4 and 8.5 (both are 6-sheeted regular covering spaces of a figure "8" curve) with those in Example 7.1.

8.6 Let F be a free group on two generators a and b. Define operations of F on the set $\{1, 2, 3\}$ as follows: a is the transposition $(1, 2)$, and b is the cyclic permutation $(1, 2, 3)$ (consider elements of F as operating on the right). Let F' be the isotropy subgroup corresponding to the element 1, i.e., the set of all elements of F which leave 1 fixed. Use Theorem 8.1 to determine a set of free generators for F', and interpret your results in terms of graphs. (HINT: Use the results of Section 2 of Appendix B.) Compare with Example 7.2.

NOTES

Most of the results on free groups in this chapter date back to the 1920's, and they are mainly the work of three men: the German mathematicians K. Reidemeister and O. Schreier, and the Danish mathematician J. Nielsen. Most of the papers of Reidemeister and Schreier were published in the journal *Hamburger Abhandlungen*, and the results were later collected by Reidemeister in a book [7].

Kuratowski's theorem on the imbedding of graphs in the plane

One of the most famous theorems on the topology of graphs was proved by the Polish topologist, K. Kuratowski, in 1930.

In Section 2 we mentioned that every finite graph can be imbedded topologically in Euclidean 3-space. However, not every such graph can be imbedded in the plane. We give two examples:

(a) Let X be a graph with five vertices and ten edges such that there is one edge connecting every pair of distinct vertices.

(b) Let Y be a graph with six vertices, $\{a, b, c, a', b', c'\}$ and nine edges such that there is one edge connecting each primed vertex to each unprimed vertex.

We leave it to the reader to verify that neither X nor Y can be imbedded in the plane (use the Jordan curve theorem). Kuratowski's theorem asserts that any finite graph which cannot be imbedded in the plane contains a subgraph homeomorphic to X or Y. For the proof, see Berge [1], Chapter 21, or a paper by G. A. Dirac and S. Schuster entitled "A Theorem of Kuratowski" (*Proc. Kon. Nederl Akad. Weten.* (ser. A.), *57,* 1954, pp. 343–348). This paper also briefly considers the problem of imbedding countable graphs in the plane. Alternative conditions for a finite graph to be imbeddable in the plane are considered in a paper by S. Mac Lane ("A Structural Characterization of Planar Combinatorial Graphs," *Duke Math. J.*, *3,* 1937, pp. 460–472). Kuratowski's theorem was generalized to a wider class of compact metric spaces by S. Claytor ("Peanian Continua Not Imbeddable in a Spherical Surface," *Ann. Math.*, *38,* 1937, pp. 631–646).

The weak topology for a graph

Our definition of a graph in Section 2 is chosen so that a graph will be a 1-dimensional CW-complex in the sense of J. H. C. Whitehead ("Combinatorial Homotopy, I." *Bull. Amer. Math. Soc.*, *55,* 1949, pp. 213–246). CW-complexes are discussed in greater detail in Chapter VII.

REFERENCES

Graph theory

1. Berge, C. *The Theory of Graphs and Its Applications.* Trans. by A. Doig. New York: Wiley, 1962.
2. Ore, O. *Theory of Graphs* (Colloquium Publications, Vol. XXXVIII). Providence: American Mathematical Society, 1962.

Topology

3. Hilton, P. J., and S. Wylie. *Homology Theory, An Introduction to Algebraic Topology.* Cambridge: The University Press, 1960. Chapter 6.
4. Seifert, H., and W. Threlfall. *A Textbook of Topology.* New York: Academic Press, 1980. Chapters 7 and 8.

Group theory

5. Hall, M. *The Theory of Groups.* New York: Macmillan, 1959. Chapter 7.
6. Kurosh, A. G. *The Theory of Groups.* 2 vols. Trans. and ed. by K. A. Hirsch. New York: Chelsea, 1955–56. Chapter IX.
7. Reidemeister, K. *Einführung in die kombinatorische Topologie.* Braunschweig: Friedr. Vieweg & Sohn, 1932. Chapters 3 and 4.
8. Rotman, J. J. *The Theory of Groups.* Boston: Allyn and Bacon, 1965. Chapter 11.

9. Schenkman, E. *Group Theory*. Princeton, N.J.: Van Nostrand, 1965. Chapter V.
10. Scott, W. R. *Group Theory*. Englewood Cliffs, N.J.: Prentice Hall, 1964. Chapter 8.
11. Specht, W. *Gruppentheorie* (Die Grundlehren der Mathematischen Wissenschaften, Band LXXXII). Berlin-Göttingen-Heidelberg: Springer-Verlag, 1956. Chapter 2.1.

CHAPTER SEVEN

The Fundamental Group of Higher-Dimensional Spaces

1 Introduction

In Chapter IV we learned how to determine the structure of the fundamental group of a compact, connected 2-manifold. Let us now look at these results in the light of the preceding chapter.

Let M be a compact, connected 2-manifold. We may look on M as obtained by identifying the edges of a certain polygon in pairs. Let X denote the subset of M which is the image of these edges under the identification. Then, X is the union of a certain number of circles, any two of which have a single point in common; thus, X may be looked on as a finite graph having a single vertex. Also, we may consider that M is constructed by "pasting" a disc onto the graph X. If we re-examine the results of Chapter IV, we see that the inclusion map of X into M induces an epimorphism $\pi(X) \to \pi(M)$, and the kernel is generated by the element of $\pi(X)$ which is "killed off" by pasting on the disc.

In this chapter, we will generalize these results as follows. Starting with an arbitrary connected graph X, we can form a new space Y by pasting any number of discs onto X in any way whatsoever. We will prove that the inclusion map of X into Y induces an epimorphism $\pi(X) \to \pi(Y)$, and the kernel is generated by precisely the elements of $\pi(X)$ which are "killed off" by pasting on the various discs. Thus, $\pi(Y)$ has a presentation with one generator for each generator of $\pi(X)$, and one relation for each attached disc.

The space Y can be enlarged further by pasting on higher dimensional balls or discs; we will show that this has no effect on the fundamental group. In a sense, this is a negative result. It shows that the fundamental group is strictly a "low-dimensional" topological invariant. It points up the need for analogous higher dimensional topological invariants.

This process leads to the important general notion of a CW-complex, due to J. H. C. Whitehead. A CW-complex is a space constructed by starting with a graph and pasting on cells of successively higher dimensions. Experience has proved that, for the purposes of algebraic topology, the category of CW-complexes is one of the most useful categories of topological spaces. CW-complexes have sufficiently nice local properties so that most pathological situations are avoided, yet they are sufficiently general to include most spaces we need to consider in the subject.

As a biproduct of these considerations, we will show that any group is isomorphic to the fundamental group of some reasonable space, in particular a space obtained by pasting a certain number of discs onto a graph (i.e., a 2-dimensional CW-complex). Finally, in Sections 5 and 6, these ideas are applied to give topological proofs of two well-known theorems of group theory.

2 Adjunction of 2-cells to a space

In this section, we assume X^* is a Hausdorff space obtained from the arcwise-connected subspace X by the adjunction or "pasting on" of a collection of ordinary 2-dimensional discs. To be precise, we assume there exists a closed subset $X \subset X^*$ such that $X^* - X$ is the disjoint union of open subsets e_λ^2, $\lambda \in \Lambda$, each of which is homeomorphic to the open disc U^2 in the plane $\mathbf{R}^2$. Each subspace e_λ^2 is called an "open 2-cell." Moreover, we assume that, corresponding to each cell e_λ^2, there exists a so-called "characteristic map," i.e., a continuous map

$$f_\lambda : E^2 \to \bar{e}_\lambda^2,$$

where $E^2 = \{x \in \mathbf{R}^2 : |x| \leqq 1\}$ as usual, f_λ is assumed to map the open disc U^2 homeomorphically onto e_λ^2, and $f_\lambda(S^1) \subset X$. Finally, we assume X^* has the "weak" topology: A subset A of X^* is closed if and only if $A \cap X$ and $f_\lambda^{-1}(A)$ are closed for all $\lambda \in \Lambda$.

As mentioned in the introduction, any compact, connected 2-manifold can be considered as obtained by the adjunction of a single 2-cell to a certain graph X. Of course, when X^* is obtained from X by the adjunction of a finite number of 2-cells, as in the case of a compact 2-manifold, it automatically has the weak topology. The condition that X^* should have the weak topology is only of interest in the case where there are an infinite number of 2-cells.

Our main object is to determine the relation between the fundamental group of X and that of X^*. For this purpose, we introduce the following notation. Choose a fixed, closed path $\varphi : I \to S^1$ which represents a generator of the fundamental group $\pi(S^1, (1, 0))$; i.e., the path φ goes around the circle S^1 exactly once. For each disc e_λ, let α_λ denote the closed path represented by the composed map $f_\lambda\varphi$. Choose a base point $x_0 \in X$; and, for each disc e_λ, a path class β_λ in X with x_0 as initial point and $f_\lambda(1, 0)$ as terminal point. Then,

$$\gamma_\lambda = \beta_\lambda \alpha_\lambda \beta_\lambda^{-1}$$

is an element of $\pi(X, x_0)$.

Theorem 2.1 *Under the above hypotheses, the homomorphism* $\pi(X, x_0) \to \pi(X^*, x_0)$ *is an epimorphism, and the kernel is the smallest normal subgroup containing the set* $\{\gamma_\lambda : \lambda \in \Lambda\}$.

PROOF: The proof is divided into three cases. In case 1, it is assumed that the set Λ consists of one element; i.e., X^* is obtained from X by the adjunction of a single cell, which, for simplicity, we denote by e^2:

$$X^* = X \cup e^2.$$

Choose a point $y \in e^2$, let $U = X^* - \{y\}$, $V = e^2$. Then, U and V are open sets, and V is contractible. Also, X is a deformation retract of U. Thus, we can apply Theorem IV.4.1 to prove the theorem in this case. The details are similar to the determination of the structure of the fundamental group of a compact surface in Section IV.5, and are left to the reader.

In case 2, it is assumed that Λ is a finite set. We can now consider that the cells $\{e_\lambda^2 : \lambda \in \Lambda\}$ are adjoined to X in succession, one after another, instead of all at once, and prove the theorem by induction on the number of elements in the set Λ. In the induction process it is necessary to use case 1.

Case 3 is the remaining case: Λ is assumed to be an infinite set. Choose a point $y_\lambda \in e_\lambda^2$ for each $\lambda \in \Lambda$. Then, $A = \{y_\lambda : \lambda \in \Lambda\}$ is a closed, discrete subset of X^*, because X^* has the weak topology. For each subset $S \subset A$ such that $A - S$ is finite or empty, let U_S denote the complement of S in X^*; i.e.,

$$U_S = X^* - S.$$

Then, U_S is an open arcwise-connected subset of X^*, and

$$\{U_S : A - S \text{ is a finite subset of } A\}$$

is an open covering of X^* which satisfies the hypothesis of the generalized Seifert-Van Kampen theorem (Theorem IV.2.2), because $U_S \cap U_T = U_{S \cup T}$. Note that, if $A - S = \{x_{\lambda_1}, x_{\lambda_2}, \ldots, x_{\lambda_k}\}$, then $U_S = X^* - S$ has as deformation retract the subspace $X \cup e_{\lambda_1}^2 \cup e_{\lambda_2}^2 \cup \cdots \cup e_{\lambda_k}^2$, which is covered by case 2. Thus, it remains to prove that the application of Theorem IV.2.2 in this situation gives the desired conclusion. This we leave to the reader. Q.E.D.

Corollary 2.2 *Given any group G, there exists an arcwise-connected space Y such that $\pi(Y)$ is isomorphic to G. If G has a presentation with a finite number of generators and relations, then we may require Y to be compact.*

PROOF: Choose a presentation of G; i.e., represent G as a quotient group of a free group F. Choose a connected graph X such that $\pi(X) \approx F$, as in Chapter VI. Adjoin 2-cells to X so as to obtain a space Y, as described in the preceding theorem. Adjoin the cells so as to "kill off" all the relations and thus obtain $\pi(Y) \approx G$. Q.E.D.

We will see in Section 4 that Y is a CW-complex, and hence has many nice properties.

3 Adjunction of higher-dimensional cells to a space

In this section, we assume that X^* is obtained from the space X by the adjunction of cells of dimension >2. The precise hypotheses are analogous to those of the preceding section: There exists an arcwise-connected, closed subspace $X \subset X^*$ such that $X^* - X$ is the disjoint union of open subsets e_λ^n, $\lambda \in \Lambda$, each of which is homeomorphic to the open n-dimensional disc U^n in $\mathbf{R}^n$. Each subspace e_λ^n is called an "open n-cell." It is also assumed that, corresponding to each n-cell e_λ^n, there exists a characteristic map

$$f_\lambda : E^n \to \bar{e}_\lambda^n$$

which is continuous, maps U^n homeomorphically onto e_λ^n, and maps S^{n-1} into X (here $E^n = \{x \in \mathbf{R}^n : |x| \leqq 1\}$). If the number of cells e_λ^n is infinite, it is assumed that X^* has the weak topology.

Theorem 3.1 *If $n > 2$, then the inclusion map of X into X^* induces an isomorphism of $\pi(X)$ onto $\pi(X^*)$.*

The proof of this theorem is analogous to that of Theorem 2.1, only now the details are easier, because all the homomorphisms involved are isomorphisms rather than epimorphisms. The reason for the difference between this theorem and Theorem 2.1 can be summarized as follows: The complement of a point in U^n has the homotopy type of an $(n-1)$-sphere, S^{n-1}. For $n > 2$, S^{n-1} is simply connected, but, for $n = 2$, its fundamental group is infinite cyclic.

4 CW-complexes

The process of adjoining cells to a space as described in Sections 2 and 3 leads naturally to the important notion of a CW-complex. Roughly speaking, a CW-complex is a space X which can be built up as follows: Start with a graph X^1 (which need not be connected) and adjoin a collection of 2-cells as described in Section 2 to obtain a space X^2. Next, adjoin a collection of 3-cells as described in Section 3 to obtain a space X^3, and so on. Then

$$X = \bigcup_{n=1}^{\infty} X^n$$

is a CW-complex.

A more precise description goes as follows. A structure of CW-complex is defined on a Hausdorff space X by the prescription of an ascending sequence

$$X^0 \subset X^1 \subset X^2 \subset \cdots$$

of closed subspaces of X which satisfy the following conditions:

(a) X^0 is a discrete space.

(b) For $n > 0$, X^n is obtained from X^{n-1} by the adjunction of a collection of n-cells as described in Section 3.

(c) $$X = \bigcup_{n=0}^{\infty} X^n.$$

(d) The space X and each of the subspaces X^n have the weak topology; i.e., a subset A of X (or X^n) is closed if and only if $A \cap \bar{e}^q$ is closed for each q-cell e^q.

The subset X^n is called the *n-skeleton* of X.

It should be emphasized that, for some values of n, there need not be any n-cells, i.e., $X^n = X^{n-1}$. It is also possible that $X = X^n$; i.e., there are no cells of dimension $> n$. If this is the case, X is called *finite dimensional*. If X is a finite-dimensional CW-complex, and n is the least integer such that $X = X^n$, then X is said to be *n-dimensional*. A CW-complex need not be connected.

In general, a Hausdorff space need not admit a structure of CW-complex. For example, a space which is not locally connected or not normal cannot be a CW-complex. Usually, if a space admits one structure of CW-complex, it will admit a great many different such structures.

A subspace Y of a CW-complex X is called a *subcomplex* if Y is a union of cells of X, and if for any q-cell e^q, if $e^q \subset Y$ then $\bar{e}^q \subset Y$. If this is the case, we define the n-skeleton Y^n by

$$Y^n = X \cap Y^n.$$

It can be shown that Y is also a CW-complex, and that it is a closed subset of X. An example of a subcomplex is the n-skeleton X^n for any integer n. The union or intersection of any collection of subcomplexes is a subcomplex.

Examples

4.1 Our first example is infinite-dimensional, real projective space. Define n-dimensional, real projective space, P_n, to be the quotient space of the sphere S^n obtained by identifying any two diametrically opposite points, x and $-x$. If we regard $\mathbf{R}^n$ as the subset of $\mathbf{R}^{n+1}$ consisting of all points $(x_1, \ldots, x_{n+1})$ such that $x_{n+1} = 0$, then S^{n-1} is a subset of S^n. This leads to a natural imbedding of quotient spaces, $P_{n-1} \subset P_n$, for all $n > 0$. Define infinite-dimensional real projective space P by

$$P = \bigcup_{n=0}^{\infty} P_n$$

with the weak topology (a subset A of P is closed if and only if $P_n \cap A$ is closed for all n). We leave it to the reader to verify that P_n is obtained from P_{n-1} by the adjunction of a single n-cell in the sense described in Section 3 (the method of obtaining P_2 from P_1 by adjunction of a disc is typical). Thus, P has been given the structure of a CW-complex, with one cell in each dimension, and the n-skeleton of P is n-dimensional, real projective space, P_n.

4.2 A triangulation of a 2-dimensional manifold (with or without boundary) gives it a structure of 2-dimensional CW-complex in an obvious way. More generally, a subdivision of a 2-manifold into polygons other than triangles, as described in Section I.8 also gives it a structure of a 2-dimensional CW-complex. Finally, the usual representation of a compact surface, as obtained by the identification of paired sides of a polygon, gives rise to a CW-complex with a single 0-cell or vertex, and a single 2-cell. Analogous considerations apply to higher dimensional manifolds which can be triangulated.

In this brief introduction to the subject we cannot give many details about CW-complexes. The interested reader can find such details in the original paper of J. H. C. Whitehead on the subject [4], or the books of Hilton [1] and S. T. Hu [2]. In these references it is proved that a CW-complex has the following properties:

(a) A CW-complex is a paracompact Hausdorff space, and hence it is normal.

(b) A CW-complex is locally contractible. This means that, given any point x and any neighborhood U of x, there exists a neighborhood V of x such that $V \subset U$ and V is contractible. This condition implies local arcwise connectivity, semilocal simple connectivity, and much more; it is a strong condition. In particular, the entire theory of covering spaces is applicable.

(c) Any covering space of a CW-complex is again a CW-complex.

(d) Any compact subset of a CW-complex meets only a finite number of cells and is contained in a finite subcomplex.

In the preceding paragraphs we have described a CW-complex as a space built up in a certain way. The usual definition is somewhat different and refers, instead, to the decomposition of X into open n-cells, e_λ^n, for $n = 0, 1, 2, \ldots$, with certain conditions imposed on this decomposition. It is then proved as a theorem that any CW-complex can be built up by the method we have described.

The results of Chapter VI and of Sections 2 and 3 are directly applicable to the fundamental group of a CW-complex. In particular, we have the following:

Theorem 4.1 *Let X be a connected* CW-*complex. The inclusion map of the* 2-*skeleton X^2 into X induces an isomorphism of $\pi(X^2)$ onto $\pi(X)$.*

This is a consequence of Theorem 3.1 and property (d) above; see Exercise II.4.11.

The 1-skeleton X^1 of a CW-complex is a graph; hence, the results of Chapter VI are applicable. Theorem 2.1 is directly applicable to determine the relation between the groups $\pi(X^1)$ and $\pi(X^2)$.

The following lemma about the fundamental group of certain CW-complexes will be needed in the following two sections.

Lemma 4.2 *Let X be a* CW-*complex which is a union of a collection of connected subcomplexes, $\{A_\lambda : \lambda \in \Lambda\}$. Assume that there exists a nonempty tree T which is a subcomplex of the* 1-*skeleton X^1 such that, for any two distinct indices λ and μ, $A_\lambda \cap A_\mu = T$. Then, for any vertex $v \in T$, the fundamental group $\pi(X, v)$ is the free product of the groups $\pi(A_\lambda, v)$ with respect to the homomorphism $\varphi_\lambda : \pi(A_\lambda, v) \to \pi(X, v)$ induced by the inclusion maps.*

PROOF: First, consider the case where X and, hence, all the A_λ's are 1-dimensional complexes. If we use the results of Section VI.5, the lemma is practically obvious in this case; it is only necessary to choose a maximal tree in X which contains the given tree T and to apply Theorem VI.5.2 to X and each A_λ.

Next, consider the case where X is 2-dimensional. We must prove that, given any group H and any collection of homomorphisms $\psi_\lambda : \pi(A_\lambda) \to H$, there exists a unique homomorphism $\sigma : \pi(X) \to H$ such that

$\sigma\varphi_\lambda = \psi_\lambda$ for all λ. Let A_λ^1 and X^1 denote the 1-skeletons of A_λ and X, and let $j_\lambda : \pi(A_\lambda^1) \to \pi(A)$ and $j : \pi(X^1) \to \pi(X)$ denote the homomorphisms induced by the inclusion maps. Then, for each index λ, we have the following commutative diagram:

$$\begin{array}{ccccc} \pi(A_\lambda^1) & \xrightarrow{j_\lambda} & \pi(A_\lambda) & \xrightarrow{\psi_\lambda} & H \\ \downarrow{\scriptstyle \varphi_\lambda^1} & & \downarrow{\scriptstyle \varphi_\lambda} & & \\ \pi(X^1) & \xrightarrow{j} & \pi(X) & & \end{array}$$

By the preceding paragraph of the proof, there exists a unique homomorphism $\sigma' : \pi(X^1) \to H$ such that

$$\psi_\lambda j_\lambda = \sigma' \varphi_\lambda^1 \tag{7.4-1}$$

for all $\lambda \in \Lambda$. By Theorem 2.1, the homomorphism j (respectively, j_λ) is an epimorphism, and the generators of the kernel are in one-to-one correspondence with the 2-cells of X (respectively, A_λ). Let e_i^2 be any 2-cell of X and let γ_i be the corresponding generator of the kernel of j. Choose an index λ such that $e_i^2 \subset A_\lambda$; then γ_i is also a generator of the kernel of j_λ. From Equation (7.4-1) and the fact that φ_λ^1 is a monomorphism it follows that $\sigma'(\gamma_i) = 0$. Because this is true for every 2-cell e_i^2, it follows that there exists a unique homomorphism $\sigma : \pi(X) \to H$ such that $\sigma' = \sigma j$. It is readily verified that σ has the required properties.

Finally, we consider the general case. We now consider the 2-dimensional skeletons of X and the A_λ and use Theorem 4.1. The details are similar to what we have just done, only somewhat simpler. We leave them to the reader. Q.E.D.

Presumably this lemma could also be proved by constructing open neighborhoods U_λ of the subcomplexes A_λ such that each A_λ is a deformation retract of U_λ, and then applying Lemma IV.3.2. However, the method of proof we have actually followed seems easier.

5 The Kurosh subgroup theorem

Assume that G is a free product of a family of groups:

$$G = \prod_{\lambda \in \Lambda}{}^* G_\lambda.$$

In the exercises to Section III.4 it was pointed out that if, for each $\lambda \in \Lambda$, we choose a subgroup G'_λ of G_λ, then the free product

$$G' = \prod_{\lambda \in \Lambda}{}^* G'_\lambda$$

can be regarded as a subgroup of G. It is natural to ask whether every subgroup of G is of this type. Simple examples show that the answer to this question is definitely *no*. However, the following well-known theorem of Kurosh shows that the answer comes close to being yes.

Theorem 5.1 *Let H be a subgroup of the free product $G = \Pi^* G_\lambda$. Then, H is itself a free product,*

$$H = F^* \left(\prod_\nu^* H_\nu\right),$$

where F is a free group and each H_ν is conjugate in G to a subgroup of one of the free factors G_λ.

PROOF: We will give a topological proof making use of the results and methods of Chapters V, VI, and VII. For each $\lambda \in \Lambda$, let X_λ be a 2-dimensional CW-complex with a single vertex v_λ such that

$$\pi(X_\lambda, v_\lambda) = G_\lambda.$$

Let v_0 be a point not belonging to any of the spaces X_λ; join the vertex v_0 to the vertex v_λ by an edge e_λ for each $\lambda \in \Lambda$. Let X denote the union of all the spaces X_λ, all the edges e_λ, and the vertex v_0; give X the weak topology. Then, X is a connected, 2-dimensional CW-complex and $\pi(X, v_0)$ can be identified with the free product G (see Lemma 4.2). Let $(\tilde{X}, p)$ be a covering space of X corresponding to the subgroup H. As mentioned in Section 4, the space $\tilde{X}$ is a CW-complex in a natural way. Choose a vertex $\tilde{v}_0 \in p^{-1}(v_0)$ such that

$$p_*\pi(\tilde{X}, \tilde{v}_0) = H.$$

The proof will be completed by using Lemma 4.2 to show that $\pi(\tilde{X}, \tilde{v}_0)$ is a free product of certain subgroups. To carry out the proof in the required generality a rather elaborate notation is needed. The reader should not let this elaborate notation obscure the essential simplicity of the proof. Perhaps the reader can get a better insight into the details that follow by taking a relatively simple special case; e.g., $\Lambda = \{1, 2, 3\}$, each G_λ is a free abelian group on 2 generators, each X_λ is a torus, H is a subgroup of finite index in G, and $(\tilde{X}, p)$ is a covering space with a finite number of sheets.

For each $\lambda \in \Lambda$, let

$$\{\tilde{X}_{\lambda\mu} : \mu \in M_\lambda\}$$

denote the set of components of $p^{-1}(X_\lambda)$. By Lemma V.2.1, $(\tilde{X}_{\lambda\mu}, p \mid \tilde{X}_{\lambda\mu})$ is a covering space of X_λ. Each $\tilde{X}_{\lambda\mu}$ is a 2-dimensional CW-complex; choose a maximal tree $T_{\lambda\mu}$ in the 1-skeleton of $\tilde{X}_{\lambda\mu}$. Let Y denote the following connected graph contained in the 1-skeleton of X : Y shall be the union of all the trees $T_{\lambda\mu}$ together with all the edges $p^{-1}(\bar{e}_\lambda)$ for all

$\lambda \in \Lambda$. Let T be a maximal tree in Y such that T contains each of the trees $T_{\lambda\mu}$; it is not difficult to prove that such maximal trees exist.[1]

We are now ready to apply Lemma 4.2 to determine the structure of the fundamental group $\pi(\tilde{X}, \tilde{v}_0)$. To do this, we consider the covering of the space $\tilde{X}$ by the following subcomplexes: Y, and $\tilde{X}_{\lambda\mu} \cup T$ for all pairs (λ, μ). Each of these subcomplexes is connected, each contains the vertex $\tilde{v}_0$, and the intersection of any two of them is the tree T. Hence, by Lemma 4.2, $\pi(\tilde{X}, \tilde{v}_0)$ is the free product of the groups $\pi(Y, v_0)$ and $\pi(\tilde{X}_{\lambda\mu} \cup T, \tilde{v}_0)$. $\pi(Y)$ is a free group by Theorem VI.5.1, and $\tilde{X}_{\lambda\mu}$ is a deformation retract of $\tilde{X}_{\lambda\mu} \cup T$ (see Theorem VI.4.1). Hence, $\pi(\tilde{X}_{\lambda\mu} \cup T) \approx \pi(\tilde{X}_{\lambda\mu})$. Under the monomorphism

$$p_* : \pi(\tilde{X}, \tilde{v}_0) \to \pi(X, v_0),$$

it is clear that $\pi(\tilde{X}_{\lambda\mu} \cup T, \tilde{v}_0)$ maps onto a conjugate of a subgroup of $\pi(X_\lambda \cup \bar{e}_\lambda, v_0) = G_\lambda$; which conjugate depends on the choice of the maximal tree T. Q.E.D.

For many purposes, it is necessary to have a more precise description of the free factors H_ν in the statement of the above theorem, and to have some idea of the extent to which they are uniquely determined by the subgroup H. To state such a refinement of Theorem 5.1, it is convenient to use the notion of a *double coset*. Recall that, for any $g \in G$, the set

$$HgG_\lambda = \{hgx : h \in H,\ x \in G_\lambda\}$$

is called a *double coset* of the subgroups H and G_λ, and any two such double cosets of H and G_λ are either disjoint or identical, just as for ordinary cosets (see M. Hall [8], p. 14, or Kurosh [9], Vol. I, p. 63).

Theorem 5.2 *Assume the hypotheses of Theorem 5.1. Then, for each index $\lambda \in \Lambda$, there exists a set of representatives*

$$\{\beta_{\lambda\mu} : \mu \in M_\lambda\},$$

one from each double coset of H and G_λ, such that

$$H = F * [\prod_{\lambda\in\Lambda}{}^{*} \prod_{\mu\in M_\lambda}{}^{*} (H \cap \beta_{\lambda\mu} G_\lambda \beta_{\lambda\mu}^{-1})],$$

where F is a free group as in Theorem 5.1.

This is clearly a more precise form of Theorem 5.2, because it gives some insight into the number of factors H_ν in Theorem 5.1, and how these factors are determined. Note that, if α and β are both elements of the

[1] One way to do this is to form a new graph Y' from Y by shrinking each of the trees $T_{\lambda\mu}$ to a vertex $v_{\lambda\mu}$; let $q : Y \to Y'$ denote the natural map. Choose a maximal tree $T' \subset Y'$, and then let $T = q^{-1}(T')$.

same double coset of H and G_λ, then the subgroups $H \cap \alpha G_\lambda \alpha^{-1}$ and $H \cap \beta G_\lambda \beta^{-1}$ are conjugate in H. For this reason, it is plausible that the $\beta_{\lambda\mu}$ should be representatives of the various double cosets.

It should be emphasized that some or all of the subgroups $H \cap \beta_{\lambda\mu} G_\lambda \beta_{\lambda\mu}^{-1}$ in Theorem 5.2 may consist of the identity element alone, even though H is a proper subgroup of G. Examples of this are given in the exercises below.

PROOF: The proof consists in looking more closely at the proof of Theorem 5.1 and reasoning more carefully. We keep the same notation as in that proof and elaborate on it somewhat.

For each index λ, let $Y_\lambda = X_\lambda \cup \bar{e}_\lambda$; then, Y_λ is a subcomplex of X which contains the base point v_0. Let $\{\tilde{Y}_{\lambda\mu} : \mu \in M_\lambda\}$ denote the set of components of $p^{-1}(Y_\lambda)$, indexed so that $\tilde{Y}_{\lambda\mu} \supset \tilde{X}_{\lambda\mu}$ for all μ. Each subcomplex $\tilde{Y}_{\lambda\mu}$ may be thought of as obtained from $\tilde{X}_{\lambda\mu}$ by adjoining a number of "whiskers"; the number of whiskers will be equal to the number of sheets in the covering space $(\tilde{X}_{\lambda\mu}, p \mid \tilde{X}_{\lambda\mu})$ of X_λ. In each subcomplex $\tilde{Y}_{\lambda\mu}$ choose a vertex $\tilde{v}_{\lambda\mu}$ such that $p(\tilde{v}_{\lambda\mu}) = v_0$; for each λ, there will exist exactly one index $\mu \in M_\lambda$ such that it is possible to choose $\tilde{v}_{\lambda\mu} = \tilde{v}_0$, but we shall not insist on such a choice. Although the $\tilde{v}_{\lambda\mu}$ having a fixed first index λ are all distinct, they need not be distinct for distinct first indices λ_1 and λ_2.

The proof of the theorem is carried out by means of the commutative diagram of groups and homomorphisms shown in Figure 7.1. Actually, there is such a diagram for each pair of indices (λ, μ). In this diagram,

$$\begin{array}{ccccc}
 & \nearrow^{j_{\lambda\mu}} & \pi(\tilde{Y}_{\lambda\mu} \cup T, \tilde{v}_{\lambda\mu}) & \xrightarrow{u_{\lambda\mu}} & \pi(\tilde{Y}_{\lambda\mu} \cup T, \tilde{v}_0) \\
 & & \downarrow \varphi_{\lambda\mu} & & \downarrow \varphi_{\lambda\mu} \\
\pi(\tilde{Y}_{\lambda\mu}, \tilde{v}_{\lambda\mu}) & \xrightarrow{i_{\lambda\mu}} & \pi(\tilde{X}, \tilde{v}_{\lambda\mu}) & \xrightarrow{u_{\lambda\mu}} & \pi(\tilde{X}, \tilde{v}_0) \\
\downarrow p_{\lambda\mu *} & & \downarrow p_* & & \downarrow p_* \\
\pi(Y_\lambda, v_0) & \xrightarrow{i_\lambda} & \pi(X, v_0) & \xrightarrow{w_{\lambda\mu}} & \pi(X, v_0)
\end{array}$$

FIGURE 7.1 Commutative diagram for the proof of Theorem 5.2.

$p_{\lambda\mu} = p \mid \tilde{Y}_{\lambda\mu}$, and the homomorphisms i_λ, $i_{\lambda\mu}$, $j_{\lambda\mu}$ and $\varphi_{\lambda\mu}$ are all induced by inclusion maps. For each vertex $\tilde{v}_{\lambda\mu}$, let $\alpha_{\lambda\mu}$ denote the unique path class in the tree T with initial vertex $\tilde{v}_0$ and terminal vertex $\tilde{v}_{\lambda\mu}$, and let

$$\beta_{\lambda\mu} = p^*(\alpha_{\lambda\mu}) \in \pi(X, v_0).$$

The isomorphisms $u_{\lambda\mu}$ and $w_{\lambda\mu}$ are defined by

$$u_{\lambda\mu}(x) = \alpha_{\lambda\mu} x \alpha_{\lambda\nu}^{-1},$$

$$w_{\lambda\mu}(y) = \beta_{\lambda\mu} y \beta_{\lambda\mu}^{-1}.$$

Note that $j_{\lambda\mu}$ is an isomorphism, $w_{\lambda\mu}$ is an inner automorphism, and that all homomorphisms in this diagram are monomorphisms.

By construction, we have

$$\pi(X, v_0) = G, \qquad p_*\pi(\tilde{X}, \tilde{v}_0) = H, \qquad i_\lambda\pi(Y_\lambda, v_0) = G_\lambda.$$

Let $p_*\varphi_{\lambda\mu}\pi(\tilde{Y}_{\lambda\mu} \cup T, v_0) = H_{\lambda\mu} \subset H$. Then, as shown in the proof of Theorem 5.1, H is the free product of F and all the groups $H_{\lambda\mu}$ (which were denoted by H_ν).

Next, we apply Proposition V.11.1 to this diagram; from this, we conclude that

$$i_\lambda p_{\lambda\mu *}\pi(\tilde{Y}_{\lambda\mu}, \tilde{v}_{\lambda\mu}) = [p_*\pi(\tilde{X}, \tilde{v}_{\lambda\mu})] \cap G_\lambda.$$

If now we apply the isomorphisms $u_{\lambda\mu}$ and $w_{\lambda\mu}$ to this relation, and make use of all the commutativity relations in Figure 7.1, we obtain

$$H_{\lambda\mu} = H \cap (\beta_{\lambda\mu}G_\lambda\beta_{\lambda\mu}^{-1}).$$

To complete the proof, we must verify that $\{\beta_{\lambda\mu} : \mu \in M_\lambda\}$ is a set of representatives for the double coset HxG_λ. For this purpose, consider the right action of $G = \pi(X, v_0)$ on the set $p^{-1}(v_0)$, as in Section V.7. The subgroup H is the isotropy subgroup corresponding to the point $\tilde{v}_0$, and we may identify the points of $p^{-1}(v_0)$ with the cosets Hx in the usual way. Consider the action of the subgroup G_λ on $p^{-1}(v_0)$ or, equivalently, on the coset space G/H. For any $\mu \in M_\lambda$, G_λ permutes the points of $Y_{\lambda\mu} \cap p^{-1}(v_0)$ transitively. Thus, the set of components $\{\tilde{Y}_{\lambda\mu} : \mu \in M_\lambda\}$ is in one-to-one correspondence with the set of double cosets HxG_λ; and any choice of paths $\beta_{\lambda\mu} \in G$ such that $\tilde{v}_0 \cdot \beta_{\lambda\mu} = \tilde{v}_{\lambda\mu} \in \tilde{Y}_{\lambda\mu}$ for all $\mu \in M_\lambda$ is a choice of representatives for these double cosets. Q.E.D.

Proposition 5.3 *Assume in Theorem 5.1 that* $\Lambda = \{1, 2, \ldots, n\}$, $G = G_1 * G_2 * \cdots * G_n$, *and that* H *is a subgroup of* G *of index* $k < \infty$. *Then, the rank of the free factor* F *in Theorem 5.1 is given by the formula*

$$\text{rank } F = 1 + k(n - 1) - \sum_{\lambda=1}^{n} c_\lambda$$

where c_λ *denotes the number of distinct double cosets* HxG_λ *for* $\lambda = 1, 2, \ldots, n$.

PROOF: Recall that, in the proof of Theorem 5.1, $F \approx \pi(Y)$, where Y is a certain graph contained in $\tilde{X}$. If $\chi(Y)$ denotes the Euler characteristic of the graph Y, then

$$\text{rank } F = 1 - \chi(Y)$$

by Theorem VI.6.2. Therefore, we must determine $\chi(Y)$. The complex X has $n + 1$ vertices; hence, $\tilde{X}$ has $k(n + 1)$ vertices. Because Y contains all vertices of $\tilde{X}$, Y has $k(n + 1)$ vertices. Next, we must count the number of edges in the graph Y. The edges of Y are of two kinds: those which project onto one of the edges e_λ in X, and those which lie in one of the trees $T_{\lambda\mu}$. Obviously, there are $k \cdot n$ edges of the first kind, which project onto some e_λ. We assert that there are

$$nk - \sum_{\lambda=1}^{n} c_\lambda$$

edges of the second kind. To see this, note that, for $1 \leqq \lambda \leqq n$, c_λ denotes the number of components $\{X_{\lambda\mu} : \mu \in M_\lambda\}$, or, alternatively, the number of trees $T_{\lambda\mu}$, $\mu \in M_\lambda$. Because the Euler characteristic of each tree $T_{\lambda\mu}$ is $+1$, the Euler characteristic of the union

$$\bigcup_{\mu \in M_\lambda} T_{\lambda\mu}$$

is c_λ. Obviously, there are k vertices in

$$\bigcup_{\mu \in M_\lambda} T_{\lambda\mu};$$

hence, there are $k - c_\lambda$ edges in

$$\bigcup_{\mu \in M_\lambda} T_{\lambda\mu}.$$

Summing over $\lambda = 1, 2, \ldots, n$ now proves the assertion. From this, it follows that

$$\chi(Y) = (1 - n)k + \sum_{\lambda=1}^{n} c_\lambda,$$

and hence the desired formula for the rank of F. Q.E.D.

Proposition 5.4 *Assume the hypotheses and notation of Theorem 5.2. For any $\beta \in G$ and $\nu \in \Lambda$, if $H \cap \beta G_\nu \beta^{-1} \neq \{1\}$, then exactly one of the subgroups $H \cap \beta_{\lambda\mu} G_\lambda \beta_{\lambda\mu}^{-1}$ mentioned in the conclusion of Theorem 5.2 is conjugate to $H \cap \beta G_\nu \beta^{-1}$ in H.*

PROOF: Because $\{\beta_{\nu\mu} : \mu \in M_\nu\}$ ranges over a complete set of representatives of the double cosets HxG_ν, there exists exactly one index $\mu \in M_\nu$ such that β and $\beta_{\nu\mu}$ belong to the same double coset. As remarked previously, this implies that $H \cap \beta G_\nu \beta^{-1}$ and $H \cap \beta_{\nu\mu} G_\nu \beta_{u\nu}^{-1}$ are conjugate subgroups of H. If $H \cap \beta G_\nu \beta^{-1}$ were conjugate to two of the subgroups $H \cap \beta_{\lambda\mu} G_\lambda \beta_{\lambda\mu}^{-1}$, then these two subgroups would be conjugate to each other. By applying Exercise III.4.8 to H, we see that this is impossible. Q.E.D.

One can reinterpret this proposition as follows: Consider the following family of subgroups of H:

$$\{H \cap \beta G_\lambda \beta^{-1} : \lambda \in \Lambda, \beta \in G, H \cap \beta G_\lambda \beta^{-1} \neq 1\}.$$

Like any family of subgroups of H, this family breaks up into various conjugacy classes. Then, among the groups $H \cap \beta_{\lambda\mu}G_\lambda\beta_{\lambda\mu}^{-1}$ occurring in the conclusion of Theorem 5.2, there is exactly one representative from each conjugacy class. Thus, although the subgroups $H \cap \beta_{\lambda\mu}G_\lambda\beta_{\lambda\mu}^{-1}$ which occur in the free product decomposition of H in Theorem 5.2 are by no means canonical or unique, the conjugacy classes of these subgroups are unique and canonical. From this, it easily follows that if one has two different decompositions of H as a free product of the type prescribed in Theorem 5.2, the free factors F which occur in these two decompositions must be isomorphic (see Exercise III.4.10). In the case where the index of H is finite, this assertion is also a consequence of Proposition 5.3.

From the facts we have just proved, it is easy to prove that any two decompositions of an arbitrary group as a free product possess isomorphic refinements, and that, if there exists a free decomposition of a given group G with indecomposable factors, then any two such decompositions are isomorphic. For the details, see Kurosh [9], Vol. II, Section 35, or Specht [10], p. 189.

Exercises

5.1 In Theorem 5.2 assume that H is a normal subgroup of G. Prove that each of the free factors $H \cap \beta_{\lambda\mu}G_\lambda\beta_{\lambda\mu}^{-1}$ is conjugate (in G) to $H \cap G_\lambda$, and that $H \cap G_\lambda$ is a normal subgroup of G_λ. If each G_λ is a *simple* group, what can be said about the free factors $H \cap \beta_{\lambda\mu}G_\lambda\beta_{\lambda\mu}^{-1}$ of H?

5.2 Let $G = G_1 * G_2$, and let N be the least normal subgroup of G containing G_2. Prove that N is the free product of the family of subgroups $\{gG_2g^{-1} : g \in G_1\}$. (See Exercise III.4.9.)

5.3 In Theorem 5.2, assume that each of the groups G_λ is an infinite group and that H is a subgroup of G of finite index. Prove that $H \cap \beta_{\lambda\mu}G_\lambda\beta_{\lambda\mu}^{-1} \neq \{1\}$ for all pairs (λ, μ).

5.4 Prove that in Theorem 5.2 it is possible to choose the representatives $\beta_{\lambda\mu}$ so that, for each $\lambda \in \Lambda$, there exists a $\mu \in M_\lambda$ such that $\beta_{\lambda\mu} = 1$. (HINT: In the proof of Theorem 5.2, if $\tilde{v}_0 \in \tilde{Y}_{\lambda\mu}$, choose $\tilde{v}_{\lambda\mu} = \tilde{v}_0$.)

5.5 In Theorem 5.2 assume that H is the commutator subgroup of G. Prove that, for any index λ, $H \cap G_\lambda$ is the commutator subgroup of G_λ. [HINT: It is obvious that the commutator subgroup of G_λ is contained in $H \cap G_\lambda$. To prove the opposite inclusion, assume that $\mu \in M_\lambda$ is such that $\tilde{v}_0 \in \tilde{Y}_{\lambda\mu}$ and $\tilde{v}_{\lambda\mu} = \tilde{v}_0$ as in the preceding exercise. Prove that $(\tilde{Y}_{\lambda\mu}, p_{\lambda\mu})$ is a regular covering of Y_λ, and that the automorphism group of $(\tilde{Y}_{\lambda\mu}, p_{\lambda\mu})$ can be considered a subgroup of the automorphism group of $(\tilde{Y}, p)$; see Example V.11.1. By hypothesis, the latter automorphism group is abelian; therefore, so is the former.]

5.6 In Theorem 5.2 assume that each of the groups G_λ is abelian, and H is the commutator subgroup of G. Prove that $H \cap \beta_{\lambda\mu}G_\lambda\beta_{\lambda\mu}^{-1} = \{1\}$ for all pairs (λ, μ) and that therefore H is a free group. Determine the rank of H in this special case where $\Lambda = \{1, 2, \ldots, n\}$ as in Proposition 5.3, and each G_λ is a finite abelian group. (HINT: To determine the index of the commutator subgroup H in this special case, use Exercise III.4.5.)

5.7 Prove that the intersection of any two free factors of a group is again a free factor; to be precise, if G_1, G_2, G_1', and G_2' are subgroups of G such that $G = G_1 * G_2 = G_1' * G_2'$, then $G_1 \cap G_1'$ is a free factor of G. (HINT: Apply Theorem 5.2 with $H = G_1'$ considered a subgroup of $G_1 * G_2$; make use of the refinement suggested by Exercise 5.4.)

6 Grushko's Theorem

One of the most important theorems concerning free groups and free products of groups is the following due to the Russian mathematician I. Grushko (1940). As motivation for this theorem, recall that, if $\{G_\lambda\}$ and $\{G_\lambda'\}$ are two families of groups indexed by the same index set Λ, and if, for each $\lambda \in \Lambda$, there is given a homomorphism $f_\lambda : G_\lambda \to G_\lambda'$, then there is a unique homomorphism

$$f : \prod_\lambda^* G_\lambda \to \prod_\lambda^* G_\lambda'$$

which is an extension of the homomorphisms f_λ (see Exercise III.4.3). For brevity, we may refer to the homomorphism f as the *free product* of the family of homomorphisms $\{f_\lambda\}$. Grushko's theorem essentially says that any epimorphism of a free group (i.e., a free product of infinite cyclic groups) onto an arbitrary free product of groups is such a free product of homomorphisms. The precise statement is as follows:

Theorem 6.1 *Let*

$$\varphi : F \to \prod_\lambda^* G_\lambda$$

be an epimorphism of a free group F onto an arbitrary free product of groups. Then, there exists a decomposition of F as a free product,

$$F = \prod_\lambda^* F_\lambda$$

such that $\varphi(F_\lambda) \subset G_\lambda$ for all $\lambda \in \Lambda$.

We will present a topological proof of this theorem due to John Stallings for the case where the index set Λ is finite and F is a free group of finite rank. Stallings' method of proof also applies in the general case, but some of the details are more complicated.

To start the proof, we construct a topological analog of the homomorphism

$$\varphi : F \to \prod_\lambda^* G_\lambda$$

as follows. For each index λ, let B_λ denote a 2-dimensional CW-complex with a single vertex v_λ such that $\pi(B_\lambda, v_\lambda) = G_\lambda$. We can assume the B_λ's are pairwise disjoint. Let Y denote the quotient space of

$$\bigcup_\lambda B_\lambda$$

obtained by identifying all the vertices v_λ to a single vertex v. Then, Y is a 2-dimensional CW-complex with a single vertex, and

$$\pi(Y, v) = \prod_\lambda{}^* G_\lambda,$$

by Lemma 4.2. Let $\{y_\tau\}$ be a basis for the free group F. For each index τ, $\varphi(y_\tau)$ has a unique representation as a reduced word in the free product $\Pi^* G_\lambda$:

$$\varphi(y_\tau) = a_1 a_2 \ldots a_n.$$

For each index τ, choose a circle S_τ. Divide the circle S_τ into n segments by n vertices. Denote these segments in order by $W_1, W_2, \ldots, W_n$. Thus, S_τ becomes a graph with n vertices and n edges. Define a continuous map $f : S_\tau \to Y$ so that $f \mid W_i$ is a closed path in some B_λ representing a_i, $i = 1, \ldots, n$. Do this for each index τ, and let X denote the union of all the circles S_τ with all the initial vertices identified to a single vertex of X. Then, X is a finite, connected graph, $\pi(X) = F$, and the maps $S_\tau \to Y$, which have been defined, determine a unique continuous map $f : X \to Y$ such that the induced homomorphism $f_* : \pi(X) \to \pi(Y)$ is equivalent to the given homomorphism φ. To each edge W_i of S_τ we can associate a unique index λ such that $f(W_i) \subset B_\lambda$. Let A_λ denote the subgraph of X consisting of all the vertices together with all edges W_i assigned the index λ (in general, A_λ is not connected). Then, f maps A_λ into B_λ, and $\bigcap_\lambda A_\lambda$ consists of the set of all vertices of X.

To prove the theorem, we construct a finite, connected, 2-dimensional CW-complex X' such that X' contains X as a deformation retract and a map $f' : X' \to Y$, which is an extension of f. It follows that $\pi(X') \approx \pi(X) = F$, and $f'_* : \pi(X') \to \pi(Y)$ is the given homomorphism φ. Moreover, X' will be the union of connected subcomplexes A'_λ, such that each $A'_\lambda \supset A_\lambda$, $f'(A'_\lambda) \subset B_\lambda$; there will exist a tree T containing all the vertices of X' such that T is contained in each A'_λ, and for any two distinct indices λ and μ,

$$A'_\lambda \cap A'_\mu = T.$$

From Lemma 4.2 it follows that $\pi(X')$ is the free product of the groups $\pi(A'_\lambda)$; therefore, we can take $F_\lambda = \pi(A'_\lambda)$, and the theorem will be proved.

Before giving Stallings' ingenious proof that such a construction is always possible, we shall illustrate it in a particular case. Let $G = G_1 * G_2$, where G_1 is cyclic of order 2 with generator a, and G_2 is cyclic of order 3 with generator b. Let F be a free group on two generators x and y, and suppose that $\varphi : F \to G$ is defined by $\varphi(x) = aba$, $\varphi(y) = ababa$. We leave it to the reader to verify that φ is an epimorphism. The space X in this example is the union of two circles, one divided into three segments, and the other divided into five segments. We can assume that X is imbedded in the plane, as shown in Figure 7.2. (The reason for this unsymmetric method of depicting X in Figure 7.2 will be clear shortly.) It is helpful to think of the edges of the complex X which are mapped into B_1 and B_2 colored differently, e.g., orange and green. Then, A_1 is the subcomplex of X consisting of all the orange edges, $a_1, a_2, \ldots, a_5$, and A_2 is the subcomplex consisting of the green edges, b_1, b_2, and b_3. The vertices, $v_0, \ldots, v_6$, must be thought of as being colored both orange and green. The arrow on each edge indicates how it is mapped to represent the element $a \in \pi(B_1) = G_1$ or $b \in \pi(B_2) = G_2$.

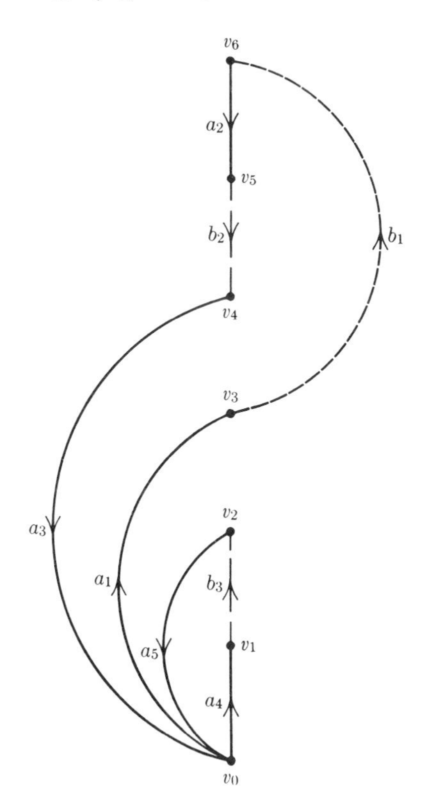

FIGURE 7.2 The complex X.

The complex X' is shown in Figure 7.3. Unfortunately, it cannot be imbedded in the plane; hence, it is shown in two pieces that must be glued together along the indicated line. X' is obtained from X by adjoining the edges $c_1, c_2, \ldots, c_6$ and the 2-cells $e_1, e_2, \ldots, e_6$; no new vertices are added. The 2-cells e_4 and e_5 are colored green, whereas e_1, e_2, e_3, and e_6 are colored orange. The new edges $c_1, \ldots, c_6$ are colored both green and orange. The map $f : X \to Y$ is extended to a continuous map $f' : X' \to Y$ as follows: Each of the edges c_i is mapped into the common base point of B_1 and B_2, the orange 2-cells are mapped into B_1, and the green 2-cells

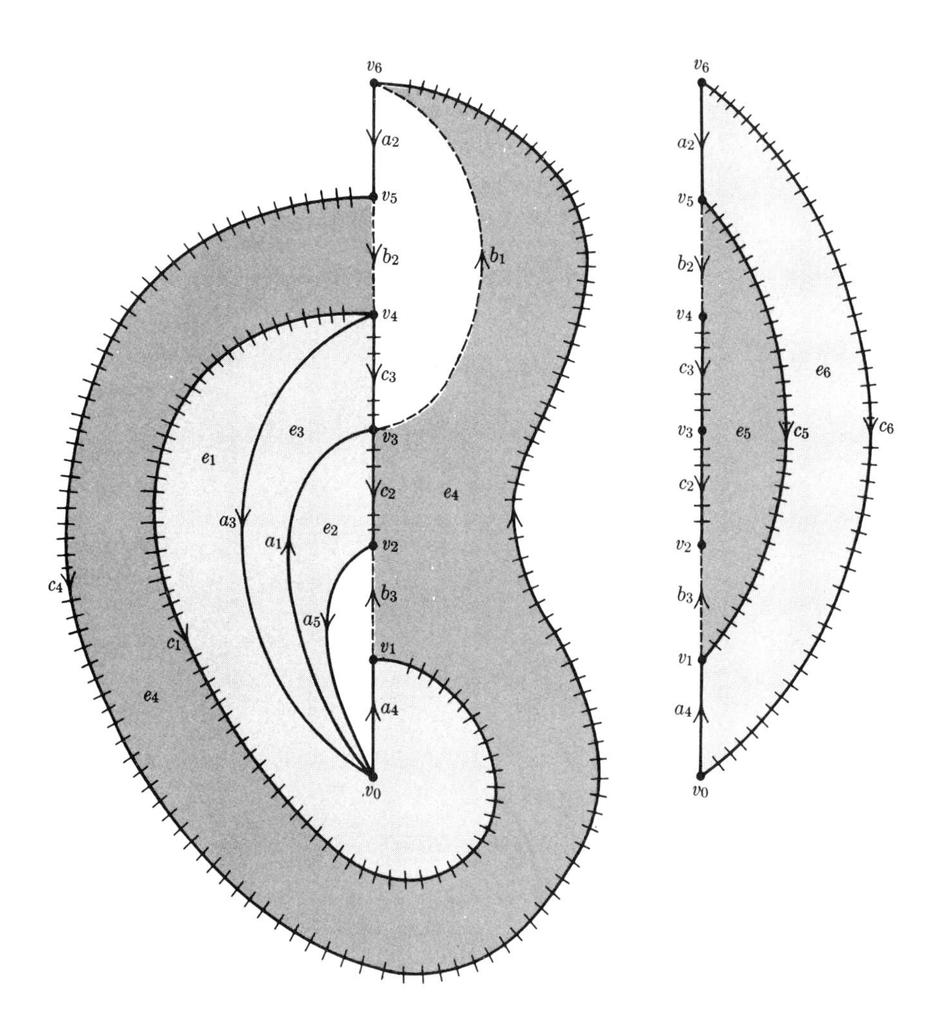

FIGURE 7.3 The complex X'. The two pieces must be glued together along the line v_0v_6.

are mapped into B_2. In each of the six cases we can verify without difficulty that the map f' can actually be extended continuously over each 2-cell (use Lemma II.8.1 and the relations $a^2 = 1$, $b^3 = 1$). It should be clear that X is a deformation retract of X', that the orange subcomplex A_1' and the green subcomplex A_2' are both connected, and that the intersection $A_1' \cap A_2'$ consists of the union of the edges $c_1, c_2, \ldots, c_6$, which is a tree containing all the vertices. Thus, by Lemma 4.2,

$$F = \pi(X) \approx \pi(X') \approx \pi(A_1') * \pi(A_2'),$$

and because $f'(A_1') \subset B_1$ and $f'(A_2') \subset B_2$, the theorem follows in this case.

This example illustrates very well the strategy of the proof in the general case. We adjoin the edges $c_1, c_2, \ldots$ to X in succession so as to connect the vertices and build a tree. Also, we must adjoin 2-cells $e_1, e_2, \ldots$ in succession so that c_i is part of the boundary of e_i, and to insure that X will be a deformation retract of X'. Finally, this entire construction must be made in such a way that the map f can be extended continuously over each 2-cell e_i to map it into one of the subcomplexes B_λ.

We now give the formal proof that such a construction can always be carried out. First, we introduce some terminology. The CW-complex Y which is the union of the subcomplexes B_λ, $\lambda \in \Lambda$, remains fixed throughout the proof. We call a system consisting of a finite, connected, 2-dimensional CW-complex K, a collection of subcomplexes C_λ, $\lambda \in \Lambda$, and a continuous map $f : K \to Y$ a *Stallings system* provided (a)

$$K = \bigcup_\lambda C_\lambda;$$

(b) for any pair of distinct indices μ and ν

$$C_\mu \cap C_\nu = \bigcap_{\lambda \in \Lambda} C_\lambda;$$

(c) for any index λ, $f(C_\lambda) \subset B_\lambda$; and (d) f maps the n-skeleton of K into the n-skeleton of Y for all n. We shall always assume that in any such Stallings system there is a definitely chosen base point for K, which is a vertex of

$$\bigcap_\lambda C_\lambda.$$

Note that we do not assume that the subcomplexes C_λ or their intersection

$$\bigcap_\lambda C_\lambda$$

are connected. It is convenient to think of the indices λ as representing different colors, as in the special example above. For this reason, we call a path in K *monochromatic* if it lies entirely in some one C_λ. A path in the 1-skeleton of K will be called a *loop* if both end points coincide in a vertex, and it will be called a *tie* if its end points are vertices in different components of the intersection

$$\bigcap_{\lambda} C_\lambda.$$

A tie $g : I \to K$ will be called a *binding tie* if there exists an index λ such that $g(I) \subset C_\lambda$, and the path $fg : I \to Y$ is equivalent in B_λ to the constant path. Note that a binding tie is always monochromatic. This definition of a binding tie could be rephrased as follows. Let η denote the equivalence class of the path g in C_λ, and let $f_\lambda : C_\lambda \to B_\lambda$ denote the restriction of f to C_λ. Then, it is required that $f_{\lambda*}(\eta) = 1$ in $\pi(B_\lambda)$.

We now describe a fundamental construction that can always be carried out in any Stallings system when given a binding tie. Let K, $\{C_\lambda : \lambda \in \Lambda\}$, and $f : K \to Y$ be a Stallings system, and let $g : I \to C_\mu$ be a binding tie of color μ, as defined above. Let D be a 2-dimensional, closed disc whose boundary circle is divided into two segments c_1 and c_2 which intersect only at their end points. Identify c_1 with the unit interval I so that g is a map of c_1 into C_μ. Let K' denote the quotient space of $K \cup D$ obtained by identifying each point $t \in c_1$ with its image $g(t) \in K$. Then, K' is a CW-complex containing K as a subcomplex, an additional edge c_2, and an additional 2-cell D. It is obvious that K is a deformation retract of K'. Let C'_μ denote the union of C_μ and the closed 2-cell D (with the given identifications), and, for any index $\lambda \neq \mu$, let C'_λ denote the union of C_λ and the edge c_2 (the end points of c_2 are identified with the end points of the path g). Then, clearly

$$\bigcap_{\lambda} C'_\lambda$$

is obtained from

$$\bigcap_{\lambda} C_\lambda$$

by adjunction of the arc c_2, which joints two distinct components of

$$\bigcap_{\lambda} C_\lambda.$$

Next, we extend the map $f : K \to Y$ to a map $f' : K' \to Y$ as follows: f' maps the arc c_2 into the unique vertex v of Y, and then f' is extended to a continuous map of the 2-cell D into B_μ. This latter extension is always possible because of the hypothesis that $fg : I \to B_\mu$ is equivalent to the constant path (see Lemma II.8.1). It is clear that the system consisting of K', $\{C'_\lambda : \lambda \in \Lambda\}$, $f : K' \to Y$ is again a Stallings system.

This construction may be used to connect two components of

$$\bigcap_{\lambda} C_\lambda$$

whenever there is a binding tie. We now take up the question of the existence of binding ties.

Lemma 6.2 *Let K, $\{C_\lambda : \lambda \in \Lambda\}, f : K \to Y$ be a Stallings system such that $f_* : \pi(K) \to \pi(Y)$ is an epimorphism. If*

$$\bigcap_\lambda C_\lambda$$

is not connected, then there exists a binding tie.

PROOF: Choose a vertex as a base point for each component of

$$\bigcap_\lambda C_\lambda.$$

Consider any loop or tie g whose initial and terminal points are such base points. It easily follows from Chapter VI that any such loop or tie is equivalent to a product of paths, each of which runs across a single edge. Thus, it is equivalent to a product of monochromatic paths. By grouping these paths into maximal monochromatic blocks, we see that

$$g \sim g_1 g_2 \cdots g_n,$$

where each g_i is monochromatic, and g_i and g_{i+1} have different colors for all i. Hence, the end points of each path g_i must belong to

$$\bigcap_\lambda C_\lambda;$$

for the initial point of g_1 and the terminal point of g_n, this is true by hypothesis. For $1 \leqq i < n$, let h_i be an edge path in

$$\bigcap_\lambda C_\lambda$$

which joins the terminal point of g_i to the base point in its component of

$$\bigcap_\lambda C_\lambda.$$

Then,

$$g \sim (g_1 h_1)(h_1^{-1} g_2 h_2) \cdots (h_{n-1}^{-1} g_n).$$

Each term here is monochromatic, and the end points of each term are among the chosen base points. Hence, each term is a loop or a tie. Thus, we have proved that *each loop or tie in a Stallings system is equivalent to a product of monochromatic loops and ties whose end points are among the set of chosen base points, and successive terms in the product have different colors.*

We now assert that there is a tie g in K whose path class η is such that $f_*(\eta) = 1$ in $\pi(Y)$. For, because K is connected and

$$\bigcap_\lambda C_\lambda$$

is not connected, there exists a tie h in K whose end points are base points in different components of

$$\bigcap_\lambda C_\lambda.$$

Let θ denote the path class of h. Because $f_* : \pi(K) \to \pi(Y)$ is an epimorphism, there is a loop k in K, based at the initial point of h, whose equivalence class $\zeta \in \pi(K)$ satisfies the equation $f_*(\zeta) = f_*(\theta)$. Then, $k^{-1}h = g$ is the desired tie, because $f_*(\zeta^{-1}\theta) = 1$.

By the assertion above, we can assume that

$$g \sim g_1 g_2 \ldots g_n$$

is a product of monochromatic loops and ties. Let η_i denote the equivalence class of the monochromatic loop or tie g_i for $i = 1, \ldots, n$ in the above equation.

Next, we assert that *there exists a tie $g \sim g_1 g_2 \ldots g_n$ representing $\eta = \eta_1 \eta_2 \ldots \eta_n$ such that the following three conditions hold:* (a) $f_*(\eta) = 1$. (b) *For all i, the monochromatic loops or ties g_i and g_{i+1} have different colors.* (c) *For any i, if g_i is a loop, then $f_*(\eta_i) \neq 1$.* We have already seen how to satisfy conditions (a) and (b); to satisfy condition (c), we simply omit from the product any loop g_i such that $f_*(\eta_i) = 1$. Because g_i is a loop, its end points coincide, and $g_1 \ldots g_{i-1} g_{i+1} \ldots g_n$ is still a well-defined product of paths. After omitting any such g_i from the product $g_1 g_2 \ldots g_n$, it may be possible to lump together the terms g_{i-1} and g_{i+1} because they have the same color. In any case, after a finite number of such reductions, we obtain the desired tie g. The number of factors n must remain $\geqq 1$, because the end points of g are distinct and remain unchanged by all these reductions.

Consider any tie $g \sim g_1 g_2 \ldots g_n$ satisfying the three conditions of the last assertion. We have the following equation in $\pi(Y)$:

$$1 = f_*(\eta) = (f_*\eta_1)(f_*\eta_2) \cdots (f_*\eta_n).$$

The terms $f_*(\eta_i)$ and $f_*(\eta_{i+1})$ belong to different free factors $\pi(B_\lambda)$ of Y for all i, and $n \geqq 1$. Therefore, there must exist an index i such that $f_*(\eta_i) = 1$; otherwise we should have a representation of the identity as a reduced word of length $\geqq 1$ in the free product

$$\prod_\lambda{}^* \pi(B_\lambda).$$

Now g_i cannot be a loop, because we have avoided such trivial loops. Hence, g_i is a tie. It is monochromatic by construction, and it is a binding tie, because $f_*(\eta_i) = 1$ in $\pi(Y)$. Because $\pi(Y)$ is a free product, we must have $f_*(\eta_i) = 1$ in the group $\pi(B_\lambda)$ for the appropriate color λ as required.

This completes the proof of Lemma 6.2.

We can now complete the proof of Theorem 6.1. Consider the Stallings system X, $\{A_\lambda : \lambda \in \Lambda\}$, $f : X \to Y$ which was constructed at the start of the proof so that $f_* : \pi(X) \to \pi(Y)$ represents the given epimorphism φ. In this Stallings system,

$$\bigcap_\lambda A_\lambda$$

consists of the vertices of the finite graph X. If the intersection is disconnected, we can apply Lemma 6.2 to conclude that there exists a binding tie. Then, we can apply the fundamental construction to obtain a Stallings system X^1, $\{A_\lambda^1\}$, $f^1 : X^1 \to Y$ such that X is a deformation retract of X^1, f^1 is an extension of f, and

$$\bigcap_\lambda A_\lambda^1$$

has one less component than

$$\bigcap_\lambda A_\lambda$$

because two vertices of

$$\bigcap_\lambda A_\lambda$$

have been joined by an arc. If

$$\bigcap_\lambda A_\lambda^1$$

is not connected, we repeat this process to obtain a new Stallings system X^2, $\{A_\lambda^2\}$, $f^2 : X^2 \to Y$, and so on. The process must come to an end after a finite number of steps, because there were only a finite number of components in

$$\bigcap_\lambda A_\lambda$$

to begin with. Thus, if

$$\bigcap_\lambda A_\lambda$$

has $n + 1$ components, after n steps we obtain a Stallings system X^n, $\{A_\lambda^n\}$, $f^n : X^n \to Y$ such that

$$\bigcap_\lambda A_\lambda^n$$

is connected; it is clear that this intersection must be a tree. If we let $X' = X^n$, $A_\lambda' = A_\lambda^n$, and $f' = f^n$, then we have carried out the construction that was promised above.

This completes the proof of Grushko's theorem.

Exercises

6.1 Assume $G = G_1 * \cdots * G_k$. Let n = minimal number of generators of G and n_i = minimal number of generators of G_i, $1 \leqq i \leqq k$. Prove that $n = n_1 + n_2 + \cdots + n_k$. Deduce that a group generated by n elements cannot be a free product of more than n nontrivial factors, and that every finitely generated group can be decomposed into the free product of a finite number of indecomposable factors.

6.2 Prove that a free group of rank n cannot be generated by fewer than n elements.

6.3 Let F and G be free groups of rank n, and $\varphi : F \to G$ an epimorphism. Prove that φ is an isomorphism. Deduce from this the following two corollaries: (a) If G is a free group of rank n, and $a_1, \ldots, a_n \in G$ are any n elements which generate G, then $\{a_1, \ldots, a_n\}$ is a basis for G. (b) If G is a free group of finite rank and N is a proper normal subgroup of G, then G/N is not isomorphic to G.

6.4 Assume that

$$G = \bigcup_{n=1}^{\infty} G_n,$$

where each G_n is a subgroup of G, G_n is a proper subgroup of G_{n+1} for all n, and G_n is generated by at most m elements for $n = 1, 2, \ldots$. If $G = H_0 * H$ with H_0 finitely generated, then H_0 is generated by less than m elements and H is not finitely generated (W. Specht [10]). (HINT: Prove the following in succession: 1. G is not finitely generated. 2. H is not finitely generated. 3. There exists an integer n_0 such that $H_0 \subset G_{n_0}$. 4. Apply the Kurosh subgroup theorem to obtain a free product decomposition of G_{n_0}. 5. Make use of Exercise 6.1.)

NOTES

CW-complexes

The introduction and systematic use of CW-complexes in topology is due to J. H. C. Whitehead [4]. Although many other kinds of complexes had been used for many years before this, it soon became apparent to workers in the field that CW-complexes had many advantages; cf. [3] in this regard.

The Kurosh subgroup theorem

In Appendix C of Volume II of [9] there are listed six different proofs of Kurosh's theorem. More recent papers on this theorem are by S. Mac Lane ("A proof of the subgroup theorem for free products," *Mathematika*, *5*, 1958, pp. 13–19), I. M. S. Dey ("Schreier Systems in Free Products," *Proc. Glasgow Math. Assoc.*, *7*, 1965, pp. 61–79), and P. J. Higgins ("Presentations of Groupoids, with Applications to Groups," *Proc. Cambridge Phil. Soc.*, *60*, 1964, pp. 7–20). The proof in the text is modeled after that of R. Baer and F. Levi ("Freie Produkte and ihre Untergruppen," *Compositio Mathematica*, *3*, 1936, pp. 391–398). Our statement of the theorem is more in line with that in some of the more recent texts and papers.

Grushko's theorem

The proof given here for Grushko's theorem is due to John Stallings ("A topological proof of Grushko's theorem on free products," *Math. Zeitschr., 90,* 1965, pp. 1–8). The text by Specht [10] derives some of the consequences of this theorem for group theory. For a purely algebraic proof, which is shorter than those in the standard textbooks, see a recent paper by R. C. Lyndon ("Grushko's Theorem," *Proc. Amer. Math. Soc., 16,* 1965, pp. 822–826). The original proof that I. Grushko gave for his theorem applied only to the case of free groups of finite rank. The first proof for the case of free groups of arbitrary rank was given in the 1951 Brown University Ph.D. thesis of Daniel Wagner; see *Trans. Amer. Math. Soc., 84,* 1957, pp. 352–378.

REFERENCES

CW-complexes

1. Hilton, P. J. *An Introduction to Homotopy Theory* (Cambridge Tracts in Mathematics and Mathematical Physics No. 43). Cambridge: The University Press, 1953. Chapter VII.
2. Hu, S. T. *Elements of General Topology.* San Francisco: Holden-Day, 1964. Chapter IV.
3. Milnor, J. "On Spaces Having the Homotopy Type of a CW-Complex." *Trans. Amer. Math. Soc., 90,* 1959, pp. 272–280.
4. Whitehead, J. H. C. "Combinatorial Homotopy I." *Bull. Amer. Math. Soc., 55,* 1949, pp. 213–245.

Fundamental group of a complex

5. Hilton, P. J., and S. Wylie. *Homology Theory, An Introduction to Algebraic Topology.* Cambridge: The University Press, 1960. Chapter 6.
6. Reidemeister, K. *Einführung in die kombinatorische Topologie.* Braunschweig: Friedr. Vieweg & Sohn, 1932. Chapters 5 and 6.
7. Seifert, H., and W. Threlfall. *A Textbook of Topology.* New York: Academic Press, 1980. Chapter 7.

Group theory

8. Hall, M. *The Theory of Groups.* New York: Macmillan, 1959. Chapter 17.
9. Kurosh, A. G. *The Theory of Groups.* 2 vols. Trans. and ed. by K. A. Hirsch. New York: Chelsea, 1955–56. Chapters IX and X.
10. Specht, W. *Gruppentheorie* (Die Grundlehren der Mathematischen Wissenschaften, Band LXXXII). Berlin-Göttingen-Heidelberg: Springer-Verlag, 1956. Chapter 2.2.

CHAPTER EIGHT

Epilogue

The preceding chapters have introduced the reader to a few topics of algebraic topology. In this chapter we give some hint of what is in store for him if he pursues the study of algebraic topology further.

In Sections II.6 and V.9 we mentioned that one of the objectives of algebraic topology is to define higher dimensional analogs of the fundamental group so as to be able to prove statements such as the Brouwer fixed-point theorem and the Borsuk-Ulam theorem. As we shall see, there are several different such higher dimensional analogs. We shall describe first the one most closely related to the fundamental group, namely, the higher dimensional homotopy groups of Hurewicz (mentioned in the Notes, Chapters II and V). These groups are defined very simply as follows: For any positive integer n, let

$$I^n = \{(x_1, \ldots, x_n) \in \mathbf{R}^n : 0 \leqq x_i \leqq 1 \text{ for } i = 1, 2, \ldots, n\}$$

denote the unit n-dimensional cube, and let $\dot{I}^n$ denote its boundary,

$$\dot{I}^n = \{(x_1, \ldots, x_n) \in I^n : x_i = 0 \text{ or } 1 \text{ for some } i\}.$$

For any topological space X and any base point $x_0 \in X$, the symbol $\pi_n(X, x_0)$ denotes the set of all relative homotopy classes of maps $f : I^n \to X$ such that $f(\dot{I}^n) = x_0$. By "relative homotopy class" we mean that all homotopies are relative to the boundary, $\dot{I}^n$ (see Section II.4 for the definition). If f and g are two maps $I^n \to X$ such that $f(\dot{I}^n) = g(\dot{I}^n) = x_0$, define their sum, $f + g$, by the formula

$$(f + g)(x_1, \ldots, x_n) = \begin{cases} f(2x_1, x_2, \ldots, x_n) & \text{if } 0 \leqq x_1 \leqq \frac{1}{2}, \\ g(2x_1 - 1, x_2, \ldots, x_n) & \text{if } \frac{1}{2} \leqq x_1 \leqq 1. \end{cases}$$

It can be shown that this defines an addition in the set of homotopy classes $\pi_n(X, x_0)$ such that $\pi_n(X, x_0)$ is a group called the *nth homotopy group of X;* the proof is almost exactly the same as that given for the fundamental group in Section II.3. In fact, for $n = 1$, the group thus defined *is* the fundamental group $\pi(X, x_0)$. It can also be shown without much difficulty that, for $n > 1$, the group $\pi_n(X, x_0)$ is abelian.

The group $\pi_n(X, x_0)$ is also analogous to the fundamental group $\pi(X, x_0)$ in several other ways. For example, any continuous map $\varphi : X \to Y$ induces a homomorphism $\varphi_* : \pi_n(X, x_0) \to \pi_n(Y, \varphi(x_0))$, defined as follows: If $f : I^n \to X$ is a representative for the homotopy class $\alpha \in \pi_n(X, x_0)$, then the composed map φf is a representative of the homotopy class $\varphi_*(\alpha) \in \pi_n(Y, \varphi(x_0))$. This induced homomorphism φ_* has the properties described in Section II.4 for the case $n = 1$. Another example of this analogy is the following property: Any path class γ in X with initial point x and terminal point y defines an isomorphism of $\pi_n(X, x)$ onto $\pi_n(X, y)$; thus, if the space X is arcwise connected, the structure of the group $\pi_n(X, x_0)$ is independent of the choice of the base point x_0.

However, this analogy between the fundamental group and the higher homotopy groups cannot be pushed too far. For example, there seems to be no reasonable analog of the Seifert-Van Kampen theorem for the higher homotopy groups. As a result, the problem of determining the structure of the groups $\pi_n(X)$ for a noncontractible space X is difficult or impossible. Most of the higher homotopy groups $\pi_n(X)$ are unknown, even in the relatively simple case where X is a k-sphere, $k > 1$; the determination of these groups is one of the important unsolved problems of algebraic topology.

Although the higher homotopy groups have been of great theoretical importance in algebraic topology ever since their introduction by Hurewicz in 1935, they are not of much help in proving theorems such as the Brouwer fixed-point theorem, or the Borsuk-Ulam theorem. For this purpose, one normally uses *homology groups* or *cohomology groups*. These groups have many properties similar to those just listed for the homotopy groups of Hurewicz, and they have one important advantage over the latter: For a wide variety of topological spaces, their algebraic structure is computable. Homology groups were introduced by Poincaré about 1895.

Unfortunately, the definition of the homology or cohomology groups of a space X is somewhat more involved than the above definition of Hurewicz homotopy groups. Actually, there are several different ways of defining the homology and cohomology groups of a space X. If X is compact and "nice" locally (e.g., a manifold or a CW-complex), these different definitions give rise to isomorphic groups. For spaces that are noncompact or are in some sense locally pathological (e.g., not locally connected or not semilocally simply connected), the different definitions may give rise to nonisomorphic groups. All the different definitions assign to each pair (X, G), consisting of a topological space X and an abelian group G (called the "coefficient group"), a sequence of *abelian* groups

$$H_n(X, G), \qquad n = 0, 1, 2, \ldots,$$

called the *homology groups* of X (with coefficient group G) and another sequence of abelian groups

$$H^n(X, G), \qquad n = 0, 1, 2, \ldots,$$

called *cohomology groups* of X (with coefficient group G). For many purposes, the most important choices for the coefficient group G are the additive group of integers, $\mathbf{Z}$, or a finite cyclic group, $\mathbf{Z}_k$. As was the case for the Hurewicz homotopy groups, a continuous map $\varphi : X \to Y$ induces homomorphisms of the homology groups

$$\varphi_* : H_n(X, G) \to H_n(Y, G), \qquad n = 0, 1, 2, \ldots,$$

and of the cohomology groups

$$\varphi^* : H^n(Y, G) \to H^n(X, G), \qquad n = 0, 1, 2, \ldots.$$

Note that these induced homomorphisms are in opposite directions for the homology and cohomology groups. The induced homomorphisms of homology and cohomology groups have properties exactly analogous to those of the induced homomorphisms of the fundamental group described in Section II.4. In particular, if two maps $\varphi_0, \varphi_1 : X \to Y$ are homotopic, they induce the *same* homomorphisms. It follows that if $\varphi : X \to Y$ is a homotopy equivalence, then the induced homomorphisms φ_* and φ^* are isomorphisms. Thus, spaces of the same homotopy type have isomorphic homology and cohomology groups.

So far, we have emphasized the similarities of the homology and cohomology groups with the Hurewicz homotopy groups (including the fundamental group). Now let us indicate the most important differences:

(a) No choice of a base point is involved in the definition of the homology and cohomology groups. This makes for greater simplicity in most considerations.

(b) The homology and cohomology groups $H_n(X, G)$ and $H^n(X, G)$ are defined for all $n \geqq 0$, whereas the homotopy groups $\pi_n(X, x_0)$ are only defined for $n > 0$. [As a matter of fact, it is often convenient and natural to define $H_n(X, G) = H^n(X, G) = \{0\}$ for all $n < 0$; it is also possible to define in a natural way a *set* $\pi_0(X, x_0)$ without any group structure, but there seems to be no reasonable way to define $\pi_n(X, x_0)$ for $n < 0$.]

(c) The homology and cohomology groups of a space are abelian, whereas the fundamental group is often non-abelian. This means that all the technique and machinery of the theory of abelian groups (especially the theory of finitely generated abelian groups) is applicable to homology and cohomology theory, and the subject has a different flavor from the theory of the fundamental group as presented in this book.

As mentioned above, the definitions of the homology and cohomology groups of a space are too involved to reproduce here. However, we can describe the structure of these groups for $n = 0$ and $n = 1$. Let X be a space having k components. Then $H_0(X, G)$ and $H^0(X, G)$ are each isomorphic to the direct sum of k copies of the coefficient group G. Thus, the structure of these 0-dimensional groups depends only on the number of connected components of the space X. If X is arcwise connected, then $H_1(X, \mathbf{Z})$ is naturally isomorphic to the abelianized fundamental group, $\pi(X)/[\pi(X), \pi(X)]$. Recall that we made use of the abelianized fundamental group of the various compact surfaces in Section IV.5; thus, we were really considering the 1-dimensional homology groups of these surfaces. If X is arcwise connected, then the 1-dimensional cohomology group $H^1(X, G)$ is naturally isomorphic to the group of all homomorphisms of $\pi(X)$ into the coefficient group G. More generally, it may be shown that the 1-dimensional groups $H_1(X, G)$ and $H^1(X, G)$ are completely determined by the fundamental groups of the various components of X.

In contrast to these general results about $H_n(X, G)$ and $H^n(X, G)$ for $n = 0$ or $n = 1$, there are various special results about these groups for larger values of n. We will now give three examples of such results. (a) If X is a space consisting of a single point, then

$$H_n(X, G) = H^n(X, G) = \begin{cases} G & \text{if } n = 0, \\ \{0\} & \text{if } n \neq 0. \end{cases}$$

From this it follows that the same formula holds for a contractible space X. (b) If M is a compact, connected, orientable n-manifold, then

$$H_n(M, G) = H^n(M, G) = G$$

and, for $q > n$,

$$H_q(M, G) = H^q(M, G) = 0.$$

In particular, this result applies if $M = S^n$, the n-sphere. This fact makes it possible to use homology groups to prove that S^n is not a retract of the $(n + 1)$-dimensional ball, E^{n+1}, by a method analogous to that used in Section II.6 to prove that S^1 is not a retract of E^2. The Brouwer fixed-point theorem follows as an easy corollary, as was pointed out in Section II.6. (c) If X is an n-dimensional CW-complex, as defined in Section VII.4, then $H_q(X, G)$ and $H^q(X, G)$ are zero for $q > n$. This result gives some hint of the connection between homology theory and dimension theory.

For homology and cohomology groups there is an analog of the Seifert-Van Kampen theorem, called the Mayer-Vietoris theorem. One can use the principles involved in this theorem to determine the homology and cohomology groups of many spaces.

For a large class of spaces (e.g., finite CW-complexes), the cohomology groups of a space are completely determined by the homology groups, and vice versa. Thus, for many problems one could equally well employ homology groups or cohomology groups to reach a solution, However, it is possible to give more structure to the cohomology groups of a space by introducing what are called "cohomology operations." This additional structure is needed for the solution of certain problems. For example, one way to prove the general Borsuk-Ulam theorem (see Section V.9 for the statement) is to use certain cohomology operations called "cup products."

We will now try to indicate some of the principal problems to which homology and cohomology theory are applicable. First, there are problems in the general field of homotopy theory:

(a) Homotopy classification of maps. Given two spaces X and Y, can we get some insight into the set of all homotopy classes of continuous maps of X into Y? Given two maps f and g of X into Y, can we associate with f and g some sort of algebraic invariants that will help us to decide whether or not f and g are homotopic?
(b) Extension of continuous maps. Let X and Y be topological spaces, let A be a closed subset of X, and let $f : A \to Y$ be a continuous map. Does there exist an extension of f over X, i.e., a continuous map $g : X \to Y$ such that $g \mid A = f$?
(c) Homotopy type of spaces. Find necessary and/or sufficient conditions for two spaces to be of the same homotopy type.

A great deal of work has been done on all of these problems and many techniques involving homology and cohomology groups have been developed for attacking them. The results obtained have found application in other branches of mathematics. The general strategy has been to try to reduce problems in homotopy theory to problems in algebra.

Another extensive group of problems arises in the study of manifolds. There are three general types of manifolds which can be considered: Topological manifolds, as defined in Chapter I, piecewise linear or combinatorial manifolds (i.e., manifolds with a preferred class of triangulations), and differentiable manifolds (i.e., manifolds with some additional structure defined so that the concept of differentiable function has meaning). One of the basic problems is the classification problem in each of these three categories of manifolds. For example, in Chapter I we discussed the classification problem for 2-dimensional manifolds. Two other basic problems are the following: Does every topological manifold admit a piecewise linear structure, i.e., a triangulation? Does every piecewise linear manifold admit a differentiable structure? The answer to the first question is unknown, except for dimensions $\leqq 4$ (see Notes, Chapter I). The answer to the second question is affirmative for dimensions < 8, and negative for dimensions $\geqq 8$. The first example of a triangulated manifold that does not admit any differentiable structure was given by M. Kervaire. If such a manifold is embedded in Euclidean space, it is impossible to smooth it out everywhere so that there are no "corners."

Some of the most striking results obtained about manifolds in recent years have been the proof of the generalized Poincaré conjecture by S. Smale (see Notes, Chapter IV) and the study of various differentiable structures on spheres by Kervaire and Milnor. These two authors show that the number of distinct differentiable structures on an n-sphere, S^n, is given by the following table:

n	5	6	7	8	9	10	11	12	13	14	15
no. of structures	1	1	28	2	8	6	992	1	3	2	16,256

They also prove that, for any integer n, the number of distinct differentiable structures on S^n is finite. It is interesting to note that the proof of these results depends heavily on homotopy theory.

Algebraic topology is a rapidly growing and expanding discipline. The exposition of most of the results obtained in recent years is rather involved and complicated; sufficient time has not yet elapsed for the research workers in the subject to polish and assemble all the details into a coherent, well-organized theory. Many definitions and proofs will seem rather unmotivated to the student. One of the great challenges for the future is to give a readable exposition of the various parts of the subject.

REFERENCES

Homology and cohomology theory

1. Artin, E. and Braun, H. *Introduction to Algebraic Topology.* Columbus: Charles E. Merrill Co., 1969.
2. Dold, A. *Lectures on Algebraic Topology.* New York: Springer-Verlag, 1972.
3. Eilenberg, S. and Steenrod, N. *Foundations of Algebraic Topology.* Princeton: Princeton University Press, 1952.
4. Massey, W. S. *Homology and Cohomology Theory: An Approach Based on Alexander-Spanier Cochains.* New York: Marcel Dekker, Inc., 1978.

5. Massey, W. S. *Singular Homology Theory* (Graduate Texts in Mathematics 70). New York: Springer-Verlag, 1980.
6. Spanier, E. H. *Algebraic Topology.* New York: McGraw-Hill, 1966.
7. Steenrod, N. E. and Epstein, D. B. A. *Cohomology Operations.* Annals of Mathematics Studies, no. 50. Princeton: Princeton University Press, 1962.
8. Vick, J. W. *Homology Theory: An Introduction to Algebraic Topology.* New York: Academic Press, 1973.

Homotopy theory

9. Hilton, P. J. *An Introduction to Homotopy Theory.* Cambridge Tracts, no. 43. Cambridge: The University Press, 1953.
10. Hilton, P. J. *Homotopy Theory and Duality.* New York: Gordon and Breach, 1965.
11. Hu, S. T. *Homotopy Theory.* New York: Academic Press, 1959.
12. Steenrod, N. E. *The Topology of Fibre Bundles.* Princeton: Princeton University Press, 1951.
13. Whitehead, G. W. *Homotopy Theory.* Cambridge, Massachusetts: The M. I. T. Press, 1966.

Differential topology

14. Milnor, J. W. "Differential Topology," in *Lectures on Modern Mathematics,* volume II, pp. 165–183. New York: John Wiley & Sons, 1964.
15. Munkres, J. R. *Elementary Differential Topology.* Annals of Mathematics Studies, No. 54. Princeton: Princeton University Press, 1963.
16. Smale, S. "A Survey of some recent developments in differential topology," *Bull. Amer. Math. Soc., 69,* 1963, pp. 131–145.
17. Wall, C. T. C. "Survey Article, Topology of Smooth Manifolds," *J. London Math. Soc., 40,* 1965, pp. 1–20.
18. Wallace, A. H. *Differential Topology: First Steps.* New York: W. A. Benjamin, 1968.
19. Milnor, J. W. *Topology from the Differentiable Viewpoint.* Charlottesville: University of Virginia Press, 1965.

APPENDIX A

The Quotient Space or Identification Space Topology

1 Definitions and basic properties

The quotient space or identification space topology makes precise our intuitive notion of the process of forming a new topological space by identifying certain points in a given topological space. It also corresponds to the idea of "pasting" or "sewing" together two or more spaces. It is an important method of forming new topological spaces from a given collection of topological spaces.

Definition Let X be a topological space, let Y be a set, and let $f : X \to Y$ be a map of X *onto* Y. The *quotient topology on* Y *determined by* f (also called the *identification topology on* Y *determined by* f) is defined as follows: A set $U \subset Y$ is open if and only if $f^{-1}(U)$ is an open subset of X.

Of course, it is necessary to verify that the open sets thus defined satisfy the axioms for a topology; this we leave to the reader. The reader should also note the following two facts:

(a) The quotient topology is the *largest* topology on Y which makes the map f continuous; in fact, this motivates the definition of the quotient topology.

(b) A subset A of Y is closed in the quotient topology if and only if $f^{-1}(A)$ is closed in X.

Examples

1.1 Let X denote the closed interval $[0, 2\pi]$ and let Y denote the unit circle, $x^2 + y^2 = 1$. Define f by $f(t) = (\cos t, \sin t)$. Then, it is readily verified that the usual topology on the circle Y is the identification topology determined by f. This example corresponds to the fact that, if we weld together the two ends of a piece of thin wire, then we obtain a circular ring.

1.2 A similar example is the following: Let Y be the cylinder

$$Y = \{(x, y, z) \in \mathbf{R}^3 : x^2 + y^2 = 1, 0 \leqq z \leqq 1\},$$

and let X be the rectangle

$$X = \{(x, y) \in \mathbf{R}^2 : 0 \leqq x \leqq 2\pi, 0 \leqq y \leqq 1\}.$$

Define $f : X \to Y$ by

$$f(x, y) = (\cos x, \sin x, y).$$

Then, the usual topology on Y is also the identification topology induced by the map f. We can think of the cylinder as obtained from the rectangle by pasting two opposite edges together.

1.3 Let R denote an equivalence relation (i.e., a relation which is reflexive, symmetric, and transitive) on the topological space X, let X/R denote the set of all equivalence classes, and let $f : X \to X/R$ denote the natural map which assigns to each point in X its equivalence class. Then, it is natural to give X/R the quotient space topology determined by f; with this topology, X/R is usually called the *quotient space of X modulo R.*

As a simple example, let $X = \mathbf{R}$, the real line, and let R be the following equivalence relation: $x \equiv y \bmod R$ if and only if $x - y$ is an integer. Then, the quotient space X/R is homeomorphic to a circle.

1.4 Another approach to the subject of equivalence relations is the following: Let $\mathfrak{U}$ be a *decomposition* or *partition* of the topological space X, i.e., a family of nonempty subsets which are pairwise disjoint and cover X. If R is an equivalence relation on X, the set of equivalence classes constitutes such a decomposition or partition. Conversely, a decomposition determines an equivalence relation: Define x and y to be equivalent if and only if they belong to the same element of the partition.

Let $f : X \to \mathfrak{U}$ be the natural map which assigns to each point $x \in X$ the unique element of $\mathfrak{U}$ which contains x. Then, it is natural to give to $\mathfrak{U}$ the quotient space topology determined by f; with this topology, $\mathfrak{U}$ is called a *decomposition space.*

1.5 Let X be a topological space, and let G be a group which operates on X on the left (see Appendix B for the definition). This action of G on X defines an equivalence relation as follows: x and y are equivalent if and only if there is an element $g \in G$ such that $g \cdot x = y$. We leave it to the reader to verify that this relation is reflexive, symmetric, and transitive. The quotient space is usually denoted by X/G. Each equivalence class is called an *orbit.* In most cases the following additional requirement is imposed: For any $g \in G$, the map $X \to X$ defined by $x \to g \cdot x$ for any $x \in X$ should be continuous. It follows as a consequence that it is a homeomorphism of X onto itself (since $x \to g^{-1} \cdot x$ is an inverse). If G is a topological group, then the following stronger condition is usually imposed: The map $G \times X \to X$ defined by $(g, x) \to g \cdot x$ shall be continuous.

Examples of groups acting on spaces may be found in Chapter V (the group of automorphisms of a covering space).

At this point, a word of caution should be interjected: If X and Y are topological spaces, and f is a continuous map of X onto Y, it does *not* follow that Y has the quotient topology determined by f. It is easy to construct counterexamples. However, if f is a *closed*[1] map, or, if f is an *open* map, then Y must have the quotient topology. The proof is easy.

If Y has the quotient topology determined by a map $f : X \to Y$, then the condition that f be a closed or an open map is often a useful hypothesis in theorems. On this account, we note the following equivalent forms of these conditions:

(a) The map f is *closed* if and only if, for any closed subset A of X, the set $f^{-1}f(A)$ is also closed.
(b) The map f is *open* if and only if, for any open subset U of X, the set $f^{-1}f(U)$ is also open.

Note that in both of these statements we are assuming that Y has the quotient topology.

For example, if G is a group which operates on the left on the space X such that, for any $g \in G$, the map $X \to X$ defined by $x \to g \cdot x$ is continuous, then the natural map $X \to X/G$ is an open map. If G is a finite group, then the natural map $X \to X/G$ is also a closed map. The proofs of these assertions are left to the reader.

Presumably the reader has already studied the subspace topology and the product space topology, and has learned many theorems about these topologies. We would hope for theorems of a similar nature about quotient spaces. Unfortunately, things do not work out this way. For example, any product of Hausdorff spaces, or any subspace of a Hausdorff space, is also a Hausdorff space. However, it is quite definitely not true that a quotient space of a Hausdorff space is Hausdorff. For example, if X is the closed interval $[0, 1]$, and $\mathcal{U}$ is the decomposition of X consisting of the three sets $\{0\}$, $(0, 1)$, and $\{1\}$, then the decomposition space $\mathcal{U}$ has only three elements, and its topology is obviously not Hausdorff. This example illustrates one of the most common problems that arise in connection with quotient spaces: Even though the space X may satisfy all the separation axioms we wish, a quotient space of X need not satisfy any separation axioms.

Proposition 1.1 *Let Y have the quotient topology determined by a map $f : X \to Y$. If X is compact, connected, or arcwise connected, then so is Y.*

This is a special case of the well-known fact that the continuous image of a space which is compact, connected, or arcwise connected has the same property.

---

[1] A map $f : X \to Y$ is *closed* if the image under f of any closed set is closed; the analogous definition holds for open maps. A continuous map may be neither open nor closed, open but not closed, closed but not open, or both open and closed. The reader should construct examples to illustrate all four possibilities. An open map is also called an *interior* map.

2 A generalization of the quotient space topology

The definition of the quotient topology is a special case of the following more general process: Let Y be a set, let $\{X_\lambda : \lambda \in \Lambda\}$ be an arbitrary family of topological spaces, and let $\{f_\lambda : X_\lambda \rightarrow Y : \lambda \in \Lambda\}$ be an arbitrary family of maps. In this situation it is natural to give Y the *largest* topology which makes all the maps f_λ continuous. This topology is defined as follows: A set $U \subset Y$ is open if and only if $f_\lambda^{-1}(U)$ is open for all $\lambda \in \Lambda$. Alternatively, a set $A \subset Y$ is closed if and only if $f_\lambda^{-1}(A)$ is closed for all $\lambda \in \Lambda$.

Another important special case of this general process is the formation of what is called (by N. Bourbaki) the "topological sum" of a collection of spaces. If in the above definition each of the maps $f_\lambda : X_\lambda \rightarrow Y$ is one-to-one, the images $f_\lambda(X_\lambda)$ are pairwise disjoint and cover Y, then we say that Y is the *topological sum* of the collection of spaces $\{X_\lambda\}$ (with respect to the maps f_λ). It is easily verified that under these hypotheses each of the spaces X_λ is mapped topologically onto its image in Y, and each such image $f_\lambda(X_\lambda)$ is an open subset of Y.

Another example of this process is the method of defining the weak topology on a CW-complex (see Chapter VII).

It should be pointed out that, in addition to the general process just described for defining a topology on Y, there is a dual process which can be defined as follows: Let X be a set, let $\{Y_\lambda : \lambda \in \Lambda\}$ be a family of topological spaces, and let $\{f_\lambda : X \rightarrow Y_\lambda : \lambda \in \Lambda\}$ be an arbitrary family of mappings. In this situation it is natural to topologize X by giving it the *smallest* topology which makes all the maps f_λ continuous. This topology has as a subbasis all sets of the form $f_\lambda^{-1}(U_\lambda)$, where U_λ is an arbitrary open subset of Y_λ. The two most common and important examples of this method of defining a topology on a space are the following:

(a) X is the Cartesian product of the spaces Y_λ,

$$X = \prod_\lambda Y_\lambda,$$

and f_λ is the projection of the product space on the factor Y_λ. In this case, this general method gives rise to the usual product space topology on X.

(b) The index set Λ has only one element; thus, there is only one space Y_λ, which we denote by Y. We assume that X is a subset of Y, and $f : X \rightarrow Y$ denotes the inclusion map. If we apply the above described general process to this case, we obtain the subspace topology on X. Thus, it is seen that the two familiar processes of forming subspaces and product spaces are special cases of a very general method of defining a topology on a space.

Not only are these two general methods of defining a topology on a space dual to each other, but in a certain sense the processes of forming subspaces and quotient spaces are dual to each other, as are the processes of forming product spaces and topological sums.

Lemma 2.1 *Let $\{X_\lambda : \lambda \in \Lambda\}$ be a family of topological spaces, let $f_\lambda : X_\lambda \to Y$ be a family of maps, and assume that Y has the largest topology which makes all the maps f_λ continuous. Then, a map $g : Y \to Z$ of Y into a topological space Z is continuous if and only if each of the composed maps $gf_\lambda : X_\lambda \to Z$ is continuous.*

The proof, which is very easy, is left to the reader.

Corollary 2.2 *Let X be a topological space, and let Y have the quotient space topology determined by an onto map $f : X \to Y$. Then, a map $g : Y \to Z$ is continuous if and only if the composed map $gf : X \to Z$ is continuous.*

Corollary 2.3 *Let Y be the topological sum of a family of spaces $\{X_\lambda : \lambda \in \Lambda\}$ with respect to maps $f_\lambda : X_\lambda \to Y$. Then, a map $g : Y \to Z$ is continuous if and only if each of the composed maps $gf_\lambda : X_\lambda \to Y$ is continuous.*

For the sake of completeness, we state the dual of Lemma 2.1.

Lemma 2.1′ *Let $\{Y_\lambda : \lambda \in \Lambda\}$ be a family of topological spaces, let $f_\lambda : X \to Y_\lambda$ be a family of maps, and assume that X has the least topology which makes all the maps f_λ continuous. Then, a function $g : W \to X$ is continuous if and only if each of the composed functions $f_\lambda g : W \to Y_\lambda$ is continuous.*

We leave it to the reader to state and prove the duals to Corollaries 2.2 and 2.3, which express well-known properties of subspaces and product spaces, respectively.

Note that, under the hypotheses of Lemma 2.1, there is no general lemma which gives a necessary and sufficient condition for a map $h : W \to Y$ to be continuous; in particular, there is no general condition for the continuity of a map into a quotient space. Similarly, there is no general condition for the continuity of a map $h : X \to Z$ under the hypotheses of Lemma 2.1′.

For the next lemma, assume we are given the following:

Z, a topological space.

$\{Y_\lambda : \lambda \in \Lambda\}$, a family of topological spaces.

$\{f_\lambda : Y_\lambda \to Z\}$, a family of continuous maps.

For each $\lambda \in \Lambda$, a family $\{X_{\lambda\mu} : \mu \in M_\lambda\}$ of topological spaces and continuous maps $\{f_{\lambda\mu} : X_{\lambda\mu} \to Y_\lambda : \mu \in M_\lambda\}$.

Lemma 2.4 *Assume that each space Y_λ has the largest topology making all the maps $f_{\lambda\mu}$, $\mu \in M_\lambda$, continuous, and that Z has the largest topology making all the maps f_λ continuous. Then, the topology on Z is the largest topology which makes all the composed maps $f_\lambda f_{\lambda\mu} : X_{\lambda\mu} \to Z$ continuous.*

The proof is trivial.

We now give three applications of this simple lemma.

(a) Let $f : X \to Y$ and $g : Y \to Z$ be onto maps of topological spaces. Assume that Y has the quotient topology determined by f, and Z has the quotient topology determined by g. Then, the topology on Z is the quotient topology determined by gf. If we think of Z as obtained from X by a process of identifying certain points, then it does not matter whether we make all the identifications at once, or make them in two stages, first obtaining Y and then Z.

(b) Assume that Z is the topological sum of a family of spaces Y_λ, and that each space Y_λ is the topological sum of a family of spaces $X_{\lambda\mu}$. Then, Z is the topological sum of all the spaces $X_{\lambda\mu}$.

(c) The processes of forming quotient spaces and topological sums are interchangeable; i.e., it does not matter in which order we perform them. To be precise, for each $\lambda \in \Lambda$ let $f_\lambda : X_\lambda \to Y_\lambda$ be an onto map of topological spaces, and assume that each space Y_λ has the quotient space topology determined by f_λ. Let X and Y be the topological sums of the families $\{X_\lambda\}$ and $\{Y_\lambda\}$, respectively, with respect to maps $\varphi_\lambda : X_\lambda \to X$ and $\psi_\lambda : Y_\lambda \to Y$, respectively. It is readily seen that there exists a unique map $f : X \to Y$ which is continuous and onto, and makes the following diagram commutative for each λ:

$$\begin{array}{ccc} X_\lambda & \xrightarrow{f_\lambda} & Y_\lambda \\ \downarrow{\scriptstyle\varphi_\lambda} & & \downarrow{\scriptstyle\psi_\lambda} \\ X & \xrightarrow{f} & Y \end{array}$$

We now assert that the quotient space topology on Y determined by f is the same as the topology Y has as a topological sum. The proof follows immediately from Lemma 2.4 and the commutativity of the above diagram.

We leave it to the reader to state the dual to Lemma 2.4. The duals of applications (a), (b), and (c) are propositions about subspaces and product spaces which are so trivial that usually they are not even stated explicitly in textbooks.

3 Quotient spaces and product spaces

The question naturally arises, is a quotient space of a product space the same as a product of quotient spaces? To be precise, suppose that, for each index $\lambda \in \Lambda$, we have a continuous, onto map $f_\lambda : X_\lambda \to Y_\lambda$, and that Y_λ has the quotient topology determined by f_λ. We can then form the product spaces

$$X = \prod_\lambda X_\lambda \quad \text{and} \quad Y = \prod_\lambda Y_\lambda,$$

and the maps f_λ determine a map $f : X \to Y$ in an obvious way: For any $x \in X$, $(fx)_\lambda = f_\lambda(x_\lambda)$. It is clear that the map f is continuous and onto. Then, the question is, does Y have the quotient topology determined by f? To put it another way, we can topologize the set Y either as a product space or a quotient space. Are the two topologies the same?

In general, the answer to this question is no, as the following example of J. L. Kelley [2] shows: Let X be a Hausdorff space which is not regular. Therefore, we can choose a closed subset A and a point b of X which do not have disjoint open neighborhoods; i.e., if U and V are any open sets containing A and b, respectively, then U and V intersect. Let Y be the quotient space of X obtained by identifying all the points of A to a single point $a \in Y$. Then, Y is not a Hausdorff space because the points a and b do not have disjoint neighborhoods. Let $f : X \to Y$ be the natural map, and consider the map $f \times f : X \times X \to Y \times Y$, where $Y \times Y$ has the product topology. We assert that $Y \times Y$ does not have the quotient topology determined by $f \times f$. For, the diagonal of $Y \times Y$ is not a closed set, because Y is not Hausdorff; but the inverse image of the diagonal under $f \times f$ is clearly a closed subset of $X \times X$.

In general, all we can say is that the quotient topology on Y is larger than the product topology (i.e., has more open sets). The following proposition gives a sufficient condition for the two topologies on Y to be the same:

Proposition 3.1 *In addition to the above hypotheses and notation, assume that each of the maps $f_\lambda : X_\lambda \to Y_\lambda$ is an open mapping. Then, the product topology and the quotient topology on*

$$Y = \prod_\lambda Y_\lambda$$

are the same.

PROOF: That each of the maps f_λ is open and onto implies that the map $f : X \to Y$ is also an open map. Hence, Y has the quotient topology determined by the map f (see the remark in Section 1). Q.E.D.

For other theorems of this type see the following article: C. J. Himmelberg, "On the Product of Quotient Spaces," *Amer. Math. Monthly*, *72*, 1965, pp. 1103–1106.

4 Subspace of a quotient space vs. quotient space of a subspace

Another natural question is the following: Is a quotient space of a subspace the same as a subspace of a quotient space? This question can be made precise as follows: Suppose that $f : X \to Y$ is a map which is onto, Y has the quotient topology determined by f, A is a subspace of X, and $B = f(A) \subset Y$. Then, we can either topologize B as a subspace of Y, or give it the quotient topology determined by the map $f \mid A : A \to B$. Are these two topologies on B the same? In general, the answer is no. The following is a simple counterexample. Let $f : X \to Y$ be as in Example 1.2; X is a rectangle and Y is a cylinder. Let

$$A = \{(x, y) \in X : 0 \leqq x < 2\pi,\ y = 0\}.$$

Then, A is a half-open interval, which is one of the edges of A. Clearly $B = f(A)$ is one of the circular boundaries of Y if we give B the subspace topology. On the other hand, since $f \mid A : A \to B$ is one-to-one and onto, the quotient topology on B makes B homeomorphic to A. But a half-open interval and a circle are not homeomorphic; one is compact and the other is noncompact. Therefore, the two topologies on B are different.

Proposition 4.1 *Under the hypotheses at the beginning of this section, the quotient topology on B is larger than the subspace topology.*

PROOF: Let $i : A \to X$ and $j : B \to Y$ denote inclusion maps. Then, the following diagram is commutative:

$$\begin{array}{ccc} A & \xrightarrow{f|A} & B \\ {\scriptstyle i}\downarrow & & \downarrow{\scriptstyle j} \\ X & \xrightarrow[f]{} & Y \end{array}.$$

If we give B the quotient topology, then it follows from Corollary 2.2 and the commutativity of the diagram that j is continuous. From this the proposition follows, because the subspace topology on B is the smallest topology which makes j continuous.

We now give a sufficient condition for the subspace and quotient space topology on B to be the same.

Proposition 4.2 *Under the same hypotheses as above, if A is a closed subset of X and $f : X \to Y$ is a closed map, or if A is an open subset of X and f is an open map, the subspace and quotient topologies on B are the same.*

PROOF: Give B the subspace topology. Consider first the case where A is an open subset and f is an open map. Then, B is an open subset of Y, and it is readily seen that $f \mid A : A \to B$ is an open map. Hence, B has the quotient space topology determined by $f \mid A$.

In the case where A is a closed subset and f is a closed map, a similar proof holds. We substitute the word "closed" for the word "open" each time it occurs in the preceding paragraph.

5 Conditions for a quotient space to be a Hausdorff space

We shall now tackle one of the most serious problems regarding quotient spaces. First, we have the following necessary condition:

Lemma 5.1 *Let $f : X \to Y$ be a continuous map. If Y is a Hausdorff space, then $\{(x_1, x_2) \in X \times X : f(x_1) = f(x_2)\}$ is a closed subset of $X \times X$.*

PROOF: We use the following easily proved fact: The space Y is Hausdorff if and only if the diagonal $D = \{(y_1, y_2) \in Y \times Y : y_1 = y_2\}$ is a closed subset of $Y \times Y$.

Consider the map $f \times f : X \times X \to Y \times Y$; it is continuous, and $(f \times f)^{-1}(D)$ is the set mentioned in the conclusion of the proposition. Because the inverse image of a closed set under a continuous map is closed, the result follows. Q.E.D.

Note that in this lemma it was not assumed that Y has the quotient topology. Thus, it gives a necessary condition for the existence of *some* topology on Y which is Hausdorff and makes f continuous. However, if there is any such topology on Y, then the quotient topology determined by f is also a Hausdorff topology; for, any topology on Y larger than a Hausdorff topology is also a Hausdorff topology.

In general, the converse of this lemma is not true, even for the quotient topology on Y. However, we do have the following partial converse.

Lemma 5.2 *Let $f : X \to Y$ be a map which is open and onto. If the set $\{(x_1, x_2) \in X \times X : f(x_1) = f(x_2)\}$ is closed, then Y is a Hausdorff space.*

PROOF: Once again, we consider the map $f \times f : X \times X \to Y \times Y$, which is also an open map. By hypothesis, the set

$$\{(x_1, x_2) \in X \times X : f(x_1) \neq f(x_2)\}$$

is open in $X \times X$; therefore, its image under $f \times f$ is open in $Y \times Y$. But this image in $Y \times Y$ is the complement of the diagonal $D \subset Y \times Y$. Therefore, D is a closed set, and Y is Hausdorff. Q.E.D.

In this lemma, we did not assume that f was continuous. We can combine the two lemmas as follows:

Proposition 5.3 (N. Bourbaki [1]) *Let $f : X \to Y$ be a map which is continuous, open, and onto. Then, Y is a Hausdorff space if and only if the set*

$$\{(x_1, x_2) \in X \times X : f(x_1) = f(x_2)\}$$

is closed in $X \times X$.

Note that the hypothesis implies that Y has the quotient topology determined by f.

The situation is somewhat nicer regarding quotient spaces of compact Hausdorff spaces. In that case, we have the following important theorem (see N. Bourbaki [1]):

Theorem 5.4 *Let X be a compact Hausdorff space and let $f : X \to Y$ be a continuous onto map. Assume that Y has the quotient topology determined by f. Then, the following three conditions are equivalent:*

(a) *Y is a Hausdorff space.*

(b) *f is a closed map.*

(c) *The set $\{(x_1, x_2) \in X \times X : f(x_1) = f(x_2)\}$ is closed in $X \times X$.*

PROOF: The theorem will be proved by demonstrating the following implications: (a) $\to$ (c), (c) $\to$ (b), and (b) $\to$ (a). The implication (a) $\to$ (c) is contained in Lemma 5.1.

Proof that (c) $\to$ (b): Let $C = \{(x_1, x_2) \in X \times X : f(x_1) = f(x_2)\}$, which is assumed to be closed in $X \times X$. Because Y has the quotient topology, to prove that f is a closed map we must prove that, for any closed set $A \subset X$, the set $f^{-1}f(A)$ is also closed. Let $p_1, p_2 : X \times X \to X$ denote the projections, $p_1(x_1, x_2) = x_1$, $p_2(x_1, x_2) = x_2$. It is readily verified that, for any subset $A \subset X$,

$$f^{-1}f(A) = p_1(C \cap p_2^{-1}A).$$

If A is closed, then so is $p_2^{-1}(A)$ and hence $C \cap p_2^{-1}A$. Therefore, $C \cap p_2^{-1}A$ is compact, because $X \times X$ is compact. Hence, $p_1(C \cap p_2^{-1}A)$is compact, and therefore closed, because X is Hausdorff.

Proof that (b) $\rightarrow$ (a): Let y_1 and y_2 be distinct points of Y; we must prove that they have disjoint neighborhoods. Note that $f^{-1}(y_1)$ and $f^{-1}(y_2)$ are disjoint closed subsets of X. Because X is compact Hausdorff, it is also normal, and hence there exist disjoint open subsets U_1 and U_2 of X with

$$f^{-1}(x_1) \subset U_1, \qquad f^{-1}(x_2) \subset U_2.$$

By hypothesis, f is a closed map; therefore, $f(X - U_1)$ and $f(X - U_2)$ are closed subsets of Y. Let V_1 and V_2 be their open complements:

$$V_1 = Y - f(X - U_1),$$
$$V_2 = Y - f(X - U_2).$$

Then, it is readily verified that $y_1 \in V_1$, $y_2 \in V_2$, and $V_1 \cap V_2 = \phi$, as required. Q.E.D.

In connection with this theorem, we should recall that a continuous map of a compact space into a Hausdorff space is always a closed map; the proof is elementary.

This theorem shows the usefulness of the hypothesis "f is a closed map." In the work of R. L. Moore, G. T. Whyburn, and their students, the point of view indicated in Example 1.4 is always adopted when dealing with quotient spaces. In the case where the natural map $f : X \rightarrow \mathfrak{U}$ is a closed map, they call the decomposition $\mathfrak{U}$ an *upper semicontinuous decomposition*. For another useful theorem concerning such decompositions, see Kelley [2], p. 148.

Exercise

5.1 Let X and Y be topological spaces, and let $f : X \rightarrow Y$, $g : Y \rightarrow X$ be continuous maps such that fg is the identity map of Y onto itself. Prove the following statements:

(a) f is onto and g is one-to-one.
(b) Y has the quotient topology determined by f.
(c) g maps Y homeomorphically onto a subspace of X (i.e., Y has the subspace topology determined by g).
(d) If X is a Hausdorff space, then so is Y.

REFERENCES

1. Bourbaki, N. *General Topology.* Reading, Massachusetts: Addison-Wesley Co., 1966. Chapter I.
2. Kelley, J. L. *General Topology.* Princeton, N.J.: Van Nostrand, 1955. Chapters 3 and 5.
3. Dugundji, J. *Topology.* Boston: Allyn and Bacon, 1966. Chapter VI.

APPENDIX B

Permutation Groups or Transformation Groups

1 Basic definitions

The reader is undoubtedly familiar with the following fact from his previous study of group theory: If E is any set (finite or infinite), then the set of all permutations of E (i.e., functions $E \to E$ which are one-to-one and onto) is a group under the operation of composition or superposition of permutations. He has undoubtedly considered examples of such a group (called the *symmetric* group of the set E), especially in the case where E is a finite set. Also, he has probably studied various subgroups of the symmetric group on a finite set.

If G is an arbitrary group, a homomorphism of G into the symmetric group on a set E is called a *representation of G by permutations of E.* If the homomorphism is an isomorphism, the representation is called *faithful.* It is an easily proved result that any group admits a faithful representation by permutations. We omit the proof, because we have no need for this theorem in this book.

We now consider another approach to this same set of ideas which occurs frequently. At first sight, this approach seems quite different, but it leads to the same result.

Definition Let E be a set and let G be a group. We say that E is a *left G-space* or that E admits G as a *group* of *operators on the left* if there is given a mapping $G \times E \to E$, denoted by $(g, x) \to g \cdot x$ for any $g \in G$ and $x \in E$, such that the following two properties hold:

(1) For any $x \in E$, $1 \cdot x = x$.
(2) For any $x \in E$ and $g_1, g_2 \in G$,

$$(g_1 g_2) \cdot x = g_1 \cdot (g_2 \cdot x).$$

For example, if G is a subgroup of the symmetric group of E, and the notation $g \cdot x$ denotes the effect of applying the permutation g to the element $x \in E$, then E is a left G-space.

Another simple example is the following: Let E denote ordinary Euclidean 3-space and let G denote the group of all rotations of E which leave the origin fixed. Let $g \cdot x$ denote the image of the point x under the rotation g. Then E is a left G-space.

Right G-spaces are defined in an analogous fashion. There is assumed given a map $E \times G \rightarrow E$, denoted by $(x, g) \rightarrow x \cdot g$, such that the following two conditions hold:

(1′) $x \cdot 1 = x$.
(2′) $x \cdot (g_1 g_2) = (x \cdot g_1) \cdot g_2$.

The essential difference between right and left G-spaces is not whether the elements of G are written on the right or left of those of E. The main point is the difference between condition (2) and condition (2′). If E is a left G-space, then the product $g_1 g_2$ operates on $x \in E$ in such a way that g_2 operates first and then g_1 operates on the result, whereas for right G-spaces, g_1 operates first, then g_2.

Exercise

1.1 Assume that E is a left G-space. For any $x \in E$ and $g \in G$, define

$$x \cdot g = (g^{-1}) \cdot x.$$

With this definition, prove that E is a right G-space.

Theorem 1.1 *Let E be a left G-space. For any $g \in G$, the map $E \rightarrow E$ defined by $x \rightarrow g \cdot x$ is a permutation of E.*

PROOF: Denote the map in question by $\varphi_g : E \rightarrow E$. Consider the map $\varphi_{g^{-1}}$. It readily follows from the axioms for a left G-space that the composed maps $\varphi_g \varphi_{g^{-1}}$ and $\varphi_{g^{-1}} \varphi_g$ are both the identity maps of E onto itself. Therefore, φ_g is one-to-one and onto, i.e., a permutation. Q.E.D.

This simple but important theorem shows that the notion of a left G-space is equivalent to the notion of a representation of G by permutations of the set E. We must not conclude, however, that such a representation is faithful; it can very well happen that there exists an element $g \neq 1$ in G such that $g \cdot x = x$ for all $x \in G$. In the case where no such element $g \in G$ exists, we say that G operates *effectively* on the set E.

If E_1 and E_2 are left G-spaces, a mapping $f : E_1 \rightarrow E_2$ is called *G-equivariant*, or simply a *mapping of left G-spaces*, in case

$$f(g \cdot x) = g \cdot (fx)$$

for any $g \in G$ and $x \in E_1$. A G-equivariant map $f : E_1 \to E_2$ is called an *isomorphism of left G-spaces* in case there exists another G-equivariant map $f' : E_2 \to E_1$ such that $f'f$ is the identity map of E_1 and ff' is the identity map of E_2. This is equivalent to the condition that f be one-to-one and onto. This definition of isomorphism is the natural one in this context. The reader should note that it is sometimes possible for a group G to operate in several different, nonisomorphic ways on a given set E. As usual, an automorphism of a G-space is a self-isomorphism.

2 Homogeneous G-spaces

Let E be a left G-space. We say that G operates *transitively* on E or that E is a *homogeneous* left G-space if the following condition holds: For any elements $x, y \in E$ there exists an element $g \in G$ such that

$$g \cdot x = y.$$

Homogeneous G-spaces are of frequent occurrence, and thus are important.

Example

2.1 Let G be a group, and let H be an arbitrary subgroup of G. We denote by G/H the set of all cosets, $g \cdot H$, $g \in G$. It is readily seen that, if we multiply all the elements in a given coset on the left by any element $g \in G$, we obtain as a result elements all of which lie in the same coset. This defines a map $G \times G/H \to G/H$, and it is readily verified that the two conditions for a left G-space hold. It is also clear that G/H is a homogeneous left G-space.

We now show that any homogeneous left G-space is isomorphic to some coset space G/H. Let E be an arbitrary homogeneous left G-space. Choose an element $x_0 \in E$, and let

$$H = \{g \in G : g \cdot x_0 = x_0\}.$$

We easily check that H is a subgroup of G. It is called the *isotropy subgroup* corresponding to x_0. Consider the map $G \to E$ defined by $g \to g \cdot x_0$. This map is onto, because E is a homogeneous G-space. Under what condition do two elements $g_1, g_2 \in G$ both map onto the same element of E? This is easily determined as follows:

$$\begin{aligned} g_1x_0 = g_2x_0 &\Leftrightarrow g_2^{-1}g_1x_0 = x_0 \\ &\Leftrightarrow g_2^{-1}g_1 \in H. \end{aligned}$$

Hence, g_1 and g_2 map onto the same element of E if and only if g_1 and g_2 belong to the same coset of H. Therefore, the map $G \to E$ induces a map $f : G/H \to E$ which is one-to-one and onto; and it is easily checked that f is G-equivariant. Thus, G/H and E are isomorphic left G-spaces, as was to be proved.

The isomorphism f and the subgroup H in the preceding argument depend on the choice of the point x_0 in E. A different choice of x_0 will give rise to a conjugate subgroup.

For the purposes of Chapter V, we need to know the structure of the group of automorphisms of a homogeneous G-space. To be consistent with the usage adopted in that chapter, we shall consider a homogeneous *right* G-space E. Let $\varphi : E \to E$ be an automorphism of E. Then, one verifies directly from the definitions that, for any point $x \in E$, the points x and $\varphi(x)$ have the same isotropy subgroup. Conversely, suppose x and y are points of E which have the same isotropy subgroup. We assert that there exists an automorphism φ of E such that $\varphi(x) = y$. We define φ in the following rather obvious way. Let $z \in E$. Then, there exists a $g \in G$ such that

$$z = x \cdot g.$$

Hence, we must have

$$\varphi(z) = \varphi(x \cdot g) = (\varphi x) \cdot g = y \cdot g.$$

Therefore, we *define* $\varphi(z) = y \cdot g$. Of course, we must check that this definition is independent of the choice of g; i.e., if $x \cdot g = x \cdot g'$, then $y \cdot g = y \cdot g'$. But this is a consequence of the assumption that x and y have the same isotropy subgroup. We must also verify that the map thus defined is G-equivariant, and that it is one-to-one and onto. It is trivial to verify the first statement, and to verify the second, we construct by the same method an inverse of φ such that $\varphi^{-1}(y) = x$.

Next, we note that if φ_1 and φ_2 are automorphisms of the homogeneous right G-space E, and, for some point $x \in E$, $\varphi_1(x) = \varphi_2(x)$, then $\varphi_1 = \varphi_2$. This is a direct consequence of the fact that G acts transitively on E.

As a consequence of these considerations, we have the following lemma:

Lemma 2.1 *A group A of automorphisms of a homogeneous G-space E is the entire group of automorphisms if and only if for any two points $x, y \in E$ which have the same isotropy subgroup, there exists an automorphism $\varphi \in A$ such that $\varphi(x) = y$.*

Next, we determine the structure of the group of automorphisms of a homogeneous G-space. First, we need a definition. Let H be a subgroup of G and

$$N(H) = \{g \in G : gHg^{-1} = H\}.$$

$N(H)$ is a subgroup of G which contains H, and it is called the *normalizer* of H. It is the largest subgroup of G which contains H as a normal subgroup.

Theorem 2.2 *Let E be a homogeneous G-space, and let H be the isotropy subgroup of G corresponding to the point $x_0 \in E$. Then the group of automorphisms of E is isomorphic to $N(H)/H$.*

PROOF: Let S denote the set of all points $x \in E$ whose isotropy subgroup is H. In view of what we have proved above, we see that the automorphism group acts transitively on S.

Next, we assert that, if $x \in S$ and $g \in G$, then $x \cdot g \in S$ if and only if $g \in N(H)$. For, the condition $xg \in S$ is equivalent to the condition

$$\{h \in G : x \cdot g \cdot h = x \cdot g\} = H.$$

But $xgh = xg$ if and only if $xghg^{-1} = x$, i.e., if and only if $ghg^{-1} \in H$, or $h \in g^{-1}Hg$. Thus, the subgroup $N(H)$ acts transitively on the subspace S, and the elements of H leave each point of S fixed. Hence, the quotient group $N(H)/H$ acts transitively on the right on S without fixed points.

We now set up an isomorphism between the automorphism group and $N(H)/H$ as follows. Let φ be an automorphism; there exists a unique element $\alpha \in N(H)/H$ such that

$$x_0 \cdot \alpha = \varphi(x_0)$$

because $N(H)/H$ operates transitively on S without fixed points. Conversely, for any element $\alpha \in N(H)/H$ there exists a unique automorphism φ such that $\varphi(x_0) = x_0 \cdot \alpha$. Thus the correspondence $\varphi \leftrightarrow \alpha$ is a one-to-one correspondence between the automorphism group and $N(H)/H$. We now check that this correspondence preserves products, as follows. Suppose that

$$\varphi(x_0) = x_0 \cdot \alpha,$$

$$\psi(x_0) = x_0 \cdot \beta.$$

Then,

$$(\varphi\psi)(x_0) = \varphi(\psi x_0) = \varphi(x_0\beta)$$
$$= (\varphi x_0)\beta = (x_0\alpha)\beta = x_0(\alpha\beta);$$

hence, $\varphi\psi$ and $\alpha\beta$ correspond. Therefore, the correspondence is an isomorphism. Q.E.D.

It should be emphasized that this isomorphism between $N(H)/H$ and the automorphism group is not natural; it depends on the choice of the point $x_0 \in E$. The student should investigate the effect of a different choice of the base point x_0 on the one-to-one correspondence which was set up.

Index

Graduate Texts in Mathematics

continued from page ii

93 Dubrovin/Fomenko/Novikov. Modern Geometry — Methods and Applications Vol. I.
94 Warner. Foundations of Differentiable Manifolds and Lie Groups.
95 Shiryayev. Probability, Statistics, and Random Processes.
96 Conway. A Course in Functional Analysis.
97 Koblitz. Introduction in Elliptic Curves and Modular Forms.
98 Bröcker/tom Dieck. Representations of Compact Lie Groups.
99 Grove/Benson. Finite Reflection Groups. 2nd ed.
100 Berg/Christensen/Ressel. Harmonic Analysis on Semigroups: Theory of Positive Definite and Related Functions.
101 Edwards. Galois Theory.
102 Varadarajan. Lie Groups, Lie Algebras and Their Representations.
103 Lang. Complex Analysis. 2nd ed.
104 Dubrovin/Fomenko/Novikov. Modern Geometry — Methods and Applications Vol. II.
105 Lang. $SL_2(\mathbf{R})$.
106 Silverman. The Arithmetic of Elliptic Curves.
107 Olver. Applications of Lie Groups to Differential Equations.
108 Range. Holomorphic Functions and Integral Representations in Several Complex Variables.
109 Lehto. Univalent Functions and Teichmüller Spaces.
110 Lang. Algebraic Number Theory.
111 Husemöller. Elliptic Curves.